COMPUTER VISION AND ROBOTICS

COMPUTER VISION AND ROBOTICS

JOHN X. LIU
EDITOR

Nova Science Publishers, Inc.
New York

For permission to use material from this book please contact us:
Telephone 631-231-7269; Fax 631-231-8175
Web Site: http://www.novapublishers.com

NOTICE TO THE READER

The Publisher has taken reasonable care in the preparation of this book, but makes no expressed or implied warranty of any kind and assumes no responsibility for any errors or omissions. No liability is assumed for incidental or consequential damages in connection with or arising out of information contained in this book. The Publisher shall not be liable for any special, consequential, or exemplary damages resulting, in whole or in part, from the readers' use of, or reliance upon, this material.

This publication is designed to provide accurate and authoritative information with regard to the subject matter covered herein. It is sold with the clear understanding that the Publisher is not engaged in rendering legal or any other professional services. If legal or any other expert assistance is required, the services of a competent person should be sought. FROM A DECLARATION OF PARTICIPANTS JOINTLY ADOPTED BY A COMMITTEE OF THE AMERICAN BAR ASSOCIATION AND A COMMITTEE OF PUBLISHERS.

LIBRARY OF CONGRESS CATALOGING-IN-PUBLICATION DATA
Computer vision and robotics / John X. Liu (editor).
p. cm.
Includes bibliographical references and index.
ISBN 1-59454-357-7
1. Robotics. 2. Computer vision. I. Liu, John X.
TJ211.C621854   2005
629.8'92637--dc22   2005011341

Published by Nova Science Publishers, Inc. ✦ *New York*

CONTENTS

PREFACE

This new book deals with control and learning in robotic systems and computers.

In Chapter 1, the authors discuss about the stereo vision and visual motion are two vision cues that allow three-dimensional (3D) information of a scene to be recovered from multiple images. When a mobile platform with two fixed camera heads is available to capture stereo pair of image streams, both cues are applicable. Yet, the cues have complementary advantages: while feature correspondence is simpler in visual motion, stereo vision offers more accurate 3D reconstruction. This paper presents an approach of integrating the two cues, that retains their advantages and removes their disadvantages. It is shown that by adopting the affine camera model for the projection model of the video cameras, the two sets of motion correspondences (on the two cameras) are actually related to the stereo correspondences (across the cameras) by a matrix rank property. The rank property is important, as it allows the inference from the more readily available motion correspondences to stereo correspondences that give more accurate 3D reconstruction. In addition, the inference process could be achieved in a time only linear with respect to the total size of the image data. With the inferred stereo correspondence, both the 3D structure of the scene as well as the motion of the mobile platform could be recovered. It is also shown that with the use of all stereo pairs of image data, not only could reconstruction accuracy be boosted, even errors in the initial motion correspondences could be detected. Experiments on real image data show that 3D reconstruction is accurate even with relatively short motion of the mobile platform.

Chapter 2 presents a new method for progressive transmission of 3D images that has four components: (1) decomposition of the image into regions using Singular Value Decomposition (SVD), (2) a reconstruction algorithm for progressive rendering that uses matrix polynomial interpolation along with approximations which are derived from SVD, (3) exploitation of a matrix norm for analyzing goodness of approximation, and (4) an optimal adaptive strategy for selecting "the next region to transmit".

SVD of matrices is used in some areas of image processing, such as restoration, but not usually in transmission. For an image (matrix) of size $m \times n$, its SVD produces an $m \times m$ matrix, an $n \times n$ matrix, and a vector of size min $\{m,n\}$. That is, the SVD generates more than double the amount of original data. Despite this fact, however, a design of an appropriate adaptive transmission strategy within this four-component procedure provides an algorithm, for lossy progressive transmission, with excellent rendering and computational performance at low percentages of data transmission.

Range image registration is a fundamental problem in range image analysis, as outlined in Chapter 3. The main task for range image registration is to establish point correspondences between overlapping range images to be registered. Since the relationship between point correspondences can be represented using motion parameters, in this chapter, we thus review the main image registration techniques from five aspects: motion representation, motion estimation, image registration using motion properties, image registration using both motion and structural properties, and the two-way constraint. In contrast with existing image registration review techniques, our review starts from the investigation of the relationship among point correspondences, motion parameters, and rigid constraints. Consequently, the review not only deepens our understanding about the relationship among motion parameters, rigid constraints, and point correspondences, but also possibly identifies the potential culprit for false matches in the process of image registration and points out future research direction to range image registration.

Given that errors in the estimates for the intrinsic and extrinsic camera parameters are inevitable, it is important to understand the behaviour of the resultant distortion in depth recovered under different motion-scene configurations. The main goal of the study in Chapter 4 is to look for a generic motion type that can render depth recovery more robust and reliable. To this end, lateral and forward motions are compared both under calibrated and uncalibrated scenarios. For lateral motion, it find that although Euclidean reconstruction is difficult, ordinal depth information is obtainable; while for forward motion, depth information (even partial one) is difficult to recover. It obtains the same conclusion in the uncalibrated case when the intrinsic camera parameters are fixed. However, when these parameters are not fixed, then lateral motion allows only a local recovery of depth order. In general, the depth distortion transformation is a Cremona transformation, and becomes a simple projective one in the case of lateral motion. It applied the above analysis to the scenario of recovering curvature of a quadric surface under lateral motion and showed that the shape estimates are recovered with varying degrees of uncertainty depending on the motion-scene configuration. Specifically, the reconstructed second order shape tends to be more distorted in the direction parallel to the translational motion than that in the orthogonal direction. It present the result of a psychophysical experiment, which confirms that in human vision, curvature estimates tend to be more erroneous and variable along the direction of lateral motion, than along its orthogonal direction.

Chapter 5 presents a stereo panoramic depth imaging system, which builds depth panoramas from multiperspective panoramas while using only one standard camera.

The basic system is mosaic-based, which means that we use a single standard rotating camera and assemble the captured images in a multiperspective panoramic image. Due to a setoff of the camera's optical center from the rotational center of the system, we are able to capture the motion parallax effect, which enables the stereo reconstruction.

The system has been comprehensively analysed. The analyses include the study of influence of different system parameters on the reconstruction accuracy, constraining the search space on the epipolar line, meaning of error in estimation of corresponding point, definition of the maximal reliable depth value, contribution of the vertical reconstruction and influence of using different cameras. They are substantiated with a number of experiments, including experiments addressing the baseline, the repeatability of results in different rooms, by using different cameras, influence of lens distortion presence on the reconstruction accuracy and evaluation of different models for estimation of system parameters. The analyses and the experiments revealed a number of interesting properties of the system.

According to the basic system accuracy we definitely can use the system for autonomous robot localization and navigation tasks.

As explained in chapter 6, the estimation of 3-D motion and structure is one of the most important function-alities of an intelligent vision system. In spite of the best efforts of a generation of computer vision researchers, we still do not have a practical and robust system for accurately estimating motion and structure from a sequence of moving imagery under all motion-scene configurations. The authour's put forth in this study a geometrically motivated 3-D motion and structure error analysis which is capable of shedding light on global effect such as inherent ambiguities. This is in contrast with the usual statistical kinds of error analyses which can only deal with local effect such as noise perturbations, and in which much of the results regarding global ambiguities are empirical in nature. The error expression that we derive allows us to predict the exact conditions likely to cause ambiguities and how these ambiguities vary with motion types such as lateral or forward motion. Such an investigation may alert us to the occurrence of ambiguities under different conditions and be more careful in picking the solution. Our formulation, though geometrically motivated, was also put to use in modeling the effect of noise and in revealing the strong influence of feature distribution. Given the erroneous 3-D motion estimates caused by the inherent ambiguities, it is also important to understand the impact such motion errors have on the structure reconstruction. In this study, various robustness issues related to the different types of second order shape recovered from motion cue are addressed. Experiments on both synthetic and real image sequences were conducted to verify the various theoretical predictions.

This study would be most beneficial for an intelligent vision system that needs to have an estimate of the robustness of the 3-D motion and structure information recovered from the world. Such information would allow the system to carry out its tasks more effectively and to seek more information if necessary.

Chapter 7 introduces multiple-view geometry for algebraic curves, with applications in both static and dynamic scenes. More precisely, it shows when and how the epipolar geometry can be recovered from algebraic curves. For that purpose, it introduce a generalization of Kruppa's equations, which express the epipolar constraint for algebraic curves. For planar curves, it shows that the homography through the plane of the curve in space can be computed. It investigates the question of three-dimensional reconstruction of an algebraic curve from two or more views. In the case of two views, it shows that for a generic situation, there are two solutions for the reconstruction, which allows extracting the right solution, provided the degree of the curve is greater or equal to 3. When more than two views are available, it shows that the reconstruction can be done by linear computations, using either the dual curve or the variety of intersecting lines. In both cases, no curve fitting is necessary in the image space.

For dynamic scenes, it is addressed the question of recovering the trajectory of a moving point, also called trajectory triangulation, from moving, non-synchronized cameras. Two cases are considered. First it address the case where the moving point itself is tracked in the images. Secondly, it focus on the case where the tangents to the motion are detected in the images. Both cases yield linear computations, using the dual curve or the variety of intersecting lines.

Eventually, it presents several experiments on both synthetic and real data, which demonstrate that our results can be used in practical situations.

In Chapter 8, a new scheme of vision based navigation was proposed for flying vehicles. In this navigation scheme, the main navigation tool is a camera, plus an altimeter. The feasibility of this navigation scheme was carefully studied both from theory and numerical analysis. Unlike most of vision based navigation approaches in which feature trajectories were utilised to compute 3D-platform motion, it was used the image geometrical transformation parameters between consecutive frames to infer 3D displacement of camera. Due to this change, the navigation process can be conducted even if there is no salient features that can be extracted from in the image sequence, for example, in the case of flying over the sea. As a result, the long-range navigation becomes possible by use EO sensor. Moreover, the way of improvement navigation accuracy was also addressed. The experiment results demonstrated that the navigation accuracy of this system is compatible to GPS (Global Positioning system), much higher than all kinds of INS (Inertial Navigation System) in terms of position estimation. It is a good alternative choice when the GPS signal is not available

In Chapter 9 an evaluation metric for calculate the behavior of a video tracking system is proposed. This metric is used for adjusting several parameters of the tracking system in order to improve the performance. The optimization procedure is based on evolutionary computation techniques. The system has been tested in an airport domain where several cameras are deployed for surveillance purposes.

In: Computer Vision and Robotics
Editor: John X. Liu, pp. 1-27

ISBN 1-59454-357-7
© 2006 Nova Science Publishers, Inc.

Chapter 1

STRUCTURE AND MOTION RECOVERY VIA STEREO-MOTION

R. Chung and *P.K. Ho*

Computer Vision Laboratory, Department of ACAE
The Chinese University of Hong Kong, Shatin, Hong Kong

Abstract

Stereo vision and visual motion are two vision cues that allow three-dimensional (3D) information of a scene to be recovered from multiple images. When a mobile platform with two fixed camera heads is available to capture stereo pair of image streams, both cues are applicable. Yet, the cues have complementary advantages: while feature correspondence is simpler in visual motion, stereo vision offers more accurate 3D reconstruction. This paper presents an approach of integrating the two cues, that retains their advantages and removes their disadvantages. It is shown that by adopting the affine camera model for the projection model of the video cameras, the two sets of motion correspondences (on the two cameras) are actually related to the stereo correspondences (across the cameras) by a matrix rank property. The rank property is important, as it allows the inference from the more readily available motion correspondences to stereo correspondences that give more accurate 3D reconstruction. In addition, the inference process could be achieved in a time only linear with respect to the total size of the image data. With the inferred stereo correspondence, both the 3D structure of the scene as well as the motion of the mobile platform could be recovered. It is also shown that with the use of all stereo pairs of image data, not only could reconstruction accuracy be boosted, even errors in the initial motion correspondences could be detected. Experiments on real image data show that 3D reconstruction is accurate even with relatively short motion of the mobile platform.

---

*E-mail address: rchung@acae.cuhk.edu.hk

1 Introduction

The capability of recovering 3D structure of a scene from visual data is important for applications like autonomous navigation and robotic manipulation. If more than one image of the scene are available the estimation problem is potentially easier because of the more information available about the imaged scene. There exists two major vision cues that employ such a multi-ocular approach. One is visual motion, in which 3D structure is recovered from an image sequence that is acquired under a relative motion between the camera and the scene. The other is stereo vision, in which 3D structure is recovered from two widely separated views of the same scene. Both the two multi-ocular cues require to solve two subproblems: the *correspondence problem*, in which image features corresponding to the same entities in 3D are to be matched across the image frames, and the *reconstruction problem*, in which 3D information is to be reconstructed from the feature correspondences.

The motion cue has the advantage that the correspondence problem is relatively easy to solve, because successive images are alike. However, it generally requires a long image sequence, up to hundreds of frames (for instance in [19]), for accurate 3D reconstruction. The reason is, 3D determination from multi-ocular vision is based upon intersecting the respective images' corresponding projection rays. To reduce the effect of disturbances like image noise etc. to the reconstruction, the physical separation between the spatial positions of the images, i.e., the baseline, must be wide enough.

In contrast, stereo vision has an easier reconstruction problem but a more difficult correspondence problem. It allows more accurate 3D reconstruction because the two views are generally widely separated. It has a more difficult correspondence problem because for each feature in one view the search distance for the correspondence in the other view is generally large, although prior knowledge of the spatial relationship of the two viewpoints could reduce the originally 2D search to 1D search along the so-called epipolar lines [12].

With the above observations, we outlined in [8] a framework of combining the two vision cues, in which the affine projection model is used for the cameras. In contrast with previous work on stereo-motion like [21, 14, 25, 24], the framework emphasizes not on how to exploit the redundancy in the image data to boost the accuracy in 3D reconstruction, but on how to couple the two vision cues in a complementary way, so that their advantages are retained and their disadvantages removed. The framework relates motion correspondences to stereo correspondences, and allows inference from the former to the latter. Accurate 3D reconstruction was demonstrated, even with relatively short platform motion.

However, several points are yet to explore. First, only one stereo pair were used for 3D reconstruction. This is not entirely reasonable, as any stereo pair is as good as the other in the image data for recovering 3D information. Second, the platform motion was not recovered. This paper presents how to compute 3D structure and motion from all stereo pairs of images. It is demonstrated that not only could more accurate reconstruction results be obtained by using all image data, even false initial motion correspondences could be detected, and thus establishment of wrong stereo correspondences could be avoided by comparing the results from different stereo pairs.

2 Previous Work

Much has been done on stereo vision; good surveys can be found in [4, 10]. Yet due to the difficulty of its correspndence problem, it hasn't been widely used in industry and society.

Visual motion has also been well-studied; classical references are listed in [13, 22]. The correspondence problem is much simpler than that in stereo vision, as consecutive images are alike, which means a feature point could not move too far between consecutive images. Very good 3D reconstruction results have been obtained, for example in [19]. One drawback is that a long image sequence is required so as to have a wide enough triangulation for accurate 3D determination. Such a drawback is not unimportant, as the longer distance the camera needs to travel, the more probable are the needed assumptions (e.g., a stationary scene) violated.

Below a few works on motion analysis that are closely related to this work are outlined.

In an elegant work, Tomasi and Kanade [19] proposed a method for reconstructing 3D from an orthographically projected image sequence. It factorizes the image measurements of object points into shape and motion matrices through singular value decomposition (SVD). Later, Poelman and Kanade [17] extended the factorization method to the case of paraperspective projection, which produces more accurate results than the original method. The factorization approach uses a large number of image measurements to counteract the noise sensitivity of structure-from-motion. However, for accurate reconstruction, a long image sequence is needed. Extensive computational time is required for processing these images. Recently, Morita and Kanade [15] presented a sequential approach for the factorization method. The sequential approach is much faster than the original one, but a long image sequence is still required.

The motion cue under an unknown motion recovers the world only up to a scale factor. One way to remove this ambiguity is to use two cameras to take stereo pair of image sequences and to combine stereo and motion analyses. The redundancy in the image data – data for both stereo and motion cues – also has the potential of allowing 3D information to be recovered more accurately. A few studies [21, 14, 25, 1, 24, 16, 7] have looked into this so-called *stereo-motion* cue.

However, the focus of the above stereo-motion work was on the exploitation of the input data's redundancy in recovering 3D information. How the two vision cues complement each other and what can be gained by combining them have not been explicitly addressed. The emphasis of this work, in contrast, is to achieve a system with the following features:

- It recovers 3D information accurately even with relatively short image sequences; this is in principle possible since widely separated views are always in the stereo-motion data regardless of how short the sequences are.

- It does not require prior knowledge of the camera motion nor the assumption of a smooth motion; this frees the system from the effect of disturbances and uncertainty in the camera motion.

- Most importantly, the stereo and motion cues are integrated in a way that they are *complementary* to each other, so that both *simple correspondence* as well as *accurate reconstruction* are possible.

3 Stereo and Motion in Complement

3.1 The Motion Model

In [19] Tomasi and Kanade proposed an elegant discrete model for the motion cue. Below the model, with some variations to pave the way for further development, is described.

Suppose F image frames observing P points in space are available. Assume an affine camera. The image position $\mathbf{p}_{fp} = (u_{fp}, v_{fp})^T$ of point p (p = 1, 2, ..., P) in image frame f (f = 1, 2, ..., F), is related to its 3D position $\mathbf{P}_p = (x_p, y_p, z_p)^T$ (with reference to the last image frame: frame F), by

$$\mathbf{p}_{fp} = J_f \underbrace{\left[\begin{array}{ccc|c} & \mathbf{R}_f & & \mathbf{t}_f \\ \hline 0 & 0 & 0 & 1 \end{array}\right]}_{\mathbf{M}_f} \left[\begin{array}{c} \mathbf{P}_p \\ 1 \end{array}\right]$$

where J_f is the affine projection matrix (a 2×4 matrix), and $(\mathbf{R}_f, \mathbf{t}_f)$ are the rotational and translational relationships between image frame f and the last image frame F. By combining the image positions of all P object points in F image frames, we have

$$\underbrace{\left[\begin{array}{c} \vdots \\ \cdots \ \mathbf{p}_{fp} \ \cdots \\ \vdots \end{array}\right]}_{\mathbf{W}} = \underbrace{\left[\begin{array}{ccc} \ddots & & \bigcirc \\ & J_f & \\ \bigcirc & & \ddots \end{array}\right]}_{\mathbf{J}} \underbrace{\left[\begin{array}{c} \vdots \\ \mathbf{M}_f \\ \vdots \end{array}\right]}_{\mathbf{M}} \underbrace{\left[\begin{array}{c} \mathbf{P}_p \\ \cdots \ 1 \ \cdots \end{array}\right]}_{\mathbf{S}}$$

Here $\mathbf{W}$, $\mathbf{J}$, $\mathbf{M}$, $\mathbf{S}$ represent the image measurements, the image projection process, the camera motion, and the scene or the object structure respectively. Each row in $\mathbf{W}$ contains the u-coordinates or v-coordinates of image points the same image frame, while each column contains the observations over the same object point. Since $\mathbf{W}$ can be factorized into matrices involving dimension four, $\mathbf{W}$ is of rank at most four (it is exactly four under general motion and general 3D structure).

3.2 The Stereo-Motion Model

The above motion model has been applied successfully to recover 3D structure [19]. However, hundreds of frames are needed. If instead a stereo pair of cameras are available to acquire a stereo pair of image sequences, potentially even with relatively short motion the 3D structure can still be estimated accurately, since widely separated views are always in the image data.

The motion model could be extended to the stereo-motion problem in the following way. Suppose a rigid stereo setup consisting of two cameras: Cameras 1 and 2, are available to capture image data as the whole setup moves in space relative to a scene. As shown in Figure 1, let $(\bar{\mathbf{R}}, \bar{\mathbf{t}})$ be the rotational and translational relationships between the stereo cameras (which are invariant with the motion of the stereo setup), in the sense that the 3D coordinates of any point with respect to the two camera coordinates frames (of Cameras 1 and 2 respectively), $\mathbf{P}$ and $\mathbf{P}'$, are related by

$$\begin{bmatrix} \mathbf{P}' \\ 1 \end{bmatrix} = \underbrace{\begin{bmatrix} \bar{\mathbf{R}} & \bar{\mathbf{t}} \\ 0\ 0\ 0 & 1 \end{bmatrix}}_{\bar{\mathbf{M}}} \begin{bmatrix} \mathbf{P} \\ 1 \end{bmatrix}$$

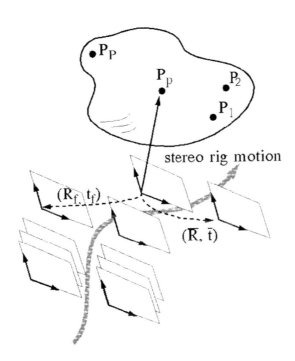

Figure 1: 3D Structure recovery from stereo-motion.

On applying Tomasi-Kanade's motion model to the two cameras separately, we have two image measurement matrices for feature points in the two cameras respectively:

$$\begin{aligned} \mathbf{W} &= \mathbf{JMS} \\ \mathbf{W}' &= \mathbf{J}'\mathbf{M}'\mathbf{S}' \end{aligned}$$

Here $\mathbf{W}', \mathbf{J}', \mathbf{S}'$ are matrices analogous to $\mathbf{W}, \mathbf{J}, \mathbf{S}$, but $\mathbf{W}', \mathbf{J}', \mathbf{S}'$ are with respect to the second camera.

Suppose stereo correspondences are established correctly across the two image sequences. This means feature points in $\mathbf{W}'$ and $\mathbf{S}'$ can be listed in the same left-to-right order of those in $\mathbf{W}$ and $\mathbf{S}$. If columns of $\mathbf{W}'$ and $\mathbf{S}'$ are so listed and $\mathbf{W}'$ is stacked beneath $\mathbf{W}$, a new image measurement matrix is obtained for the stereo-motion data:

$$\underbrace{\left[\begin{array}{c} \mathbf{W} \\ \hline \mathbf{W}' \end{array}\right]}_{\mathcal{W}} = \left[\begin{array}{c} \mathbf{JMS} \\ \hline \mathbf{J}'\mathbf{M}'\mathbf{S}' \end{array}\right] = \left[\begin{array}{c} \mathbf{JMS} \\ \hline \mathbf{J}'\widetilde{\mathbf{M}}\mathbf{MS} \end{array}\right]$$

$$= (\underbrace{\left[\begin{array}{c|c} \mathbf{J} & \bigcirc \\ \hline \bigcirc & \mathbf{J}' \end{array}\right]}_{\mathcal{J}} \underbrace{\left[\begin{array}{c} \mathbf{I}_{4F} \\ \hline \widetilde{\mathbf{M}} \end{array}\right]}_{\widetilde{\mathcal{M}}} \mathbf{M})\mathbf{S} \qquad (1)$$

where $\widetilde{\mathbf{M}}$ is a $4F \times 4F$ matrix representing the stereo camera geometry :

$$\widetilde{\mathbf{M}} = \left[\begin{array}{ccc} \ddots & & \bigcirc \\ & \bar{\mathbf{M}} & \\ \bigcirc & & \ddots \end{array}\right]$$

The matrices $\mathcal{W}$ and $\mathcal{J}$ are analogues of $\mathbf{W}$ and $\mathbf{J}$ (which are for single-camera motion) in stereo-motion. The matrix $\mathcal{W}$ (size: $4F \times P$) represents the image measurements from the stereo cameras with the stereo correspondences correctly established, and $\mathcal{J}$ (size: $4F \times 8F$) represents the image projection parameters of the stereo cameras. $\widetilde{\mathcal{M}}$, a term not present in the original motion model, is a $8F \times 4F$ matrix representing the geometry of the stereo camera setup. Notice that although here we have a stereo pair of cameras not one camera, the 3D structure term $\mathbf{S}$, like the counterpart in the motion model, is with reference to the camera coordinate frame of Camera 1 over the last image frame (i.e., Image F).

Since the factorization in Equation (1) involves matrices with dimension four, $\mathcal{W}$ in stereo-motion, like $\mathbf{W}$ or $\mathbf{W}'$ in single-camera motion, is of rank at most four and in general four (under general 3D structure and motion). Such a property is unlikely to be satisfied accidentally, as $\mathcal{W}$ is $4F \times P$ large; it is however satisfied when $\mathcal{W}$ is constructed under fully correct stereo matching. As will be discussed in the next section, the property allows stereo correspondences to be inferred from motion correspondences which are easier to obtain.

3.3 Inferring Stereo Correspondences from Motion Correspondences

Our stereo-motion system proceeds in the following way. A rigid stereo rig of cameras is constructed, and it undergoes a motion during which F pairs of images are taken from the cameras. Distinct feature points are then extracted independently from the two image sequences, and tracked in the two sequences separately.

We assume most of the estimated motion correspondences are correct since the image frames are dense and thus adjacent images are very much alike (however, we do allow mistakes in the motion correspondences, which are to be addressed in Section 3.6). With such motion correspondences the image measurement matrices $\mathbf{W}^*$ and $\mathbf{W}'^*$ for the two image sequences can be constructed. $\mathbf{W}^*$ and $\mathbf{W}'^*$ are in the same form as $\mathbf{W}$ and $\mathbf{W}'$, except that their columns are not necessarily properly ordered, i.e., stereo correspondences are not established yet. They may also have different number of columns, as feature points observable in one image sequence may not be observable in the other.

Our idea is to transfer the motion correspondences, which are easier to obtain, to stereo correspondences, which allows more accurate 3D reconstruction. Establishing stereo correspondences across the two images sequences is equivalent to matching columns of $\mathbf{W}^*$ with columns of $\mathbf{W}'^*$, so as to have matched pairs of columns to form the matrix $\mathcal{W}$ in Equation (1).

As $\mathcal{W}$ is of rank four, its column space is only a 4D subspace in a (4F)-dimension vector space, and all columns in $\mathcal{W}$ are linear combinations of 4 independent vectors. Suppose four basis vectors of $\mathcal{W}$ are available as $\mathbf{b}_1, \mathbf{b}_2, \mathbf{b}_3, \mathbf{b}_4$, and let $\mathbf{B}$ be $[\mathbf{b}_1, \mathbf{b}_2, \mathbf{b}_3, \mathbf{b}_4]$. Since $\mathbf{W}$ and $\mathbf{W}'$ are sub-matrices of $\mathcal{W}$ and of rank 4, the upper and lower sub-matrices of $\mathbf{B}$ — $\mathbf{B}_W$ and $\mathbf{B}_{W'}$ (size: $2F \times 4$) — are also matrices consisting of basis vectors for $\mathbf{W}$ and $\mathbf{W}'$ respectively.

Take any column in $\mathcal{W}$, which has $\mathbf{h}_W$ as its upper sub-column and $\mathbf{h}_{W'}$ as its lower sub-column. $\mathbf{h}_W$ represents a column of $\mathbf{W}$ that corresponds to the motion correspondence of a feature point in Image Sequence 1, and $\mathbf{h}_{W'}$ represents a column of $\mathbf{W}'$ that corresponds to the motion correspondence of the same feature point in Image Sequence 2. If there is a way to predict $\mathbf{h}_{W'}$ for every $\mathbf{h}_W$, the problem of inferring stereo correspondences is essentially solved.

It turns out if Basis $\mathbf{B}$ of $\mathcal{W}$ is available (and thus Basis $\mathbf{B}_W$ of $\mathbf{W}$ and Basis $\mathbf{B}_{W'}$ of $\mathbf{W}'$ as well), $\mathbf{h}_{W'}$ of $\mathbf{W}'$ could indeed be predicted for every $\mathbf{h}_W$ of $\mathbf{W}$ as:

$$\mathbf{h}_{W'} = \mathbf{B}_{W'}(\mathbf{B}_W^T \mathbf{B}_W)^{-1} \mathbf{B}_W^T \mathbf{h}_W \tag{2}$$

Derivation of the above formula is simply based upon the fact that the set of linear combination coefficients that generate $[\mathbf{h}_W^T, \mathbf{h}_{W'}^T]^T$ from Basis $\mathbf{B}$ also generate $\mathbf{h}_W$ from Basis $\mathbf{B}_W$, and $\mathbf{h}_{W'}$ from Basis $\mathbf{B}_{W'}$ as well.

In other words, given any column of $\mathbf{W}^*$, the corresponding column in $\mathbf{W}'^*$ can be predicted, provided that the basis vectors of $\mathcal{W}$ are known. The basis vectors can be formed if four linearly independent columns of $\mathcal{W}$ are available, which are equivalent to a minimum of four features matched across any stereo pair in the image data. Such initial correspondences may be obtained by epipolar constraint of stereo cameras. If more than 4 matches are available, a more accurate basis can be determine by the SVD technique [19].

Thus, the stereo-motion framework could use Equation (2) for inferring stereo correspondences from motion correspondences. With noise, the estimated column $\mathbf{h}_{W'}$ may not be exactly that in $\mathbf{W}'^*$, but should be quite close to it. A column is then selected from $\mathbf{W}'^*$ that has least-squares-error with it. This way stereo correspondences can be fully established, and $\mathbf{W}^*$ and $\mathbf{W}'^*$ can be organized to form $\mathbf{W}$ and $\mathbf{W}'$ and also $\mathcal{W}$. An input-output description of the inference mechanism is summarized in Figures 2.

Notice that once $\mathbf{B}$ is known, for each feature point whose motion correspondence in

Image Sequence 1 is $\mathbf{h}_W$, Equation (2) takes only an effort linear with respect to the length of $\mathbf{h}_W$, i.e., the total number of image frames, to compute. Thus, to compute the stereo correspondences of all feature points the inference process requires a computational complexity only linear with respect to the total number of feature points and image frames, which could not exceed the total size of the spatial-temporal volume of image data.

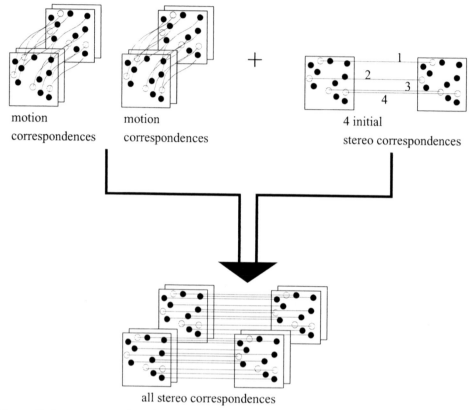

Figure 2: Input-output description of the stereo correspondence-inference mechanism.

To increase the accuracy of inference process, an iterative approach is employed in this work. At the end of each iteration, the matrix $\mathbf{B}$ is updated using the newly acquired stereo matches. The accuracy of both $\mathbf{B}$ and $\mathbf{h}_{W'}$ is thus improved. This makes the inferred vectors be closer to the actual ones and inferring more stereo correspondences be possible.

3.4   Computational Complexity of the Inference Process

Here a simple analysis of the computational complexity that it takes to infer stereo correspondences from motion correspondences is given. Computations for extracting features from images and tracking the features along motion frames are not considered, as they are not the contribution of this work, and any feature extraction and motion correspondence systems in the literature are just the same to be used in this work.

Suppose P features are extracted and matched in the two image sequences, and suppose each image sequence is of F image frames. The system first uses the epipolar geometry to extract a few initial stereo matches, spending $O(P)$ time. Most likely a little over four initial stereo correspondences, say I of them, are obtained. The system then extracts the optimal basis of a 4D space (the column space of $\mathcal{W}$) from the initial stereo correspondences, by applying SVD to a $4F \times I$ matrix. The SVD process takes $O(FI^2)$ (if $I \leqslant 4F$) or $O(IF^2)$ (if $I > 4F$) time, which can be assumed constant and negligible since both I and F are most likely small. Once the basis is available, the system infers columns of $\mathcal{W}$ one by one using a linear algorithm. Since each column has 4F entries, it takes $O(F)$ time to infer a column. Since there are altogether P columns, it takes $O(PF)$ time to infer all columns of $\mathcal{W}$.

In total, it takes $O(P) + O(PF) = O(PF)$ time to infer stereo correspondences from motion correspondences. The inference process is therefore linear with respect to the total number of features in each image and the total number of image frames in each image sequence.

3.5 Determining Motion and Structure

Once stereo correspondences are available, the motion of the mobile platform and the 3D structure of the scene are to be determined. Let $\mathbf{P}_{fp}$ be the 3D position of the p-th object point ($p = 1, 2, \ldots, P$) as determined by stereo pair f ($f = 1, 2, \ldots, F$) and with respect to the reference camera coordinate frame there (in this work the reference coordinate frame of any particular stereo pair is taken as the camera coordinate frame of Camera 1 over that stereo pair of images). Notice that $\mathbf{P}_{fp}$'s for all P points and all F stereo pairs are obtainable by applying the triangulation process to the stereo pairs respectively as outlined below. Suppose the extrinsic parameters of the stereo cameras are $(\bar{\mathbf{t}}, \bar{\mathbf{R}})$ (in the convention that $\mathbf{P}' = \bar{\mathbf{R}}\mathbf{P} + \bar{\mathbf{t}}$ for all 3D coordinates $\mathbf{P}$ and $\mathbf{P}'$ of the same spatial position with respect to the camera coordinate frames of Camera 1 and Camera 2 respectively), which are invariant with the stereo rig motion and could be estimated in an off-line process (e.g., [3]) or on-line process (e.g., [5]). Then for all corresponding image positions $\mathbf{p}_{fp}$ and $\mathbf{p}'_{fp}$ of the p-th feature point in the two images at time frame f, the 3D position of the spatial point with respect to the reference coordinate frame (the camera coordinate frame of Camera 1 here) at time frame f could be determined as

$$\mathbf{P}_{fp} = \frac{-(\bar{\mathbf{t}} \times \begin{bmatrix} \mathbf{p}'_{fp} \\ f' \end{bmatrix}) \cdot (\bar{\mathbf{R}} \begin{bmatrix} \mathbf{p}_{fp} \\ f \end{bmatrix} \times \begin{bmatrix} \mathbf{p}'_{fp} \\ f' \end{bmatrix})}{\|\bar{\mathbf{R}} \begin{bmatrix} \mathbf{p}_{fp} \\ f \end{bmatrix} \times \begin{bmatrix} \mathbf{p}'_{fp} \\ f' \end{bmatrix}\|^2} \begin{bmatrix} \mathbf{p}_{fp} \\ f \end{bmatrix}$$

where f and f' are the focal lengths of Cameras 1 and 2 respectively.

Notice that even though affine cameras are assumed in the course of inferring stereo correspondences, the above 3D determination assumes a full perspective projection model for the cameras. 3D determination in this work could thus still be of high accuracy.

There is however this issue. $\mathbf{P}_{fp}$'s from different stereo pairs may not be consistent with one another because of the different image noise and disturbances they contain. We could choose to use for simplicity only one stereo pair. We could also choose to integrate information from all these stereo pairs to determine the set of 3D positions $\mathbf{P}_p$'s (with respect to the last stereo pair, i.e., stereo pair F here) which best fit all the image data. Such $\mathbf{P}_p$'s are the optimal structure we want for the scene. Yet, to recover $\mathbf{P}_p$'s, we have to register the $\mathbf{P}_{fp}$'s from different stereo pairs by putting them under the same coordinate system for reference. This means we have to estimate the rigid transformations between the stereo pairs, which are actually the 3D motion of the mobile platform we also desire.

There are a number of ways to formulate the problem. Bundle adjustment [6, 20] is one. Below is a variant of bundle adjustment we used.

Suppose $\mathbf{R}_f^\star$ and $\mathbf{t}_f^\star$ represent the rotational and translational relationships between the reference coordinate frame of the stereo pair f and the reference coordinate frame of the last stereo pair F, such that $\mathbf{R}_f^\star \mathbf{P}_{fp} + \mathbf{t}_f^\star$ transfers the 3D coordinates of object point p from the coordinate frame in stereo pair f to the coordinate frame in the last stereo pair F. $\mathbf{R}_f^\star$'s and $\mathbf{t}_f^\star$'s for all stereo pairs collectively represent the platform motion. The optimal object positions $\mathbf{P}_p$'s ($p = 1, 2, \ldots, P$) and the motion $(\mathbf{R}_f^\star, \mathbf{t}_f^\star)$'s ($f = 1, 2, \ldots, (F-1)$) (it is known that $\mathbf{R}_F^\star = \mathbf{I}_3$ and $\mathbf{t}_F^\star = [0, 0, 0]^T$) are naturally the values that minimize the function

$$\sum_{f=1}^{F} \sum_{p=1}^{P} \| \mathbf{P}_p - (\mathbf{R}_f^\star \mathbf{P}_{fp} + \mathbf{t}_f^\star) \|^2$$

The optimization function is quadratic with respect to the unknowns in $\mathbf{P}_p$'s and $(\mathbf{R}_f^\star, \mathbf{t}_f^\star)$'s, and unique solution for them are available through a linear method. However, $\mathbf{R}_f^\star$'s so estimated are not necessarily orthonormal. To enforce that condition, a second non-linear optimization process is necessary which can employ the linear solution as the starting point for an iterative solution.

In this work, we choose to use a simpler way of estimating the 3D structure and motion. Experimental results show that the method does give very accurate results. The rationale behind the method is the following. The structure and the motion are related by

$$\mathbf{S}_f = \underbrace{\left[\begin{array}{ccc|c} & \mathbf{R}_f & & \mathbf{t}_f \\ \hline 0 & 0 & 0 & 1 \end{array} \right]}_{\mathbf{M}_f} \mathbf{S}^*$$

where $\mathbf{S}_f = \left[\cdots \begin{array}{c} \mathbf{P}_{fp} \\ 1 \end{array} \cdots \right]$ is the 3D structure computed from the f-th stereo pair, $\mathbf{S}^* = \left[\cdots \begin{array}{c} \mathbf{P}_p^* \\ 1 \end{array} \cdots \right]$ is the optimal object structure to determine (with respect to the last stereo pair F), and $\mathbf{M}_f$ represents the rigid transformation between the f-th stereo pair and the last stereo pair. Again, note that $\mathbf{S}_f$'s for all stereo pairs are available once stereo correspondences are obtained. Seeing that $\mathbf{M}_f$'s are readily recoverable from the above equation if $\mathbf{S}^*$ is known, we adopt the following iterative scheme for estimating the platform motion $\mathbf{M}_f$'s and the optimal structure $\mathbf{S}^*$ in an alternate fashion.

Initially, the optimal structure $\mathbf{S}^*$ is set equal to $\mathbf{S}_F$. The points sets $\mathbf{S}_f$ and $\mathbf{S}^*$ are then translated so that their geometric centers are located at the origin of the same coordinate system (in a way taking away the relative translation $\mathbf{t}_f$ and leaving only the relative rotation $\mathbf{R}_f$ of the two point sets). Using a unit quaternion representation [2] for $\mathbf{R}_f$, which assures the orthonormal property of the rotation matrix, we determine the quaternion by minimizing the squared distances between the two sets of points [11] using a least-squares-error method which has a closed-form solution [9]. The rotation matrix $\mathbf{R}_f$ is then reconstructed from the quaternion vector. Once $\mathbf{R}_f$ is determined, the translation vector $\mathbf{t}_f$ can be determined from

$$\mathbf{t}_f = \mathbf{C}_f - \mathbf{R}_f \mathbf{C}^*$$

where $\mathbf{C}^*$ and $\mathbf{C}_f$ are the centroids of $\mathbf{S}^*$ and $\mathbf{S}_f$ respectively. The above is done for every stereo pair f. This way we can estimate $\mathbf{M}_f$ for all stereo pair f when the optimal structure $\mathbf{S}^*$ is assumed to be $\mathbf{S}_F$.

In the next iteration we first determine a more accurate estimate for $\mathbf{S}^*$ by combining all structures $\mathbf{S}_f$'s from different stereo pairs. To do this, all $\mathbf{S}_f$'s have to be referenced under the same coordinate system, which can be done by multiplying them with the corresponding motion matrix $\mathbf{M}_f$ just estimated. We refine the optimal structure as the least-median-of-squared-error (LMedS) value of all these sets of points [23], i.e.,

$$\mathbf{S}^* = \mathbf{LMedS}\,(\mathbf{M}_1\mathbf{S}_1, \mathbf{M}_2\mathbf{S}_2, \ldots, \mathbf{M}_F\mathbf{S}_F)$$

The least-median-of-squared-error estimator finds a structure that yields the smallest median of squared error computed for the entire data set. Unlike the least-squares-error estimator, the LMedS approach is robust toward outliers caused by limited image resolution or occasional errors in motion correspondences and thus stereo correspondences inherited from them. With the refined value of $\mathbf{S}^*$, the motion matrices $\mathbf{M}_f$'s for all stereo pairs are once again estimated using the quaternion method described above.

The motion-and-structure-recovery procedure is repeated, having the optimal structure $\mathbf{S}^*$ and the motion matrices $\mathbf{M}_f$'s refined alternately, until the values of both $\mathbf{S}^*$ and $\mathbf{M}_f$'s are stable enough. In our experiments, typically less than three iterations are enough to generate accurate results. We conjecture that this is because the initial estimate of $\mathbf{S}^*$, $\mathbf{S}_F$, is quite close to the true value already.

3.6   Identifying Incorrect Motion Correspondences

The above framework assumes that prior motion correspondences are correctly established in the two image streams. However, feature tracker does give incorrect results occasionally, especially when the image feature is not very distinct or when there are repetitive patterns in the neighborhood. If the motion correspondences over a feature are wrong, the inferred position of the stereo correspondence will also be incorrect, and the system may establish a false stereo correspondence for this feature. So features with faulty motion correspondences should be discarded.

With the use of all stereo pairs faulty motion correspondences in isolated image frames can be easily identified. If the motion correspondence over a feature in a particular image frame is faulty, the stereo correspondence in the corresponding stereo pair will not be consistent with the stereo correspondences over the same feature in the other stereo pairs. To put it more precisely, the 3D reconstructions from the different stereo pairs simply would not align, and the variance of the 3D positions would be large. In our system, we simply discard features which have large variance in their 3D positions estimated from the various stereo pairs.

4   Experiments

The framework proposed in this paper has been implemented and tested with synthetic and real image data. The synthetic data experiments, in which ground truth about the 3D structure and motion is available, allow the performance of the system to be evaluated objectively. The real image data experiments on a variety of indoor and outdoor scenes, on the other hand, allows the robustness of the system (against all kinds of natural disturbances) to be examined.

4.1   Synthetic Image Data

A scene with a synthetic sphere initially 1.2 m away from the stereo cameras was simulated. In the scene the stereo cameras were 40 cm apart, each with a focal length of 60mm and a resolution of 500×500. The diameter of the sphere was 10 cm long, with 500 dots randomly distributed on it (Figure 3). In a duration of 15 image frames, the cameras moved along a seesaw-shape trajectory on a plane parallel to the image plane of Camera 1 at the first time instant. The cameras translated 5 mm along the path over each frame, and rotated a small angle to keep the object always in sight.

The random dots were treated as feature points and tracked along the image streams, with Gaussian noise of zero mean and 0.7 pixel variance added to them. Totally 204 points were successfully tracked in the two streams. The measurement matrices $\mathbf{W}^*$ and $\mathbf{W}'^*$ (size: 30×204) were then constructed. Under the epipolar constraint 33 unique matches were located across the two streams. The matrix basis was determined from these initial matches and other stereo correspondences were inferred and refined. In total 160 point correspondences were established.

The computed structure is shown in Figure 4. The recovered structure was very accurate. The root-mean-square distance error, which was mainly due to additive Gaussian noise to the image positions, was only 2 mm in Euclidean distance. The computed motion is shown in Figure 5. The recovered motion was close to the true motion, despite the challenge that for small motion the effect of camera rotation could appear very similar to that of camera translation.

In Figure 6, the stereo correspondence positions $\mathbf{h}_{W'}$ estimated from Equation (2) were compared with their true values under different lengths of the image streams. The differ-

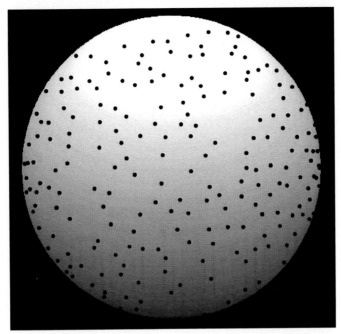

Figure 3: First image frame of synthetic data (a spherical surface).

ence was measured in average Euclidean distance between the predicted and true image positions. As expected, the prediction error decreased as the total number of image frames increased, and was less than 5 pixels already even if the image streams were only 11 frames long. This shows that the inference of stereo correspondences is satisfactory even with relatively short image sequences.

4.2   Real Image Data

The proposed method was also tested with various sets of indoor and outdoor image data.

In one experiment, a house model was to be reconstructed. Figure 7 shows the first and the last image pairs of the house model image sequences. The images were taken from a laboratory using a stereo rig of two CCD cameras, both with 50 mm lens, which translated approximately 1 cm sideways and rotated a small angle over each image frame. The house model was about 2m away from the cameras. The baseline of the cameras was about 58 cm. The camera optical axes were convergent, forming an angle of 17 degrees. The stereo cameras were first calibrated using the method described in [3]. Objects with orthogonal trihedral vertices were placed in front of the cameras and the relative geometry of the cameras was estimated off-linely from the image projections of the vertices.

Each camera captured 9 images in the entire duration. Image features were selected and tracked individually in both sequences, using a publicly available feature tracker [18]. The feature tracker automatically selected and tracked 300 features throughout both image sequences. A total of 15 unique stereo matches were located using the epipolar constraint and a simple correlation-based matching algorithm, which evaluates match candidates by the

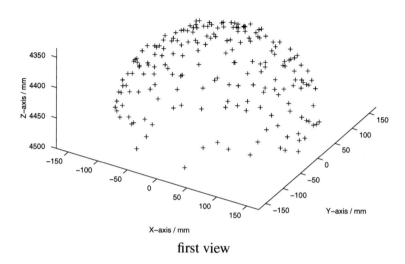

first view

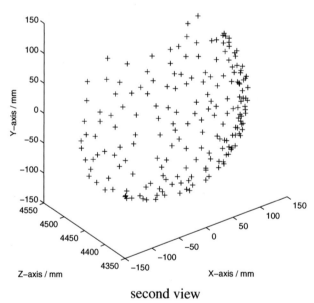

second view

Figure 4: 3D structure recovered from synthetic image sequences (a spherical surface).

mean square difference between the corresponding feature windows in two images. Other stereo correspondences of features were then predicted and refined. Altogether 167 features were successfully matched across the stereo image sequences in three iterations. The image positions of these matched features are overlaid on the first image pair in Figure 7. Figure 8

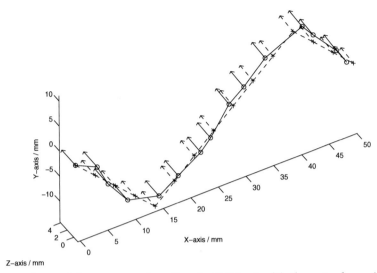

Figure 5: Comparison of the computed motion (solid line) with the actual one (dashed line) in synthetic images. The arrows show the directions of the camera optical axes.

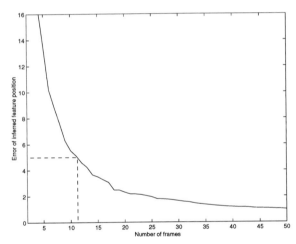

Figure 6: Deviation of inferred feature positions from their true values.

shows two different views of the computed 3D structure, as well as the actual house model for comparison. The positions of the roof, the chimneys, and the walls were all correctly recovered.

In another experiment, two disjointed objects, an oscilloscope and a soda can, were observed by the cameras (Figure 9). As the stereo-motion framework does not make any assumption on the object's shape, it is applicable to even disjointed objects, as long as the affine projection approximation is valid, i.e. the object's depth range is small compared with its average depth from the cameras.

The feature tracker established 300 motion correspondences in both sequences. A total of 19 initial stereo matches were found by the epipolar constraint. After three iterations, the system was able to infer 141 stereo correspondences. The overhead view of the reconstructed 3D structure is illustrated in Figure 10. The positions of the two objects are clearly separated. Points in the lower-left corner belong to the curved surface of the soda can. Points in the upper-half belong to the frontal surface and a side-wall of the oscilloscope. The five points in the center belong to the cable connected to the oscilloscope.

Figure 11 shows another set of image data. A set of 400 features were tracked in both sequences and 35 initial matches were established by the epipolar constraint. In total, 210 stereo correspondences were found in three iterations. Figure 12 shows two different views of the reconstructed bowl surface. To make reprojection possible, the object surface was approximated by a set of triangular patches, which were generated from the 3D object points by the Delaunay Triangulation method. The pixel values of the last image pairs were then mapped onto the corresponding patches in the new image. It can be seen that the reconstructed surface is smooth, as very little distortion is observed in the views. This shows that the 3D reconstruction is accurate, though relatively short image sequences were used in the experiment.

Figure 13 shows the images of an outdoor scene. The image streams were taken with two hand-held digital camcorders mounted on a metal bar. A person held the camcorders and walked sideways slowly in front of a Red-Cross building. Images were extracted from the two video streams every 0.25 seconds. Twenty frames were obtained in each sequence in a duration of 5 seconds.

A total of 500 features were tracked and 49 initial matches were found. The system was able to infer 282 stereo correspondences in three iterations. Over 80% of them were on the Red-Cross building. The inference was accurate. The deviation between the inferred position and the true one was less than 1 pixel for most of the feature points on the building. The top-view of the recovered 3D structure is shown in Figure 14. It can be seen that the position and the size of the Red-Cross building and the cars are correctly recovered. In the reconstruction, the angle between the two major walls of the building is about 115 degree, which is consistent with the physical measurement. There are a number of other buildings in the background, which were all far away (about 50-60m) from the cameras. Due to the limited resolution of the cameras for such long distance, the computed 3D structure there appear to be a bit noisy.

Table 1 shows the computational time requirements in inferring stereo correspondences from motion correspondences when the inference mechanism was run on an UltraSPARC-II machine. It should be noted that the time for detecting features in the images and tracking them along the motion sequences is not included; those are not the focuses of this research, and any feature detection and motion tracking algorithm will fit just as well to the proposed stereo-motion framework. It can be seen that the computational time was roughly proportional to the number of features to infer and the number of image frames involved.

Table 1: Computation time requirement of the Correspondence Inference Mechanism.

Experiment	No. Stereo Pairs	No. Point Corr. Established	No. Iterations	Computational Time (sec)
house model	9	167	3	1.67
Oscilloscope & soda can	9	141	3	1.60
bowl	9	210	3	2.32

first image (Camera 1)   first image (Camera 2)

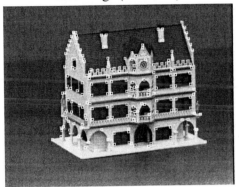

last image (Camera 1)   last image (Camera 2)

Figure 7: The first and the last image pairs in the "house model" image sequences. The sequences consist of 9 image pairs. The house model was about 2 m away from the cameras, which had a baseline of about 58 cm. The matched features are overlaid on the last image pairs, with the background darkened.

4.3 Motion Recovery

In this experiment the algorithm was tested with image sequences of a snack bag (Figure 15).

Each camera captured 9 images in the entire duration. 300 image features were detected

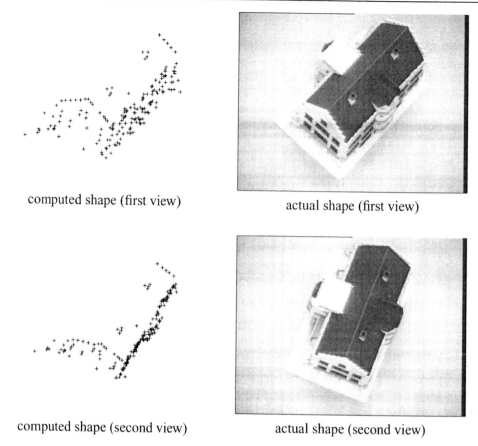

computed shape (first view)   actual shape (first view)

computed shape (second view)   actual shape (second view)

Figure 8: Reconstruction results of the house model.

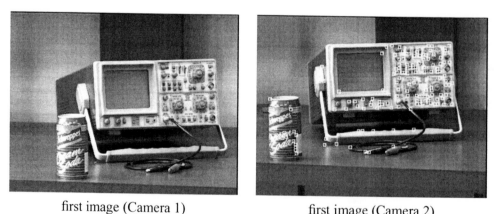

first image (Camera 1)   first image (Camera 2)

Figure 9: The first image pair in the "oscilloscope and soda can" image sequences. Each sequence consists of 9 frames. The objects were about 4 m from the cameras. The separation between cameras was about 60 cm. The stereo-rig motion was similar to that in the house model experiment.

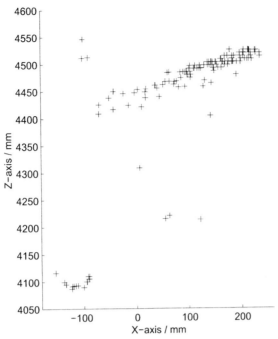

Figure 10: Top view of 3D structure recovered for the oscilloscope and soda can.

first image (Camera 1)   first image (Camera 2)

Figure 11: The first image pair in the "bowl" image sequences, which consist of 9 pairs of images. The bowl was about 1.2 m from the cameras, which had a baseline of about 40 cm. The stereo-rig motion was similar to that of the house model sequences.

and tracked in the two sequences independently, using the publicly available feature tracker [18]. A total of 19 unique stereo matches were located using the epipolar constraint and a simple correlation-based matching algorithm. Altogether 143 features were successfully matched across the two sequences. The computed motion is shown in Figure 16. No ground truth was available for the motion, but the direction and magnitude of the recovered motion were roughly as expected.

Several features were poorly tracked, and as a result the deviations of their computed

Figure 12: Two different views of the reconstructed bowl surface.

first image (Camera 1)   first image (Camera 2)

Figure 13: The first image pair in the "outdoor" image streams. The streams consist of 20 pairs of images. The separation between the cameras was about 60 cm.

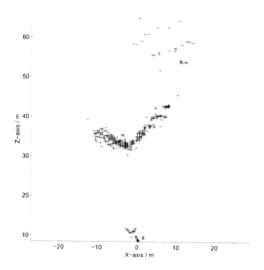

Figure 14: The top view of the computed 3D structure of the outdoor scene.

first image (Camera 1)   first image (Camera 2)

Figure 15: The first image pairs in the "snack bag" image sequences, which consist of 9 pairs of images. The snack bag was about 2 m away from the cameras, which had a baseline of about 58 cm. The camera optical axes were convergent, forming an angle of 8 degrees. The matched features are overlaid on the second image.

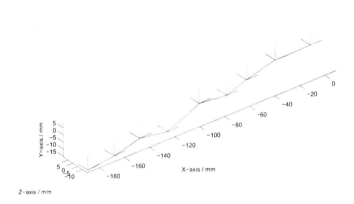

Figure 16: Recovered motion for the snack bag image sequences.

3D positions were rather large (Figure 17(a)). These features were discarded from the measurement matrix, because results obtained from these features would not be reliable. The object structure was then reconstructed from the remaining 137 features. The reconstruction was accurate, as can be seen in the reprojected images shown in Figure 18.

4.4   One-Stereo-Pair vs. All-Stereo-Pairs Reconstruction

To illustrate the improvement made by utilizing all stereo pairs (all-pairs-method), both the one-stereo-pair recovery and the all-stereo-pairs recovery methods were applied to the same synthetic image sequences used in Section 4.1. The shaded image of the spherical surface reconstructed from the two methods are shown in Figure 4. The surface recovered from the one-pair-method was rather rough; the error was due to the additive Gaussian noise in

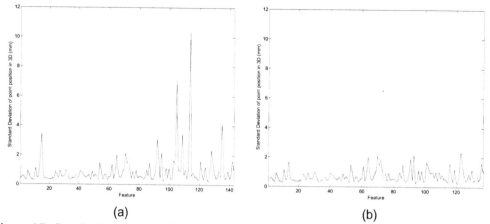

(a)   (b)

Figure 17: Standard deviation of 3D point position in the snack bag data: (a) before filtering, (b) after discarding unwanted features.

Figure 18: Reprojections of the snack bag's surface to viewpoints different from the original ones.

the image data. The all-pairs-method is more robust toward noise and produced a more accurate result. The root-mean-square error was 2mm for the all-pairs-method, which was about one third of the error from the one-pair-method.

Although the all-pairs-method produces more accurate results, it requires more computational effort. For the 'snack bag' experiment in Section 4.3, the all-pairs-method spent 54 seconds to compute the results, while the one-pair-method only required 1.7 seconds. The reason is that the all-pairs-method has to spend much time on calculating the quaternion vectors and the least-median-square-fit solution. The all-pairs-method is particularly suitable for applications which require high accuracy in the results and which allow off-line processing; an example application is scene reprojection.

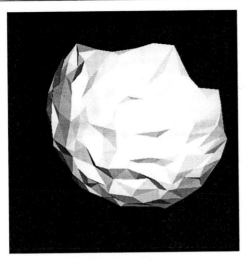 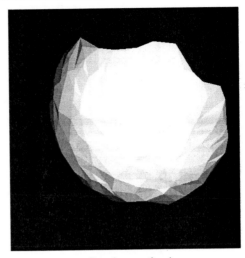

one-pair method   all-pairs method

Figure 19: 3D reconstruction results under the use of (a) one stereo pair, and (b) all stereo pairs of image data.

5   Conclusion

A framework of combining visual motion and stereo vision for reconstructing 3D structure has been described. The framework offers the advantages of both vision cues: simple correspondence, and accurate 3D reconstruction. Using the affine projection model for the cameras, the framework allows stereo correspondences to be inferred directly from motion correspondences in a computational effort linear with respect to the total size of the image data. The framework also recovers structure and motion at the same time. Notice the 3D recovery are based upon the full perspective camera model not the affine camera model, and thus the reconstruction accuracy is not compromised; the affine camera model is used only to trim down the search of stereo correspondences to small neighborhoods around certain predicted positions. The use of all stereo pairs of image data for the reconstruction could improve the accuracy further. The use of all stereo pairs in a robust estimation shell also allows errors in the initial motion correspondences to be detected.

In contrast with previous work, the approach does not use the smoothness assumption, but the affine camera assumption instead, to capture the inter-relationship of feature correspondences for establishing stereo correspondences.

Experiments show that the framework does not require long image sequences in the input for accurate 3D reconstruction. Typically, for objects of about 2m away from the cameras and a 3D reconstruction accuracy of 2mm root-mean-square error, only 9 image frames over a 9cm travel of the stereo rig are needed. This is an advantage, since with shorter image sequences there would be less chance for independent motion in the scene during the imaging period, and complications in the motion analysis would be reduced. The problem of occlusions and disocclusions during the movement of the cameras is also less severe with shorter image sequences.

Acknowledgment

The work described in this paper was substantially supported by a grant from the Research Grants Council of Hong Kong Special Administrative Region, China, under a project with Reference Number CUHK4114/97/E.

References

[1] P. Balasubramanyam and M. A. Snyder. The P-Field: A Computational Model for Binocular Motion Processing. In *Proceedings of the IEEE Conference on Computer Vision and Pattern Recognition*, pages 115–120, Maui, Hawaii, June 1991.

[2] O. Bottema and B. Roth. *Theoretical Kinematics*. North-Holland Pub. Co., New York, 1979.

[3] R. Chung and S.-k. Wong. Stereo Calibration from Correspondences of OTV projections. *IEE Proceedings: Vision, Image and Signal Processing*, **142**(5):289–296, October 1995.

[4] U. R. Dhond and J. K. Aggarwal. Structure from stereo–A review. *IEEE Transactions on Systems, Man & Cybernetics*, **19**(6):1489–1510, November/December 1989.

[5] F. Dornaika and R. Chung. Stereo Geometry from 3D Ego-motion Streams. *IEEE Transactions on Systems, Man & Cybernetics: Part B*, 2003. To appear in 2003. A preliminary version was published as Self-Calibration of a Stereo Rig without Stereo Correspondence, in Proceedings of the 1999 Conference on Vision Interface, May 1999, Quebec, pp. 264-271.

[6] Richard Hartley. Euclidean Reconstruction from Uncalibrated Views. In Joseph L. Mundy, Andrew Zisserman, and David Forsyth, editors, *Applications of Invariance in Computer Vision*, pages 237–256. Springer-Verlag, LNCS 825, 1994.

[7] A. Ho and T. Pong. Cooperative Fusion of Stereo and Motion. *Pattern Recognition*, January 1996.

[8] P.K. Ho and R. Chung. Stereo-Motion with Stereo and Motion in Complement. *IEEE Transactions on Pattern Analysis and Machine Intelligence*, **22**(2):215–220, February 2000.

[9] B. K. P. Horn. Closed-form solution of absolute orientation using unit quaternions. *Journal of the Optical Society of America*, **4**(4):629–642, 1987.

[10] G. A. Jones. Constraint, Optimization, and Hierarchy: Reviewing Stereoscopic Correspondence of Complex Features. *Computer Vision and Image Understanding*, **65**(1):57–78, January 1997.

[11] Simon K. Kearsley. An algorithm for the simultaneous superposition of a structural series. *Journal of Computational Chemistry*, **11**(10):1187–1192, 1990.

[12] T. J. Keating, P. R. Wolf, and F. L. Scarpace. An Improved Method of Digital Image Correlation. *Photogrammetric Engineering and Remote Sensing*, **41**(8):993–1002, August 1975.

[13] Stephen Maybank. *Theory of reconstruction from image motion*. Springer series in information sciences 28. Springer-Verlag, Berlin, 1993.

[14] A. Mitiche. A Computational Approach to the Fusion of Stereo and Kineopsis. In W. N. Martin and J. K. Aggarwal, editors, *Motion Understanding: Robot and Human Vision*, pages 81–95. Kluwer Academic Publishers, 1988.

[15] T. Morita and T. Kanade. A Sequential Factorization Method for Recovering Shape and Motion From Image Streams. *IEEE Transactions on Pattern Analysis and Machine Intelligence*, **19**(8):858–867, August 1997.

[16] M. Okutomi and T. Kanade. A Multiple-baseline Stereo. *IEEE Transactions on Pattern Analysis and Machine Intelligence*, **15**:353–363, 1993.

[17] C. J. Poelman and T. Kanade. A Paraperspective Factorization Method for Shape and Motion Recovery. In *Proceedings of the European Conference on Computer Vision*, pages 97–108, Stockholm, Sweden, May 1994. Springer-Verlag.

[18] J. Shi and C. Tomasi. Good Features to Track. In *Proceedings of the IEEE Conference on Computer Vision and Pattern Recognition*, pages 593–600, Seattle, Washington, June 1994.

[19] C. Tomasi and T. Kanade. Shape and Motion from image streams under orthography: A Factorization Method. *International Journal of Computer Vision*, **9**(2):137–154, 1992.

[20] Bill Triggs, Philip McLauchlan, Richard Hartley, and Andrew Fitzgibbon. Bundle Adjustment - A Modern Synthesis. In Bill Triggs, Andrew Zisserman, and Richard Szeliski, editors, *Vision Algorithms: Theory and Practice*, pages 298–375. Springer-Verlag, LNCS 1883, 2000.

[21] A. M. Waxman and J. H. Duncan. Binocular Image Flows: Steps toward stereo-motion fusion. *IEEE Transactions on Pattern Analysis and Machine Intelligence*, **8**:715–729, November 1986.

[22] J. Weng, T. S. Huang, and N. Ahuja. *Motion and structure from image sequences*. Springer series in information sciences 29. Springer-Verlag, Berlin, 1993.

[23] Gang Xu and Zhengyou Zhang. *Epipolar Geometry in Stereo, Motion and Object Recognition : a unified approach*. Kluwer Academic Publishers, Boston, 1996.

[24] Z. Zhang and O. D. Faugeras. Three-Dimensional Motion Computation and Object Segmentation in a Long Sequence of Stereo Frames. *International Journal of Computer Vision*, 7(3):211–241, 1992.

[25] Z. Zhang, O. D. Faugeras, and N. Ayache. Analysis of a sequence of stereo scenes containing multiple moving objects using ridigity constraints. In *Proceedings of the IEEE International Conference on Computer Vision*, Tampa, Florida, 1988.

In: Computer Vision and Robotics
Editor: John X. Liu, pp. 27-47

ISBN 1-59454-357-7
© 2006 Nova Science Publishers, Inc.

Chapter 2

SVD AND MATRIX POLYNOMIAL INTERPOLATION FOR LOSSY PROGRESSIVE TRANSMISSION OF 3D IMAGES*

I. Baeza, J.A. Verdoy and R.J. Villanueva[†]
Instituto de Matemática Multidisciplinar,
Universidad Politécnica de Valencia, Valencia, Spain
A. Law
Department of Clinical Neurosciences,
University of Calgary, Calgary, Canada

Abstract

This paper presents a new method for progressive transmission of 3D images that has four components: (1) decomposition of the image into regions using Singular Value Decomposition (SVD), (2) a reconstruction algorithm for progressive rendering that uses matrix polynomial interpolation along with approximations which are derived from SVD, (3) exploitation of a matrix norm for analyzing goodness of approximation, and (4) an optimal adaptive strategy for selecting "the next region to transmit".

SVD of matrices is used in some areas of image processing, such as restoration, but not usually in transmission. For an image (matrix) of size $m \times n$, its SVD produces an $m \times m$ matrix, an $n \times n$ matrix, and a vector of size $\min\{m,n\}$. That is, the SVD generates more than double the amount of original data. Despite this fact, however, a design of an appropriate adaptive transmission strategy within this four-component procedure provides an algorithm, for lossy progressive transmission, with excellent rendering and computational performance at low percentages of data transmission.

---

*This research has been supported by the Spanish Ministerio de Ciencia y Tecnología and FEDER grant TIC2002-02249, the Generalitat Valenciana grant GV04B/483, the Natural Sciences and Engineering Research Council and the Multiple Sclerosis Society of Canada, and the Seaman Family MR Research Centre in Calgary, Canada.

[†]To whom correspondence should be addressed (E-mail address: rjvillan@mat.upv.es.), J. Villanueva-Oller

1 Introduction

Traditional data transmission methods work in a sequential way; that is, data are compressed (or at least packed in some way) before being sent, and once received, data are decompressed (or unpacked). This approach has a main drawback: the data are only available once the whole data set has been received. See for example [15] and [17] for a wavelet approach, or [16] and [25] for further mathematical structures such as interpolation.

An alternative option which solves this limitation is the so-called *progressive transmission* [24]. Under this approach the data set is split in some way into many subsets, and the subsets are sent step by step. The receiver uses these incoming subsets (in fact, data regions) to perform a *progressive reconstruction*, that is, a partial reconstruction using only the available data at each step. Of course the evolving reconstruction will not be precisely the same as the original data set until all data are received, but a continually-improving approximation emerges since incoming data are added to those already received for on-going reconstruction. This process stops once the whole data set has been completely transmitted, or the receiver stops the process due to having achieved a suitable reconstruction quality.

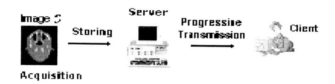

Figure 1: Simple client-server scheme for progressive transmission of digital data.

This progressive transmission pattern makes sense in environments where low bandwidth channels must deal with a large amounts of image data, or even when high bandwidth is available, with a number of massive images that are transmitting to completion [24]. A good example of the latter situation is the radiology department of a large hospital, where the amount of digital data (CT, MR, and digital RX) being processed for diagnosis is so huge that the hospital's entire local network can slow down. In such an environment, it may be beneficial to exploit resources at the receiver node and execute a progressive reconstruction process each time new data are received, rather than waiting the time needed to collect the entire image data set.

Several techniques for progressive transmission have been developed recently for 2-dimensional (2D) and 3D images: for instance, decorrelation techniques [14] or wavelets [15] and other approaches currently under investigation may be found in [21], [26] and [27]. Matrix polynomial structures are also being used for 3D progressive imaging [2, 4, 6, 7, 8].

Early strategies developed in the overall progressive transmission process were often static in the sense that they were independent of internal details, and hence the order of data transmission was fixed and determined once and for all. But, more recently adaptive strategies have been proposed, being adaptive in the sense that changes in data within the image might produce differently-ordered data subsets sequences for transmission. In precisely, such cases, the main goal is to select the most relevant data, previously manipulated if needed, in order to send first this information, then, the client generates good quality reconstructions with few, but relevant, data transmitted [3, 4]. As well, an adaptive strategy might be structured in such a way that its outcome changes if the rendering mechanism to be employed at the recipient node is changed: the evolving image the recipient observes could be responsive to what feature(s) might be being sought or investigated.

Singular Value Decomposition (SVD) of matrices is used in a wide variety of areas of image processing, such as restoration, noise reduction, computational tomography, image deblurring, and hidden transmission of digital images (steganography), [5, 11, 12, 13], where it is useful to remove the small singular values and obtain a matrix approximation whose rank equals the number of remaining singular values. SVD is also used as a part of the process of producing an image, but it is not usual to include it in the transmission.

The SVD transformation, just like Discrete Cosine Transform (DCT) for example, provides a lossy image compression, but it can achieve a higher compression rate. Although full SVD generates more than double the amount of original data, the range of 19% to 21% SVD matrix rank can yield a nondeteriorated reconstructible image, with no significant visual improvement observed beyond 21% [18]. Thus, despite the growth of the amount of data handled, it is possible to design an algorithm, for lossy progressive transmission of images, that has excellent rendering and computational performance at low percentages of data transmission: this is the main thrust of the development below. SVD will, in fact, provide a type of data image which is suitable for subsequent processing using matrix polynomial interpolation – mainly piecewise Newton matrix interpolation, a technique in 3D image reconstruction oriented to 2D slice transmission that has been tested in several recent papers [6, 7, 8].

This chapter is organized as follows. In Section 2 we recall some results about SVD, norms and how to reconstruct a matrix from its SVD. Section 3 discusses the general principles and what is needed to determine and define a process for progressive transmission of data. In Section 4 we apply the ideas developed in Section 3 to design a process for progressive transmission where SVD and piecewise Newton matrix polynomials play fundamental roles. A complete experiment using a data set composed of 93 CT slices of a human head is presented in Section 5. Finally, in Section 6, some conclusions and future developments are discussed.

It is convenient to note that one of our main goals is to define the process of progressive transmission by means of algorithms that allow for design of automatic implementations.

Throughout, we use the nouns "image" and "matrix" interchangeably when the latter is a numerical description whose visualization would be the former.

As well, $\mathbb{R}^{m \times n}$ shall denote the space of $m \times n$ real matrices.

2 On SVD

In this section, we recall some results about matrix theory and SVD. Details can be found in [9].

Definition 1 *If $A = (A_{ij})$ is an $m \times n$ matrix, then $rank(A) = p$, with $p \leq \min\{m, n\}$, is the number of rows (columns) of A that are linearly independent.*

Definition 2 *$U \in \mathbb{R}^{n \times n}$ is an orthogonal matrix if $U^T U = I_n$, where I_n is the identity matrix of order n.*

Theorem 3 (Singular Value Decomposition (SVD)) *[9] If $A \in \mathbb{R}^{m \times n}$ then there exist orthogonal matrices*

$$U = [u_1, \ldots, u_m] \in \mathbb{R}^{m \times m},$$

and

$$V = [v_1, \ldots, v_n] \in \mathbb{R}^{n \times n},$$

such that

$$U^T A V = diag(\sigma_1, \ldots, \sigma_p), \quad p = \min\{m, n\},$$

where

$$\sigma_1 \geq \sigma_2 \geq \cdots \geq \sigma_p \geq 0.$$

The σ_i are the singular values *of A and the column vectors u_i and v_i are, respectively, the i-th* left singular vector *and the i-th* right singular vector *of A.*

Let us define the 2-norm and the Frobenius norm. The first one will give bounds of certain useful approximations, and the second one will be used as a measure of goodness in the progressive transmission process. They are related.

Definition 4 *If $A = (A_{ij})$ is an $m \times n$ matrix and*

$$U^T A V = diag(\sigma_1, \ldots, \sigma_p), \quad p = \min\{m, n\},$$

is its SVD given by Theorem 3, we define the 2-norm of A by

$$\|A\|_2 = \sigma_1. \tag{1}$$

Definition 5 *If $A = (A_{ij})$ is an $m \times n$ matrix, then the Frobenius norm of A is [9] given by*

$$\|A\|_F = \sqrt{\sum_{i=1}^{m} \sum_{j=1}^{n} |A_{ij}|^2}. \tag{2}$$

It can be shown [9] that, if $A \in \mathbb{R}^{m \times n}$, and

$$U^T A V = diag(\sigma_1, \ldots, \sigma_p), \quad p = \min\{m, n\},$$

is its SVD given by Theorem 3, then

$$\|A\|_F^2 = \sigma_1^2 + \sigma_2^2 + \cdots + \sigma_p^2, \tag{3}$$

and, furthermore, from (1) and (3), the following relationship between the 2-norm and Frobenius norm holds:

$$\|A\|_2 \leq \|A\|_F \leq \sqrt{n} \|A\|_2. \tag{4}$$

The underlying idea to extract from expression (4) is that, if two matrices (images) are close in 2-norm, they are close in Frobenius norm, and vice-versa.

Corollary 6 *[9] Let the SVD of $A \in \mathbb{R}^{m \times n}$ be given by Theorem 3. If $k < r = rank(A)$ and*

$$A_k = \sigma_1 u_1 v_1^T + \sigma_2 u_2 v_2^T + \cdots + \sigma_k u_k v_k^T,$$

then

$$\min_{rank(B)=k} \|A - B\|_2 = \|A - A_k\|_2 = \sigma_{k+1}.$$

The above corollary implies that the best approximation to A using matrices of rank k is given by A_k, and that the approximation error is known, it is the $k+1$ singular value. Moreover, when k increases, the approximation is improved. SVD is usually used for approximating large matrices with a number of zero singular values, or under a certain, pre-fixed, threshold. The threshold is useful for removing irrelevant singular values and, hence, saving computations. These facts lead us to the following algorithm for iterative optimal approximation of a matrix from its SVD.

Algorithm 7 (Reconstruction of a matrix from its SVD) *Let $A \in \mathbb{R}^{m \times n}$ be a matrix with its SVD given by Theorem 3. Let $\sigma_1 \geq \cdots \geq \sigma_q \geq 0$ be the first q singular values of A and $\{u_1, \ldots, u_q\}$, $\{v_1, \ldots, v_q\}$ be their corresponding singular vector sets.*

1 *Approximate A by $A_1 = \sigma_1 u_1 v_1^T$.*

2 *FOR $i = 2, 3, \ldots, q$*

    2.1 *Improve the approximation, A_i, to A by $A_i = A_{i-1} + \sigma_i u_i v_i^T$.*

3 *END FOR*

Note that $\|A - A_q\|_2 = \sigma_{q+1}$, and if $q = p$, then $A_p = A$.

So far, a mathematical and algorithmic description of SVD have been summarized, but it is important to observe that SVD is a way for organizing the information of a matrix (image), from the most relevant, higher singular values and their associated singular vectors, to the lesser relevant, lower singular values and their associated singular vectors. In this sense, the first q singular values of A can be chosen so that

$$\sigma_1 \geq \cdots \geq \sigma_q \geq \varepsilon > \sigma_{q+1} \geq \cdots \geq \sigma_p \geq 0,$$

where $\varepsilon \geq 0$ is a prefixed threshold, and the singular values smaller than the threshold and their corresponding singular vectors are removed − they are assumed to not provide substantial improvement in the image reconstruction.

3   What is Needed to Define a Process for Progressive Transmission? General Method for Progressive Transmission of Data with Four Components (4CPT)

Let us suppose we have

- a sender, which has the data stored in a convenient way, and is ready to send data regions upon demand,

- a receiver, which requires a given image, and so performs a request to the sender,

- a channel for transmission, which is the way the data must be transmitted across,

- and a data set (image) to be sent from the sender node, through the channel, to the recipient node.

In order to develop a general process for progressive transmission, some of the constituent steps are considered first.

Progressive transmission is founded upon a decomposition of the image. The decomposition could be direct in the sense that it consist simply of (disjoint) subimages, or it could, instead, be based on previous manipulations of the image − in which case a decomposition using this latter structure might be better said to consist of "regions" of the image. For instance, consider a 2D image O : after applying the Haar wavelet transform to O, the regions are the corresponding wavelet coefficients. The segmentation method is not trivial, many are possible, and some segmentation methods can have advantages (or, on the other hand, disadvantages) compared with others. Moreover, a segmentation method working well with an image type might not be as good with another image type. Thus, the decision about what segmentation method should be used must be made before the transmission step. This component is called *decomposition* or *segmentation*.

A reconstruction scheme must be chosen as part of an overall progressive transmission process. Such a scheme should be appropriate to the receiver, in terms of, e.g., quality of images and computation time requirements. A more powerful receiver (in terms of CPU power) can use reconstruction algorithms a weaker receiver could not use. This component is called *reconstruction algorithm*.

It is also necessary to define a measure that allows us to know if the evolution of the progressive transmission is, or would be, adequate. Dealing with the "goodness" of a reconstruction method entails the difficult task of measuring the image quality. To measure the difference between images (the original and a reconstructed one), some comparative methods will be required. This component is called *measure*.

Finally, a progressive transmission not only requires splitting the data into regions, but also an order in which to send them, prefixed or determined by some strategy for selecting "the next region to transmit". A desirable transmission pattern is one capable of sending first the more relevant data region, so the receiver can perform a more accurate reconstruction in the first transmission steps, or perhaps even initiate an optimal transmission sequence. Determination of such an optimal sequence will be directly related to the accuracy level of the progressive reconstruction. This component is called *strategy for selecting the next region to transmit*.

The consideration of the above four components, jointly, is defined as a *General Method for Progressive Transmission of Data with Four Components (4CPT)*.

Summarizing, to define completely a specific process for progressive transmission, it is necessary to determine the four components of 4CPT, i.e., we need a method to decompose the image into regions, a reconstruction algorithm implemented at the receiver node, a measure for analyzing goodness of approximation, and a strategy for selecting the next region to transmit, preferably in some optimal (non-static) way.

Most of the known methods for progressive transmission of images [7, 14, 15, 21, 24, 27] can be defined as a 4CPT, with static strategies.

4 A 4CTP Based on SVD and Linear Piecewise Newton Matrix Interpolation

Following the components discussed in the above section, a process for progressive transmission of 3D digital images using SVD and linear piecewise Newton matrix interpolation is developed in this section.

4.1 Decomposition

The algorithm for decomposition presented below is based on the following idea: if O is a 3D image of size $r \times s \times t$, it can be decomposed into r 2D parallel slices of size $s \times t$ that form the 3D data set

$$\{O_1, O_2, \ldots, O_r\},$$

where O_i corresponds to slice i. Then, an SVD with a given threshold can be applied to each slice, thereby providing a decomposition of each slice into regions.

Algorithm 8 (Decomposition of a 3D image) *Let O be a 3D image of size $r \times s \times t$ and $\varepsilon \geq 0$ be a given threshold.*

1. *Divide O into r 2D parallel slices (matrices) O_i, $i = 1, 2, \ldots, r$, of size $s \times t$.*

2. *Apply SVD given by Theorem 3 to each O_i, $i = 1, 2, \ldots, r$. Let q_i be the index such that*

$$\sigma_1^i \geq \cdots \geq \sigma_{q_i}^i \geq \varepsilon > \sigma_{q_i+1}^i \geq \cdots \geq \sigma_p^i \geq 0, \quad p = \min\{s, t\},$$

and consider $\left\{u_1^i, \ldots, u_{q_i}^i\right\}$, $\left\{v_1^i, \ldots, v_{q_i}^i\right\}$ the corresponding singular vector sets.

3. *Define the region (i, j) as $reg(i, j) = \left\{u_j^i, \sigma_j^i, v_j^i\right\}$, $i = 1, 2, \ldots, r$, $j = 1, 2, \ldots, q_i$.*

As can be seen, the regions defined by Algorithm 8 are very different from the more usual subsets of a given image, as indicated above in the definition of region.

Now, two considerations are in order here. First, starting with a 3D image with $r \times s \times t$ data, the decomposition following Algorithm 8 generates $\sum_{i=1}^{r} q_i$ regions with $s + t + 1$ data, i.e.,

$$\left(\sum_{i=1}^{r} q_i\right)(s + t + 1), \tag{5}$$

and in the most unfavorable case where $\varepsilon = 0$ ($q_i = p$ for all i) and $p = s = t$ (square matrices), expression (5) is transformed into $2(r \times s \times t) + s \times t$, more than double the amount of original data. Second, the election of the threshold ε depends on the use of SVD or the type of matrices. A small threshold may, or may not, involve the removal of a large number of singular values. Therefore, depending on the number of removed singular values, the number of regions is reduced, and consequently so is the amount of data generated by Algorithm 8.

It is interesting to note that the SVD is usually computed numerically and therefore is affected by roundoff errors of computation. This implies that the use of thresholds in order to remove singular values smaller than the aforementioned threshold is very usual, and leads to methods of progressive transmission, where SVD is used, that are of *lossy progressive transmission* type: the corresponding information removed will not be sent and the image will not usually be reconstructed at the recipient node to be identical with original image.

4.2   Reconstruction Algorithm

The reconstruction algorithm $\mathcal{R}$ selected here is based on use of SVD and a linear piecewise matrix Newton interpolation method for progressive transmission, as developed recently in [8].

Algorithm 9 (Reconstruction algorithm $\mathcal{R}$) *Let O be a 3D image of size $r \times s \times t$ and $reg(i, j) = \left\{u_j^i, \sigma_j^i, v_j^i\right\}$, $i = 1, 2, \ldots, r$, $j = 1, 2, \ldots, q_i$, be the regions of O obtained from Algorithm 8 with a given threshold $\varepsilon \geq 0$. Suppose that in some stage $reg(i^*, j^*)$ is transmitted.*

1 *Compute a better approximation to O_{i^*} as*

$$\left[\sigma_1^i u_1^i \left(v_1^i\right)^T + \cdots + \sigma_{j^*-1}^i u_{j^*-1}^i \left(v_{j^*-1}^i\right)^T\right] + \sigma_{j^*}^i u_{j^*}^i \left(v_{j^*}^i\right)^T,$$

where the $j^* - 1$ first regions were already transmitted in previous stages (see Algorithm 7).

2 *Let $O_{i_1}^*, O_{i_2}^*, \ldots, O_{i_k}^*$, $i_1 < i_2 < \ldots < i_k$, be the approximate slices obtained in previous stages and in STEP 1.*

3 *If $\{i_1, i_2, \ldots, i_k\} = \{1, 2 \ldots, r\}$ no interpolation is needed. GOTO STEP 6.*

4 *Construct a matrix linear Newton interpolatory polynomial as follows*

$$
\begin{aligned}
P_j(x) &= O_{i_j}^* + \frac{O_{i_{j+1}}^* - O_{i_j}^*}{i_{j+1} - i_j} (x - i_j), \\
j &= 1, 2, \ldots, k-1,
\end{aligned}
\tag{6}
$$

and the piecewise matrix Newton polynomial for the jth segment in this transmission step by

$$P(x) = P_j(x) \text{ if } x \in [i_j, i_{j+1}].$$

5 *Compute $P(i)$, $i = 1, 2, \ldots, r$, to obtain a full set of r slices, k of them will be the original data $O_{i_1}^*, O_{i_2}^*, \ldots, O_{i_k}^*$, and $r - k$ of them are generated by the matrix polynomial $P(x)$.*

6 *END*

It is clear that the piecewise matrix Newton polynomial $P(x)$ (step 4 of Algorithm 9) changes in each step, but only in the "pieces" where a new region of the new transmission is inserted. So, steps 1, 2, 4 and 5 can be reduced by computing only with those slices where there have been changes from the last transmission.

4.3 Measure

Examples of measures which have been used in the literature include signal measures (entropy, signal-to-noise-ratio [10], peak signal noise ratio [23]), statistical errors (mean, standard deviation, variance, mean square error), and scalar, vector or matrix norms [9].

Using the Frobenius norm (2), let us define a measure μ that allows for analyzing the goodness of approximation: let O be a 3D image of size $r \times s \times t$, decomposed into r parallel slices O_i, $i = 1, 2, \ldots, r$, of size $s \times t$. Then,

$$\mu(O) = \sum_{i=1}^{r} \|O_i\|_F.
\tag{7}$$

This selection may seem strange at first, taking into account that SVD and the 2-norm have, jointly, interesting properties (see Section 2). But the Frobenius norm is easier and quicker to compute, and the expression (4) ensures that both perform in a similar way. Moreover, the Frobenius norm has been used effectively in recent papers [7, 8], and is independent on the image size.

The reconstruction algorithm $\mathcal{R}$ builds, from data of transmitted regions, a 3D image, approximating to the original one. But, it will be necessary to extract and compare slices of this 3D approximation. For this, the matrix operator determined simply through element-by-corresponding-element multiplication plays a basic role in the sequel.

Definition 10 *If $A = \left(a_{ijk}\right)$ and $B = \left(b_{ijk}\right)$ are two $r \times s \times t$ matrices, then $A \otimes B$ is the $r \times s \times t$ matrix given by*

$$A \otimes B = \left(a_{ijk}\, b_{ijk}\right).$$

Corresponding to an $r \times s \times t$ matrix, we now define an $r \times s \times t$ matrix companion to it. The companion matrix provides the tool to determine optimal strategies for progressive transmission.

Definition 11 *If $A = \left(a_{ijk}\right)$ is an $r \times s \times t$ matrix, then the companion matrix to A, $\delta(A)$, is the $r \times s \times t$ matrix $B = \left(b_{ijk}\right)$ for which*

$$b_{ijk} = \begin{cases} 0, & \text{if } a_{ijk} = 0 \\ 1, & \text{otherwise} \end{cases} \tag{8}$$

for all i, j, k.

It is important to note that if P is a reconstruction (i.e., some approximating matrix) for O, then

$$\mu\left(\delta(O_a) \otimes O - \delta(O_a) \otimes P\right),$$

estimates how well P approximates O "at" the slice identified by $\delta(O_a)$. In other words, this expression provides a numerical estimate of how close the reconstruction's slice, $\delta(O_a) \otimes P$, is to the corresponding slice, $\delta(O_a) \otimes O$, of the original image O.

4.4 Strategy for Selecting "the Next Region to Transmit" and Design of a Process for Progressive Transmission

The strategy for selecting "the next region to transmit" is a crucial decision in order to ensure effectiveness in the transmission and the reconstruction algorithm. An efficient strategy design depends on what is being sought.

The next algorithm introduces a process for progressive transmission based on decomposition of 3D images as given by Algorithm 8 (slices and SVD), and the reconstruction algorithm 9 (SVD and piecewise matrix Newton interpolation), where an adaptive strategy for selecting "the next region to transmit" is included in key steps. The main idea of this adaptive strategy is to transmit regions related to the worst approximated slices, to improve them. Notice that the order of region transmission is not prefixed, and it is constructed at each step, depending on the internal details of each image transmitted progressively.

The algorithm starts by sending initially the first region of slice 1, $reg(1,1)$, and the first region of the last slice, $reg(r,1)$. Hence, we use information about the two boundary regions as initialization, because of the poor approximation of slices corresponding to data that are outside of the interpolation interval.

Algorithm 12 (Progressive transmission with adaptive strategy) *Consider the 3D data set O of size $r \times s \times t$, $S = \{O_i\}_{i=1}^r$, its r 2D parallel slices of size $s \times t$, and its decomposition into regions*

$$\{reg(i,j), \quad i = 1, 2, \ldots, r, \quad j = 1, 2, \ldots, q_i\},$$

obtained by Algorithm 8. Let $\mathcal{R}$ be the reconstruction Algorithm 9, and consider the measure μ defined by expression (7).

1 *Set $\tilde{N} = (\sum_{i=1}^r q_i) \times r$ to the maximum possible number of transmissions.*

2 *Let $T_1 = \{reg(1,1), reg(r,1)\}$ (transmitted regions) and*

$$U_1 = \{reg(i,j), \quad i = 1, 2, \ldots, r, \quad j = 1, 2, \ldots, q_i\} \sim T_1,$$

(untransmitted regions).

3 *Let $V_O = ((1,1), (r,1))$ (the vector where the order of region transmission is stored).*

4 *FOR $k = 1, 2, 3, \ldots, \tilde{N}$ DO*

    4.1 *Find a slice O_{i^*} of S for which*

$$\max_{O_a \in S} \mu(\delta(O_a) \otimes \mathcal{R}(T_k) - \delta(O_a) \otimes O),$$

    is achieved.

    4.2 *Select for transmission the region $reg(i^*, j^*)$, the first untransmitted region of slice O_{i^*}.*

    4.2 *Set $T_{k+1} = T_k \cup \{reg(i^*, j^*)\}$ (transmit the next region $reg(i^*, j^*)$).*

    4.3 *Set $U_{k+1} = U_k \sim \{reg(i^*, j^*)\}$ (remove the transmitted region from the set of untransmitted).*

    4.4 *Append (i^*, j^*) at the end of V_O.*

    4.5 *Compute $\mathcal{R}_O^k = \mathcal{R}(T_{k+1})$ the approximate reconstruction of O at stage k.*

5 *END DO*

Although we consider $\tilde{N}$ in step 1 above as the maximum possible number of transmissions, it is not necessarily a prefixed quantity, because the process of progressive transmission may be terminated before all the data have been received.

The adaptive strategy is defined in steps 4.1 and 4.2 of Algorithm 12. Note that this adaptive strategy implies that the region $reg(i,j)$ cannot be transmitted if $reg(i,j-1)$ has not already been transmitted. This restriction allows a simple design of the strategy and avoids a large amount of computation.

The progressive reconstruction is done in step 4.5.

5 Experiment and Results

The previous section develops the general process for progressive transmission of images that has four components: 4CPT's constituents are decomposition, reconstruction, measure and strategy for transmission. These are now illustrated and discussed using a benchmark, CT head data set.

Consider the 3D data set O of Figure 2, a reconstruction of a human head using VTK [19, 20] from a CT scan consisting of 93 parallel slices. Each slice is a 2D CT image of size 256×256 whose pixels are 16-bit gray level.

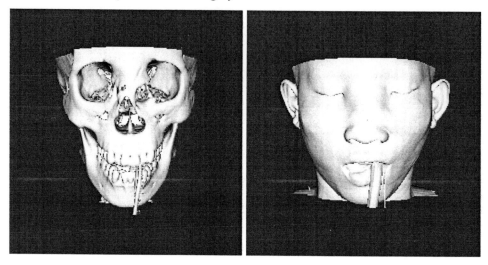

Figure 2: Bone- and skin-level renderings of the full, 93 CT slice, test data set.

The adaptive process for progressive transmission defined by Algorithms 8, 9 and 12 has been implemented in MATLAB 6.5.1 code in a PC Pentium IV, speed 2.4 Ghz, 512 MB of RAM, under Windows XP.

For decomposition, we apply Algorithm 8 to O, with threshold $\varepsilon = 10^{-9}$. After step 1 we obtain O_i, $i = 1, 2, \ldots, 93$, each 2D slice being a 256×256 matrix to which the SVD is applied in step 2. The CPU time for computing the SVD of the 93 slices is 100.4 seconds. After step 3, the obtained regions are

$$
\begin{array}{|c|c|c|c|}
\hline
O_1 & O_2 & \cdots & O_{93} \\
\hline
reg(1,1) & reg(2,1) & & reg(93,1) \\
reg(1,2) & reg(2,2) & & reg(93,2) \\
\vdots & \vdots & & \vdots \\
reg(1,q_1) & reg(2,q_2) & & reg(93,q_{93}) \\
\hline
\end{array}
\tag{9}
$$

where, in this experiment, $q_1 = q_2 = \ldots = q_{93} = 240$. This means that the original data consists of

$$93 \times 256 \times 256 = 6094848 \text{ values,}$$

and after Algorithm 8, the data consists of

$$93 \times 240 \times (256 + 1 + 256) = 11450160 \text{ values,}$$
$$93 \times 240 = 22320 \text{ region transmissions,}$$

(each region has $256 + 1 + 256 = 513$ values) which represents an increment of 187.86%. The threshold $\varepsilon = 10^{-9}$ saves for transmission

$$[93 \times 256 \times (256 + 1 + 256)] - [93 \times 240 \times (256 + 1 + 256)] = 763344 \text{ values,}$$
$$\text{or } 93 \times 16 = 1488 \text{ region transmissions.}$$

For all graphs presented below, we consider 6094848 as 100% of data (11881 region transmissions, approximately). Therefore, the horizontal axis in the graphs shows the interval $[0, 187.86]$, and it is easier to compare the goodness of the reconstruction and the amount of data transmitted.

The execution, Algorithm 12, starts by transmitting the first region of slice 1, $reg(1,1)$, and the first region of the last slice, $reg(93,1)$.

To illustrate the obtained results, we present 3D renderings using VTK [19, 20], at bone level (Figure 3) and skin level (Figure 4). Note that, despite the increasing of data due to SVD, we get good quality reconstructions at low percentages of data transmission – for instance the "15% data" image in Figure 3 compares well, visually, with the original bone level rendering in Figure 2.

The ordering transmission vector V_O is not included here, because of its length (22320 region transmissions), but it can be found via the web address *http://adesur.mat.upv.es/w3/ Mateo2004/*.

Graphic errors are presented in Figures 5 and 6. In Figure 5, the evolution of the error through to transmission of all data is shown. It is interesting to note the decreasing speed of the error when 0% to 30% of the data are transmitted.

Figure 6(A), shows the zoom of the beginning (from 0% to 25% of data transmitted), and Figure 6(B), the end (from 100% to 187.86% of data transmitted) of the graph in Figure 5. It should be noted in Figure 6(B) that the magnitude of the error is small compared with that corresponding to the interval when 0% to 75% of data transmitted. The evolution of the accumulated CPU time is showed in Figure 7. Note that the increasing of the CPU time is of linear order. Although the total CPU time is approximately 4 hours and half, it is worth noticing that the reconstruction algorithm creates a new reconstruction each time a new region of the 22319 is received. On the other hand, recalling that the effect of this 4CPT is essentially lossy, the transmission of the 30% of data transmission corresponds to 3564 regions, and the CPU time of reconstruction is 2722.44 seconds (45 minutes approximately). Time can be saved if some intermediate reconstructions are avoided. Each reconstruction is carried out in 0.77 seconds, approximately.

Other interesting analysis is to determine where is located the most relevant information of the image, i.e., the first to be transmitted. Figures 8, 9 and 10 address this for the data set being used: they give the frequency of transmitted regions belonging to corresponding slices, when 1%, 5% and 10% of the data are transmitted, respectively.

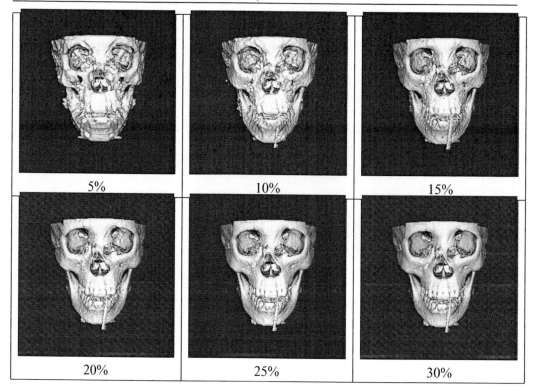

Figure 3: 3D renderings using VTK at bone level at several percentages of data transmitted.

The code and specific, detailed instructions to reproduce this experiment, results, time computation graphs, error graphs, regions, order transmission vector and visualization of 3D renderings at several percentages of data transmitted, can be found in the we address *http://adesur.mat.upv.es/w3/Mateo2004/*.

6   Conclusion

Section 3 provides a *General Process for Progressive Transmission of Data with Four Components (4CPT)*,

1. decomposition,

2. reconstruction algorithm,

3. measure, and

4. strategy for selecting the next region to transmit.

It presents an overall framework for designing processes for progressive transmission of data. The selection or decision taken in any one of the four affects the others. As well, in component 4, adaptive strategies may be introduced naturally as a part of the process.

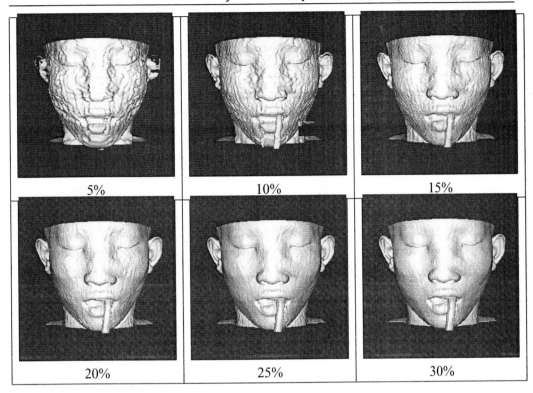

Figure 4: 3D renderings using VTK at skin level at several percentages of data transmitted.

Following 4CPT, Section 4 develops a specific, new scheme, for progressive transmission, wherein decomposition is based on the common planar slicing of the data set, but SVD of each slice along with reconstruction based on SVD plus linear piecewise matrix Newton interpolation is introduced. As well, the idea of adaptive strategies[1] is discussed, and then the four components are defined using algorithms to permit automating the overall process. A basic aim is to create a transparent process for future users in related implementations.

The experimental results, illustrated and discussed in Section 5, show promise despite the increase in data which accompanies application of SVD: good progressive approximations are obtained even when only low percentages of data are transmitted.

This general 4CPT method, and the specific new process developed and illustrated based on SVD, could be applied to other areas in image processing in which relevant information within the image can be detected and extracted.

Finally, as can be seen in [18], SVD may be applied for video compression, and it might be possible to reach higher compression rates using piecewise matrix Newton interpolation. Moreover, if the component *strategy for selecting the next region to transmit* is generalized to *strategy for selecting the next subcollection of regions to transmit*, given a prefixed compression rate, the 4CPT designed process selects the most relevant data, i.e., subcollection

---

[1]Adaptive strategies for progressive transmission of images have appeared recently in the literature [2, 3, 4] and they are providing an effective structure for evolving research in this field.

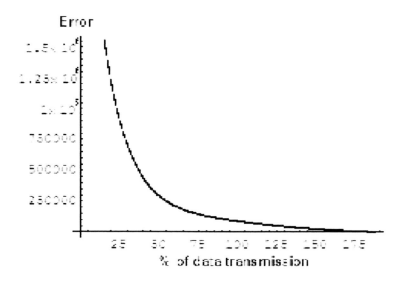

Figure 5: Graphic evolution of the error. Error is computed using the measure μ defined in (7).

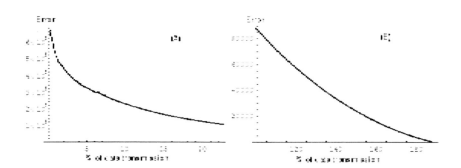

Figure 6: These graphs represent the first and the last part of Figure 5, where it can be seen the upper bound of the approximation error after sending $reg(1,1)$ and $reg(93,1)$, and the error after 100% of data transmission, can be seen.

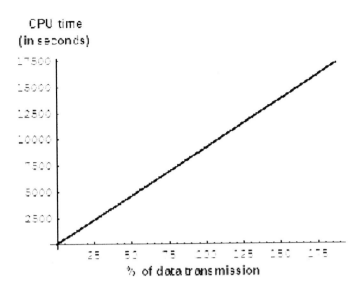

Figure 7: Evolution of the accumulated CPU time (in seconds), approximately 4 hours and half, of the whole process of reconstruction. Note that the reconstruction algorithm builds a new approximation for each of the 22319 region transmissions (in the first transmission, two regions are transmitted).

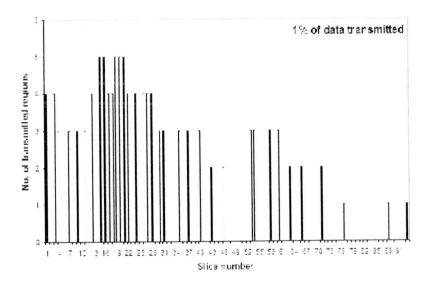

Figure 8: Frequency of regions transmitted, corresponding to slices used, when 1% of the total data (187.86% of the original amount of data) is transmitted. Note that the slices with more regions transmitted are in the interval between slice 15 and slice 28. We can say, therefore, that the most relevant information of the 93 CT head used in this experiment, at 1% of data transmitted, is located in the interval of slices 15-28.

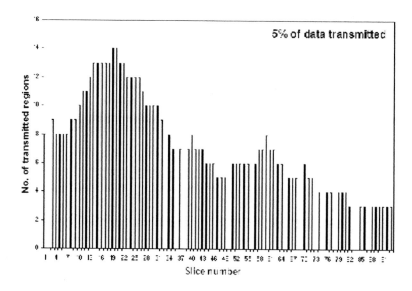

Figure 9: Frequency of regions transmitted, corresponding to slices used, when 5% of the total data (187.86% of the original amount of data) is transmitted. The interval where is located the most relevant information at 5%, has increased, compared with that of Figure 8, from slice 7 to slice 31. This is not unexpected since the interval of slices corresponds to a thicker part of the head, close to the the top.

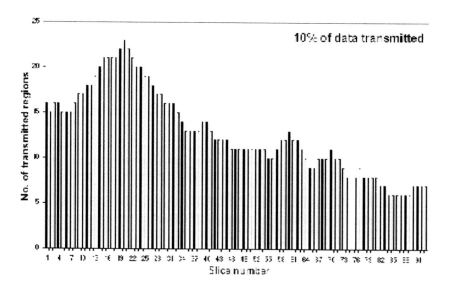

Figure 10: Consolidation of the fact that the most relevant information is located between slices 7 and 31. Note that, still, no regions of slice 75 have been sent – the first region of slice 75 is sent in transmission #1189, and after this latter, matrix Newton piecewise interpolation is not needed.

of regions corresponding at the prefixed compression rate. Preliminary approaches have been studied with adaptive strategies for progressive transmission of images at a prefixed compression rate [1], but, currently, computing needs are expensive, in a serial processing environment.

References

[1] Baeza, I. (July 2004). *Polinomios matriciales y wavelets para la transmisión progresiva adaptativa de imágenes digitales.* Ph.D. dissertation. Universidad Politécnica de Valencia.

[2] Baeza, I., Law, A.G., Villanueva-Oller, J., & Villanueva, R.J. (To appear 2004). Progressive transmission of images: Adaptive best strategies. *Mathematical and Computer Modelling.*

[3] Baeza, I., Villanueva, R.J., Law, A.G., Mitchell, J.R., Zhu, H., & Brown, R.A. (Submitted) Progressive Transmission of Digital Images: Four-step, Multidimensional Procedures. *Journal of Mathematical Imaging and Vision.*

[4] Baeza, I., Villanueva-Oller, J., Villanueva, R.J., & Díez, S. (2004) Envío y reconstrucción progresiva de imágenes digitales 3D utilizando interpolación matricial. *III Congreso Internacional sobre Métodos Numéricos en Ingeniería y Ciencias Aplicadas.* Monterrey (México).

[5] Banham, M.R., & Katsaggelos, A.K. (1997) Digital image restoration. *IEEE Signal Proc. Magazine*, Vol. 14, pp. 24-41.

[6] Defez, E., Hervás, A., Law, A.G., Villanueva-Oller, J., & Villanueva, R.J. (2000) Progressive transmission of images: PC-based computations using orthogonal matrix polynomials. *Mathematical and Computer Modelling*, Vol. 32, pp. 1125-1140.

[7] Defez, E., Law, A.G., Villanueva-Oller, J., & Villanueva, R.J. (2002) Matrix cubic splines for progressive transmission of images. *Journal of Mathematical Imaging and Vision,* Vol. 23, No. 1, pp. 41-53.

[8] Defez, E., Law, A.G., Villanueva-Oller, J., & Villanueva, R.J. (2002) Matrix Newton interpolation and progressive 3D imaging: PC-based computation. *Mathematical and Computer Modelling*, Vol. 35, No. 3-4, pp. 303-322.

[9] Golub, G.H., & van Loan, C.F. (1989) *Matrix Computations.* Baltimore, MA: Johns Hopkins Univ. Press.

[10] González, R.C., & Woods, R.E. (1993) *Digital Image Processing.* New York: Addison-Wesley.

[11] Hansen, P. C., Jacobsen, M., Rasmussen, J. M., & Sørensen, H. (2000) The PP-TSVD algorithm for image reconstruction problems. *In P. C. Hansen, B. H. Jacobsen and K. Mosegaard (Eds.), Methods and Applications of Inversion, Lecture Notes in Earth Science*, Vol. 92, pp. 171-186. Berlin: Springer.

[12] Hansen, P. C., & Jensen, S.H. (1998) FIR filter representation of reduced-rank noise reduction. *IEEE Trans. Signal Proc.*, Vol. 46, pp. 1737-1741.

[13] Johnson, N., & Jagodia, S. (February 1998) Exploring Steganography: Seeing the Unseen. *Computer*, pp.26-34.

[14] Kim, Y.-S., & Kim, W.-Y. (1998) Reversible decorrelation method for progressive transmission of 3D medical image. *IEEE Trans. Medical Imaging*, Vol. 17, No. 3, pp. 383-394.

[15] Kofidis, E., Kolokotronis, N., Vassilarakou, A., Theodoridis, S., & Cavouras, D. (1999) Wavelet-based medical image compression. *Future Generation Computer Systems,* Vol. 15, pp. 223-243.

[16] Lehmann, T., Gönner, C., & Spitzer, K. (1999) Interpolation Methods in Medical Image Processing. *IEEE Transactions on Medical Image Processing*, Vol. 18, No. 11, pp. 1049-1075.

[17] Mallat, S.G. (1980) A theory for multiresolution signal decomposition: The wavelet representation. *IEEE Trans. PAMI*, Vol. 11, No. 7, pp. 84-95.

[18] Sharifinejad, A. (2003) Utilizing Autoregressive Truncated Singular Value Decomposition algorithm for obtaining more efficiently Compressed Images. *Image and Vision Computing New Zealand 2003*, Massey University, Palmerston North, New Zealand.

[19] Schroeder, W., & Martin, K. (1999) *The VTK user's guide*, Kitware Inc.

[20] Schroeder, W., Martin, K., & Lorensen, B. (1997) *The visualization toolkit. An object-oriented approach to graphics*, Prentice-Hall.

[21] Sigitani, T., Iiguni, Y., & Maeda, H. (1999) Progressive cross-section display of 3D medical images. *Phys. Med. Biol.*, Vol. 44, No. 6, pp. 1565-1577.

[22] Stollnitz, E.J., DeRose, T.D., & Salesin, D.H. (1995) Wavelets for computer graphics: A primer, Part I. *IEEE Computer Graphics and Applications*, Vol. 15, No. 3, pp. 76-84.

[23] Strang, G., & Nguyen, T., (1996) *Wavelets and Filter Banks*, Wellesey-Cambridge Press.

[24] Tzou, K. (1987) Progressive image transmission: A review and comparison of techniques. *Opt. Eng.*, Vol. 26, pp. 581-589.

[25] Unser, M. (1999) A perfect fit for signal and image processing. *IEEE Signal Processing Magazine*, Vol. 16, No. 6, pp. 22-38.

[26] Wrazidlo, W., Brambs, H.J., Lederer, W., Schneider, S., Geiger, B., & Fischer, C. (2000) An alternative method of three-dimensional reconstruction from two-dimensional CT and MR data sets. *Med. Biol. Eng. Comput.*, Vol. 38, No. 2, pp. 140-149.

[27] Zhu, H., Brown, R.A., Villanueva, R.J., Villanueva-Oller, J., Lauzon, M.L., Mitchell, J.R., & Law, A.G. (2004) Progressive imaging: S-transform order. *ANZIAM J.*, Vol. 45(E), pp. C1002-C1016.

In: Computer Vision and Robotics
Editor: John X. Liu, pp. 49-97

ISBN 1-59454-357-7
© 2006 Nova Science Publishers, Inc.

Chapter 3

RANGE IMAGE REGISTRATION: A SURVEY

Yonghuai Liu[*]
Department of Computer Science, University of Wales,
Aberystwyth Ceredigion SY23 3DB, Wales, UK
Baogang Wei[†]
School of Computer Science, Zhejiang University
Huangzhou 310027, P. R. China

Abstract

Range image registration is a fundamental problem in range image analysis. The main task for range image registration is to establish point correspondences between overlapping range images to be registered. Since the relationship between point correspondences can be represented using motion parameters, in this chapter, we thus review the main image registration techniques from five aspects: motion representation, motion estimation, image registration using motion properties, image registration using both motion and structural properties, and the two-way constraint. In contrast with existing image registration review techniques, our review starts from the investigation of the relationship among point correspondences, motion parameters, and rigid constraints. Consequently, the review not only deepens our understanding about the relationship among motion parameters, rigid constraints, and point correspondences, but also possibly identifies the potential culprit for false matches in the process of image registration and points out future research direction to range image registration.

1 Introduction

1.1 Problem Statement

Traditional TV and modern digital cameras simulate single human eyes, outputting projective images. In contrast, laser scanning systems (range camera, laser range finder) (Figure 1)

---

[*]E-mail address: yyl@aber.ac.uk
[†]E-mail address: wbg@zju.edu.cn

simulate not only both human eyes, but human brains as well. The 3D imaging geometry of laser scanning systems essentially creates a stereo vision and/or performs some function of human brains through performing and post-processing some measures of interest. As a result, laser scanning systems output range images, which depict 3D information of object with free form surface. Since the laser scanning systems have limited field of view, a number of images have to be taken from different viewpoints so that a full coverage of the object surface can be obtained. All these images are depicted in local laser scanning system centred coordinate frames. For the construction of a full model of the object, all these images have to be aligned in a single global coordinate frame. This process is called registration.

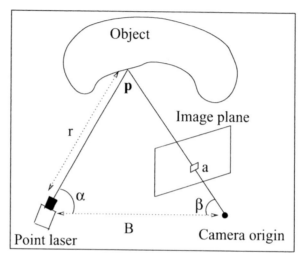

Figure 1: Diagram for principles of a structured light system operation [100].

It is likely in the future that laser scanning systems become essential components of robotic systems due to the fact that laser scanning systems directly capture depth information and the recovery of depth information from projective images is often sensitive to noise caused by sensor mechanics, instrument electronics, surface orientation, or reflective properties. However, before this conjecture comes true, we still have a number of challenges to tackle. The automatic registration of overlapping range images is one of these challenges due to partial overlapping between the images to be registered and the number of degrees of freedom with respect to the registration parameters [68].

Range image registration is a fundamental problem for range image analysis, as it finds applications in many areas such as, for instance, object modelling [51], 3D map construction [57], and simultaneous localization and map building (SLAM) problem [128]. Range image registration has two goals: one is to determine correspondences between different data sets representing the same object surface from different viewpoints, the other is to estimate the motion parameters bringing one data set into alignment with the other. In practice, these two goals are often interwoven, thus complicating range image registration.

1.2 Milestones in the Quest of Machine Vision

Of the five senses – vision, hearing, smell, taste, and touch – vision is undoubtedly the one that man has come to depend upon above all others, and indeed the one that provides most of that data he receives [25]. Thus to endow machines with the ability to understand and interpret images taken from different viewpoints, or from different time instants, or even using different modalities and to recognise objects in the images has been the goal of the machine vision community during the last five decades. Because human beings depend on the eye to capture images, machine vision research naturally has followed a similar method to implement simulated vision systems. This leads to 2D projective images being the most widely used representation to simulate human vision especially in the early stages of development of machine vision systems. During the last two decades, significant progress and important milestones have been achieved towards the goal of developing machine's ability to see and interpret the world.

This chapter has identified five milestones in the quest of machine vision. The **first** milestone in machine vision analysis can be considered the work of Longuet-Higgins in 1981 [86] and Tsai and Huang in 1984 [130] who proposed a perfect linear algorithm for motion estimation from 2D projective images and the interpretation of 2D projective images based on epipolar geometry. This algorithm is linear and mathematically perfect and efficient for real applications assuming that the correspondences between the two overlapping projective images have been established. Unfortunately, such 2D-2D motion estimation and structural analysis is sensitive to noise and prone to multiplicity of solutions [56].

Research has since changed the focus of attention from interpretation of 2D projective images to interpretation of range images. Range images explicitly provide depth information which greatly reduces the complexity of image understanding and interpretation. Moreover, at this time, the development of laser, sound, electronics, and other technologies make it realistic to directly capture the depth information about the scene studied [35, 108]. The **second** milestone can be considered the work of Arun et al. in 1987 [2] who proposed the constraint least squares (CLS) method for motion estimation from range images with available correspondence knowledge. This algorithm is also mathematically perfect, accurate, robust, and efficient in applications [31, 93]. However, this algorithm assumes that knowledge about correspondences between the two range images have been established. The actual situation especially for hand-held cameras is that even though the two images have been taken with some overlapping, the actual quantities of overlapping and the knowledge about the distribution of points, occlusion, appearance and disappearance of points are often not available. Even though the motion information can be approximately acquired through imaging devices, this, however, often limits the application of range cameras. Thus, the establishment of correspondences between the range images plays a key role in understanding and interpreting range images.

The **third** milestone can be considered the work of Besl and McKay [8], Zhang [141], Chen and Medioni [16], and Menq, Yau, and Lai [96] who in 1992 independently proposed an efficient algorithm called iterative closest point (ICP) algorithm for the registration of range images. Since the ICP criterion is efficient, of general use, and easy to implement for the establishment of possible point correspondences between overlapping range images to be registered, it has become a *de facto* standard approach to image registration. However, unless initial motion parameters are accurate and the image points are well distributed, the ICP criterion inevitably introduces false matches, since a single distance constraint generally cannot completely determine the position of a 3D point. Clearly, the assumption of the ICP criterion is strict and often violated in practice. Thus the establishment of correspondences is still a challenging task for the machine vision community [15].

While motion estimation from 2D projective images has been proven to be more difficult and that from range images has been proven to be easier, there is a mixture between them which is motion estimation from the correspondences between a 3D model and its 2D projective image (3D-2D problem). The **fourth** milestone can be considered the work of Quan and Lan in 1998 and 1999 [104, 105] who proposed a linear algorithm for the 3D-2D problem. This algorithm is based on distance constraints and it is accurate and efficient in real applications.

With the technological development in optics and electronics in the last two decades, range image acquisition systems has become more affordable, accurate, and acceptable [108] and thus, a significant amount of attention from the machine vision communities has been attracted. A breakthrough has been eventually made for the establishment of correspondences between two overlapping range images to be registered. Thus, the **fifth** milestone can be considered the work of Gold and Rangarajan [43, 44] who proposed the softassign algorithm for range image registration. The softassign algorithm explicitly models outliers in the process of image registration, applies the Sinkhorn iterative alternate row and column normalisation procedure [124] to enforce the two-way constraint, and employs the deterministic annealing scheme to ensure that good results can be obtained. Thus, the softassign algorithm is theoretically elegant and practically of general use.

The above analysis has shown that the establishment of correspondences between two 2D projective images, two range images, and a 3D model and a 2D projective image is of vital importance for the understanding and interpretation of images and the recognition of objects from their images. Once the correspondences have been established, the motion estimation and image interpretation problem is relatively easier to be solved by the methods mentioned above.

1.3   Classification of Registration Methods

An overall analysis reveals that logically, there are two kinds of information that can be used to register two overlapping range images: one is the information extracted from each image and the other is the information connecting points in different images. But these kinds of information cannot guarantee that the established point correspondences between different images are real. Thus, the key to successfully apply this two kinds of information to register overlapping range images is to eliminate false point matches. However, under special conditions, the point correspondences between different images can be directly determined without the need to evaluate them. Thus the existing registration methods can be classified into the following three basic categories, or a combination of them (Figure 2) [81, 82]:

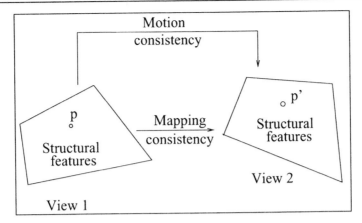

Figure 2: The classification of existing registration methods.

1. Structure consistency based methods [58, 61, 62, 134, 116, 138, 139]. This class of methods have dominated image registration and matching techniques. The characteristics of this class of methods lie in that they assume that satisfying structural constraints is a necessary condition for a pair of points to represent a real correspondence. Thus, they first extract some structural features from each image and then examine the consistency between these features from different images to establish or evaluate the possible point matches between different images to be registered. The structural features extracted are independent of rigid motion parameters. The methods in this class have to deal with four primary problems [134]: (1) the features to be extracted from images are expressive in representing different views of objects; (2) the features to be extracted from images are robust to noise, occlusion, and appearance and disappearance of points; (3) the similarity metric is powerful in discriminating different features; and (4) since the features attached to a point in one image have to match those attached to each candidate in another, the established point matches could be completely wrong. Solving any of these problems is a challenging task;

2. Motion consistency based methods [84, 76, 107]. This class of methods is relatively new and needs to be further explored. The characteristics of this class of methods lie in that they assume that satisfying rigid motion constraints is a necessary condition for a pair of points to represent a real correspondence. Thus, a set of possible point matches is first established by the traditional ICP criterion, then the quality of point matches is measured based on rigid motion constraints and finally false matches resulted from occlusion and appearance and disappearance of points are rejected based on their quality measures. Rigid motion constraints describe the relationship between each point correspondence and the motion parameters of interest and bridge the points described in two different coordinate frames before and after a rigid motion. Since the algorithms based on global rigid motion constraints can make full use of redundant data points for the estimation of the motion parameters of interest, they are in

general accurate and robust. However, some parameters necessary for rejecting false point matches must be properly defined; and finally

3. Mapping consistency based methods [8, 70]. The characteristics of this class of methods lie in that they directly use a certain mapping to establish point correspondences between the images to be registered without the need to evaluate them. In order to make sure that the established correspondences are feasible, the existing methods in this class often impose some constraints on image data acquisition or require image pre-processing. For example, the traditional ICP algorithm described in [8] assumes that one set of data points is a subset of another in 3D space. The disadvantage of the existing methods in this class is that their assumptions are often restrictive.

To our knowledge, while existing structural consistency based methods never ensure that the established point correspondences between different images to be registered represent real correspondences, satisfying exactly global rigid motion constraints is also a sufficient condition for a pair of points to represent a real correspondence. All these methods have their own advantages and disadvantages and can succeed in one situation, but degrade in another. Thus, it is always attractive and desirable to develop methods that are robust for general use.

1.4   Related Work

Sabata and Aggarwal in 1991 [111] published an excellent review paper on the motion estimation from images with available correspondences. They classify the techniques into two categories: local features and global features.

In 1992, Brown reviewed the 2D projective image registration [11]. He evaluated the registration techniques from four aspects: feature space, a similarity metric, search space, and search strategy and classified the techniques into four categories: correlation and sequential methods, Fourier methods, point mapping methods, and finally elastic model-based matching.

Huang and Netravall in 1994 [56] published another excellent review paper on the motion estimation from images with available correspondences. They evaluated the techniques based on point and line correspondences for motion estimation in problem formulation, efficient algorithms for solution, and sensitivity to noise.

Maurer and Fitzpatrick in 1993 [94] presented an excellent review about medical image registration. They first clarify the difference between registration, reformatting, and rendering, then classify the registration techniques into: (1) stereotactic frame systems; (2) point methods; (3) curve and surface methods; (4) moment and principal axes methods; (5) correlation methods; (6) interactive methods; and (7) atlas methods.

Elsen et al. in 1993 [33] presented another extensive classification scheme for medical image registration. They classify techniques according to a number of criteria: (1) dimensionality (1D vs. 2D vs. 3D vs. 4D); (2) type of features used for registration (intrinsic or extrinsic); (3) domain of the transformation (local or global); (4) type of transformation (rigid vs. affine vs. projective vs. curved); (5) parameters determination (search vs. closed-form solution); and (6) interaction (interactive vs. semi-automatic vs. automatic).

Recently, Maintz and Viergever [92] review medical image registration techniques from the following aspects: dimensionality, nature of registration basis, nature of transformation, domain of transformation, interaction, optimisation procedure, modalities involved, subject, and object.

1.5   Organization of this Chapter

In this chapter, we review the range image registration techniques from the following five aspects: (1) motion representation, (2) the application of motion representation for motion estimation from given correspondences, (3) the application of motion representation for image registration, (4) the combination of structural constraints with motion constraints for image registration, and (5) the two-way constraint for range image registration.

Our framework for the review of range image registration techniques is inspired by the following facts. Since point correspondences in different coordinate systems are connected by the motion parameters, the motion representation must be able to reveal the inherent relationship among point correspondences, distance constraint, rotation constraint, and projection constraint. Consequently, once the point correspondences are given, then the motion representation techniques can be used to estimate the intermediate motion parameters of interest, leading to the final motion estimation. Since point correspondences must satisfy the motion constraints, the motion constraints can then be used to reject false matches that do not satisfy the motion constraints. The image registration algorithms based on motion representation techniques do not make use of information in the structured image data. Research has shown that structural information can be used to enhance the performance of the registration algorithms [22] that do not use neighbouring information of points. All these existing techniques did not consider the two-way constraint for image registration. Finally, we review the registration techniques based on the two-way constraint and relative optimisation techniques.

Through the review of range image registration techniques, the audience are expected to have a clear idea about the state of the art of motion representation, motion estimation and image registration, which help them to choose appropriate techniques for practical robotic applications.

The rest of this chapter is organised as follows: for completeness, we first review the four classical algorithms for the motion estimation from given correspondence data. Then Section 3 reviews the motion representation approaches, Section 4 reviews approaches for motion estimation based on motion properties, Section 5 reviews the approaches for range image registration using the motion properties, Section 6 reviews approaches for the improvement of performance of range image registration algorithms using both motion and structural properties, while Section 7 reviews approaches for range image registration based on the two-way constraint. Finally, Section 8 draws some conclusions.

2 Classical Algorithms for Motion Estimation from Given Correspondences

Image registration can be broadly classified into two categories: one is based on image correspondences and the other is not. Generally, once knowledge about correspondences is available (that is, the positions of feature points are known in both images before and after the motion), then the motion estimation from these correspondences is relatively easy. In this section, we review four classical algorithms for this purpose: constraint least squares, quaternion method, orthonormal matrices method, and dual quaternion method.

2.1 The Constraint Least Squares (CLS) Method

Schonemann in 1966 [113] proposed a general solution to the orthonormal procrustes problem which transforms a matrix into another by an orthonormal matrix. His solution is based on eigenstructure analysis which actually is the earliest explicit version of the CLS method assuming that the correspondences between images are available.

In 1975, Gower investigated how to compare a number of configurations [48]. In order to make the comparison sensible, the relationships between different configurations are modelled as rotation, translation, and scaling. Then he transformed the global comparison between different configurations to be a pair-wise comparison between each configuration and their consensus configuration. His work also implicitly gave an early version of the CLS method for motion estimation from two images with available correspondences.

Sibson in 1978 [121] published the first in a series of papers about multidimensional scaling. Multidimensional scaling investigates the difference between two configurations assuming that the point correspondences between these two configurations have been pre-assigned. In order to do so, optimal matching of the two configurations under the operations such as translation, rotation/reflection, and scale change has to be made. The process of constructing such a matching is known as a procrustes analysis. The analysis of sum of squared fitting error is a procrustes statistics. When Sibson analysed the procrustes statistics, he implicitly provided the early version of the CLS method as well as orthonormal matrices method for the fit of two configurations.

Arun et al. [2] were the first to propose the singular value decomposition method for the estimation of motion parameters in computer vision. Later Umeyama in 1991 [132], Kanatani in 1994 [65], Goryn et al. in 1995 [46] refined the method so that a proper rotation matrix is always returned. This method is outlined as follows: first, centring all correspondence points $(\mathbf{p}_i, \mathbf{p}'_i)$ $(i = 1, 2, \cdots, n)$ where $\mathbf{p}_i$ is a point described in a coordinate frame before the motion and $\mathbf{p}'_i$ is its correspondent described in another coordinate frame after the motion, yielding $(\tilde{\mathbf{p}}_i, \tilde{\mathbf{p}}'_i)$:

$$\tilde{\mathbf{p}}_i = \mathbf{p}_i - \bar{\mathbf{p}}, \ \tilde{\mathbf{p}}'_i = \mathbf{p}'_i - \bar{\mathbf{p}}',$$

where $\bar{\mathbf{p}} = \frac{1}{n} \Sigma_{i=1}^n \mathbf{p}_i$ and $\bar{\mathbf{p}}' = \frac{1}{n} \Sigma_{i=1}^n \mathbf{p}'_i$.

The singular value decomposition of the cross-correlation matrix $\mathbf{A}$ of the centred data is then defined as:

$$\mathbf{A} = \sum_{i=1}^{n} \tilde{\mathbf{p}}'_i \tilde{\mathbf{p}}_i^T = \mathbf{U}\mathbf{W}\mathbf{V}^T \qquad (1)$$

Finally, the rotation matrix $\hat{\mathbf{R}}$ can be estimated as:

$$\hat{\mathbf{R}} = \mathbf{V} \begin{pmatrix} 1 & 0 & 0 \\ 0 & 1 & 0 \\ 0 & 0 & det(\mathbf{V}\mathbf{U}^T) \end{pmatrix} \mathbf{U}^T$$

where $det(\cdot)$ denotes the determinant of a matrix. The translation vector $\hat{\mathbf{t}}$ can be estimated in the least squares sense as:

$$\hat{\mathbf{t}} = \bar{\mathbf{p}}' - \hat{\mathbf{R}}\bar{\mathbf{p}}$$

This algorithm is called the constraint least squares (CLS) method. Haralick et al. [52] proposed a weighted version of this algorithm considering the uncertainty of correspondence data. They derived the same solution through Lagrangian multipliers so that the rigid constraints can be imposed.

Chaudhuri, Sharma, and Chatterjee in 1996 [15] developed the recursive CLS method for motion estimation from a long sequence of range images assuming that the correspondences between successive frames have been established. In order to deal with mismatches, a least median of squares method has been applied. For the sake of improving computational efficiency, the Monte Carlo resampling technique has been adopted. A number of experiments based on both synthetic data and real images have shown that the proposed algorithm is accurate and robust.

In summary, the CLS method is mathematically perfect, accurate, robust, and also efficient in applications [87, 31, 93]. Thus, it is widely used for motion estimation from two sets of correspondence data.

2.2 Quaternion Method

In [34, 35, 36, 53, 55] it was proposed using the unit quaternion to represent the rotation matrix leading to a concise and efficient estimation of rotation parameters. Using the formulation of Kanatani, the ith measurement a 3×4 matrix $\mathbf{M}_i$ is defined as:

$$\mathbf{M}_i = [\tilde{\mathbf{p}}'_i - \tilde{\mathbf{p}}_i, \ (\tilde{\mathbf{p}}'_i + \tilde{\mathbf{p}}_i) \times \mathbf{I}]$$

where the vector product between a vector $\mathbf{a}$ and a matrix $\mathbf{B}$ is a matrix with columns given by the vector products between $\mathbf{a}$ and the columns of $\mathbf{B}$. Let $\mathbf{q} = [q_0 \ q_1 \ q_2 \ q_3]^T$ be the quaternion representing the unknown rotation. Then $\hat{\mathbf{q}}$ can be estimated as the eigenvector corresponding to the minimum eigenvalue of matrix $\mathbf{M}$:

$$\mathbf{M} = \sum_{i=1}^{n} \mathbf{M}_i^T \mathbf{M}_i$$

Since the quaternion method is a different formalisation of motion estimation from range images, the obtained solution is identical with the CLS method up to the processor precision when the image data are not degenerate [87, 31].

2.3   Orthonormal Matrices Method

In 1952, Green [50] investigated the general procrustes problem where he proposed an earliest version of orthonormal matrices method for approximating one matrix by another under an orthonormal transformation. But he assumed that the given matrices must be of full column rank. Since then, the orthonomal matrices method has been mentioned by a number of researchers in their work (e.g. [48, 121, 69]).

In 1988, Horn [54] explicitly proposed using the orthonormal matrices method to estimate the motion parameters from given correspondence data. When matrix $\mathbf{A}$ is non-singular, it can be decomposed as a product of an orthonormal matrix $\mathbf{W}$ and a positive definite matrix $\mathbf{S}$:

$$\mathbf{A} = \mathbf{W}\mathbf{S}$$

where

$$\mathbf{S} = (\mathbf{A}^T\mathbf{A})^{1/2}$$

is the square root of positive definite matrix $\mathbf{A}^T\mathbf{A}$ while

$$\mathbf{W} = \mathbf{A}(\mathbf{A}^T\mathbf{A})^{-1/2}$$

Horn et al. show that Trace($\mathbf{R}\mathbf{A}$) is maximised for $\mathbf{R} = \mathbf{W}$, which is the nearest orthonormal matrix to the non-singular matrix $\mathbf{A}$. $\mathbf{W}$ is computed by decomposing $\mathbf{A}^T\mathbf{A}$ in terms of its eigenvalues and eigenvectors as:

$$\mathbf{A}^T\mathbf{A} = \lambda_1\mathbf{u}_1\mathbf{u}_1^T + \lambda_2\mathbf{u}_2\mathbf{u}_2^T + \lambda_3\mathbf{u}_3\mathbf{u}_3^T$$

where $\{\lambda_i\}$ are eigenvalues of matrix $\mathbf{A}^T\mathbf{A}$ and $\{\mathbf{u}_i\}$ are corresponding eigenvectors. Now

$$\mathbf{S}^{-1} = \frac{1}{\sqrt{\lambda_1}}\mathbf{u}_1\mathbf{u}_1^T + \frac{1}{\sqrt{\lambda_2}}\mathbf{u}_2\mathbf{u}_2^T + \frac{1}{\sqrt{\lambda_3}}\mathbf{u}_3\mathbf{u}_3^T$$

and

$$\mathbf{R} = \mathbf{W} = \mathbf{A}\mathbf{S}^{-1}$$

The translation vector $\mathbf{t}$ is estimated as the difference of the centroids between two sets of points in the least squares sense. The orthonormal matrices method is still another formulation of the same problem and it provides exactly the same motion estimation solution as the CLS method when the image data are not degenerate. But when the image data are limited, the orthonormal matrices method is more efficient than the CLS method [87].

2.4   Dual Quaternion Method

In 1991, Walker et al. [133] proposed a novel method for motion estimation from two sets of correspondence data. Originally, the algorithm was designed to minimise a cost function:

$$J = \sum_{i=1}^{L} \alpha_i ||\mathbf{n}_{1i} - \mathbf{R}\mathbf{n}_{2i}||^2 + \sum_{i=1}^{N} \beta_i ||\mathbf{p}'_i - \mathbf{R}\mathbf{p}_i - \mathbf{t}||^2$$

where $\mathbf{n}_{1i}$ and $\mathbf{n}_{2i}$ are two sets of L corresponding unit normal vectors, and α_i and β_i are weighting factors reflecting data reliability. The least squares minimisation equation is a special case of this cost function if $\alpha_i = 0$ and $\beta_i = 1$ are set.

In this method, the rotation and translation are represented together using a dual quaternion, $\mathbf{q}_d = [\mathbf{r}, \mathbf{s}]$. Here motion is modelled as a simultaneous rotation and translation along a particular line, with direction $\mathbf{n} = [n_x, n_y, n_z]$, passing through a point $\mathbf{a} = [a_x, a_y, a_z]$. The amount of motion is given by an angle θ and distance d, defining the components of $\mathbf{q}_d$ as:

$$\mathbf{r} = \begin{pmatrix} \sin(\theta/2)\mathbf{n} \\ \cos(\theta/2) \end{pmatrix}, \mathbf{s} = \begin{pmatrix} \frac{d}{2}\cos(\theta/2)\mathbf{n} + \sin(\theta/2)\mathbf{a} \times \mathbf{n} \\ -d/2\sin(\theta/2) \end{pmatrix} \tag{2}$$

Again the least squares minimisation equation can be rewritten in this new framework, resulting in an equation involving $\mathbf{r}$ and $\mathbf{s}$:

$$J = \frac{1}{N}(\mathbf{r}^T \mathbf{C}_1 \mathbf{r} + N\mathbf{s}^T \mathbf{s} + \mathbf{s}^T \mathbf{C}_2 \mathbf{r} + const.)$$

where

$$\mathbf{C}_1 = -2\sum_{i=1}^{N} \begin{pmatrix} \mathbf{K}(\mathbf{p}')\mathbf{K}(\mathbf{p}) + \mathbf{p}'\mathbf{p}^T & -\mathbf{K}(\mathbf{p}')\mathbf{p} \\ -\mathbf{p}'^T\mathbf{K}(\mathbf{p}) & \mathbf{p}'^T\mathbf{p} \end{pmatrix}$$

$$\mathbf{C}_2 = 2\sum_{i=1}^{N} \begin{pmatrix} -\mathbf{K}(\mathbf{p}) - \mathbf{K}(\mathbf{p}') & \mathbf{p} - \mathbf{p}' \\ -(\mathbf{p} - \mathbf{p}')^T & 0 \end{pmatrix}$$

$$\mathbf{K}(\mathbf{n}) = \begin{pmatrix} 0 & -n_z & n_y \\ n_z & 0 & -n_x \\ -n_y & n_x & 0 \end{pmatrix}$$

and $const = \sum_{i=1}^{N}(\mathbf{p}_i^T \mathbf{p}_i + \mathbf{p}'_i{}^T \mathbf{p}'_i)$.

From Equation 2, it is known that $\mathbf{r}^T \mathbf{r} = 1$ and $\mathbf{r}^T \mathbf{s} = 0$, thus, the optimal dual quaternion is obtained by minimising:

$$J = \frac{1}{N}(\mathbf{r}^T \mathbf{C}_1 \mathbf{r} + N\mathbf{s}^T \mathbf{s} + \mathbf{s}^T \mathbf{C}_2 \mathbf{r} + const. + \lambda_1(\mathbf{r}^T \mathbf{r} - 1) + \lambda_2(\mathbf{r}^T \mathbf{s})) \tag{3}$$

where λ_1 and λ_2 are Lagrange multipliers. Taking the partial derivatives about $\mathbf{r}$ and $\mathbf{s}$ gives:

$$\frac{\partial J}{\partial \mathbf{r}} = \frac{1}{N}[(\mathbf{C}_1 + \mathbf{C}_1^T)\mathbf{r} + \mathbf{C}_2^T \mathbf{s} + 2\lambda_1 \mathbf{r} + \lambda_2 \mathbf{s}] = 0 \tag{4}$$

$$\frac{\partial J}{\partial \mathbf{s}} = \frac{1}{N}[2N\mathbf{s} + \mathbf{C}_2\mathbf{r} + \lambda_2\mathbf{r}] = 0 \tag{5}$$

Multiplying Equation 5 by $\mathbf{r}^T$ gives $\lambda_2 = -\mathbf{r}^T\mathbf{C}_2\mathbf{r} = 0$, since $\mathbf{C}_2$ is skew symmetric. Thus $\mathbf{s}$ is given by:

$$\mathbf{s} = -\frac{1}{2N}\mathbf{C}_2\mathbf{r}$$

Substituting these into Equation 4 yields:

$$\mathbf{A}\mathbf{r} = \lambda_1\mathbf{r}$$

where

$$\mathbf{A} = \frac{1}{2}(\frac{1}{2N}\mathbf{C}_2^T\mathbf{C}_2 - \mathbf{C}_1 - \mathbf{C}_1^T)$$

Thus the quaternion $\mathbf{r}$ is an eigenvector of matrix $\mathbf{A}$ and λ_1 is the corresponding eigenvalue. Substituting the above results back into Equastion 3 gives:

$$J = \frac{1}{N}(const. - \lambda_1)$$

The error is thus minimised if we select the eigenvector corresponding to the largest eigenvalue.

Having computed $\mathbf{r}$ and $\mathbf{s}$, it is easy to compute the motion parameters rotation matrix $\hat{\mathbf{R}}$ and translation vector $\hat{\mathbf{t}}$ through the following procedure. Let the vector and scalar parts of quaternions $\mathbf{r}$ and $\mathbf{s}$ be $\mathbf{r}_1$ and r_2, $\mathbf{s}_1$ and s_2, respectively, then the rotation matrix $\hat{\mathbf{R}}$ can be estimated as:

$$\hat{\mathbf{R}} = (r_2^2 - \mathbf{r}_1^T\mathbf{r}_1)\mathbf{I} + 2\mathbf{r}_1\mathbf{r}_1^T + 2r_2\mathbf{K}(\mathbf{r}_1)$$

and the translation vector $\hat{\mathbf{t}}$ can be estimated as the vector part of the quaternion $\mathbf{q}$ given by:

$$\mathbf{q} = \begin{pmatrix} r_2 - \mathbf{K}(\mathbf{r}_1) & \mathbf{r}_1 \\ -\mathbf{r}_1^T & r_2 \end{pmatrix}^T \mathbf{s}$$

This method also provides exactly the same motion estimation results as the CLS method when the image data are not degenerate, and has the advantage that matrices $\mathbf{C}_1$ and $\mathbf{C}_2$ can be computed incrementally [141].

3   Motion Representation

While many methods that calibrate rigid body transformation parameters mainly consider the relationships between object feature points and their corresponding image points in different coordinate frames based on analytic geometry, perspective geometry, or epipolar geometry, we here mainly review the relationships between correspondences that have been synthesised into a single coordinate frame. The general assumptions and constraints of these methods are given as follows:

1. All transformations must be rigid body transformations;

2. All correspondences must undergo the same rigid body transformation;

3. All correspondences must be synthesised into just a single coordinate frame without changing their coordinates;

4. The rigid body transformation must include rotational part, in other words, the rotation angle θ of the rigid body transformation must be defined as: $0 < \theta < \pi$.

Essentially, these assumptions normalise the relationship between objects and range cameras where objects are in motion and the range camera is static. The relationship among point correspondence $(\mathbf{p}, \mathbf{p}')$, rigid motion parameters rotation matrix $\mathbf{R}$ and translation vector $\mathbf{t}$, and rigid constraints about distance, angle and projection is outlined as follows.

3.1 Motion Representation Based on Correspondence Vectors

In order to design accurate, robust, and efficient motion estimation and image registration algorithms, it is necessary to obtain a deep understanding of 3D rigid non-pure translational transformations which are often used in 3D motion estimation and image registration. While Chasles' theory [4] gives the sufficient representation of rigid non-pure translational transformations, in machine vision the necessary representation of rigid non-pure translational transformations is often more desired for the calibration of rigid non-pure translational transformation parameters. As a result, a general sufficient and necessary representation theorem regarding 3D rigid non-pure translational transformations based on the explicit expressions of distance, angle, and projection measurements is put forward in [72] (see Figure 3):

Theorem 1 *A non-pure translational transformation* $(\mathbf{R}, \mathbf{t})$ *is a 3D rigid non-pure translational transformation if and only if there exists a fixed axis* $\mathbf{h}$ *and a fixed point* $\mathbf{c}$ *with* $\mathbf{h}^T \mathbf{c} = 0$, *such that*

$$||\underline{\mathbf{p}} - \mathbf{c}|| = ||\underline{\mathbf{p}}' - \mathbf{c}|| \tag{6}$$

$$\frac{(\underline{\mathbf{p}} - \mathbf{c})^T (\underline{\mathbf{p}}' - \mathbf{c})}{(\underline{\mathbf{p}} - \mathbf{c})^T (\underline{\mathbf{p}} - \mathbf{c})} = \cos\theta \tag{7}$$

$$\mathbf{h}^T (\mathbf{p} - \mathbf{p}') = -\mathbf{h}^T \mathbf{t} \tag{8}$$

where $(\underline{\mathbf{p}}, \underline{\mathbf{p}}')$ *are the projections of an arbitrary correspondence* $(\mathbf{p}, \mathbf{p}')$ *on the plane perpendicular to the fixed axis* $\mathbf{h}$ *and* $\cos\theta$ *and vector* $\mathbf{t}$ *are all constants uniquely determined by the rigid non-pure translational transformation.*

The point $\mathbf{c}$ in this theorem is called the critical point $\mathbf{c}$ in 3D. Any line in the plane perpendicular to the rotation axis

$\mathbf{h}$ and passing through the critical point $\mathbf{c}$ in 3D is called a critical line in 3D. The line with unit length parallel to the rotation axis $\mathbf{h}$ and passing through the critical point $\mathbf{c}$

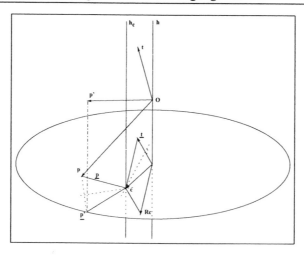

Figure 3: The projections $(\underline{\mathbf{p}}, \underline{\mathbf{p}'})$ of correspondence $(\mathbf{p}, \mathbf{p}')$ on the plane perpendicular the rotation axis $\mathbf{h}$ is equidistant to the critical point $\mathbf{c}$ in 3D, the including angle between $\underline{\mathbf{p}} - \mathbf{c}$ and $\underline{\mathbf{p}'} - \mathbf{c}$ is equal to the rotation angle θ of the rigid non-pure translational transformation, and the projection of any correspondence vector $\mathbf{p} - \mathbf{p}'$ along the rotation axis $\mathbf{h}$ is equal to a constant uniquely determined by the rigid non-pure translational transformation.

is called the critical axis $\mathbf{h}_c$. Any cylinder with axis along the critical axis $\mathbf{h}_c$ is called a critical cylinder. Any plane passing through the critical axis $\mathbf{h}_c$ is called a critical plane.

This theorem has shown that any correspondence $(\mathbf{p}, \mathbf{p}')$ can be effected by a rotation around the critical axis $\mathbf{h}_c$ and a translation along this critical axis $\mathbf{h}_c$. Thus this theorem formalises the Chasles' screw motion theory [4] about spatial displacement based on original image point data. The screw rotation is defined as a rotation around the critical axis $\mathbf{h}_c$ and the screw translation is defined as the projection of the translation vector along the critical axis $\mathbf{h}_c$. This theorem is called the representation theorem of 3D rigid non-pure translational transformations based on correspondence vectors with a correspondence vector defined as: $\mathbf{CV} = \mathbf{p} - \mathbf{p}'$. Since Equation 6 encapsulates distance information, Equation 7 encapsulates angle information, and Equation 8 encapsulates projection information, thus the representation of 3D rigid non-pure translational transformation does reflect exactly the three mutually independent rigid constraints about distance, angle, and projection measurements in 3D. In addition, the constraints represented by Equations 6, 7, and 8 also physically define the position of the correspondent: Equation 6 shows that the correspondent $\mathbf{p}'$ must lie on the critical cylinder passing through the original point $\mathbf{p}$. Equation 7 shows that the correspondent $\mathbf{p}'$ must lie on the critical plane. The including angle between this critical plane and another passing through the original point $\mathbf{p}$ is equal to a constant. Equation 8 shows that the correspondent $\mathbf{p}'$ must lie on the plane perpendicular to the critical axis. The distance between this plane and another perpendicular to the critical axis $\mathbf{h}_c$ and passing through the original point $\mathbf{p}$ is equal to a constant. Any constraint represented by one of Equations 6, 7, and 8 confines the position of the correspondent $\mathbf{p}'$ from three to two dimensions. Any two constraints represented by two of Equations 6, 7, and 8 confine the position of the correspondent $\mathbf{p}'$ from three to one dimension. The three constraints represented by Equations 6, 7, and 8 together uniquely define the position of the correspondent to be the intersection point of two planes and a semicylinder. Such useful property provides a practical and efficient method for searching correspondences in 3D space.

3.2 Motion Representation Based on Reflected Correspondence Vectors

The more deeply we understand rigid body transformations, the larger the probability of developing efficient motion estimation and image registration algorithms. Thus, the following sufficient and necessary representation theorem regarding 3D rigid transformations based on reflected correspondence $(\mathbf{p}, \mathbf{p}'')=(\mathbf{p}, -\mathbf{p}')$ was proposed in [75, 74] (see Figure 4):

Theorem 2 *A transformation $(\mathbf{R}, \mathbf{t})$ is a 3D rigid transformation if and only if there exists a fixed axis $\mathbf{h}$ and a fixed point $\mathbf{e}$ such that one of the following three equations:*

$$||\mathbf{p} - \mathbf{e}|| = ||\mathbf{p}'' - \mathbf{e}|| \tag{9}$$

or

$$||\underline{\mathbf{p}} - \mathbf{e}|| = ||\underline{\mathbf{p}}'' - \mathbf{e}|| \tag{10}$$

or

$$||\underline{\mathbf{p}} - \underline{\mathbf{e}}|| = ||\underline{\mathbf{p}}'' - \underline{\mathbf{e}}|| \tag{11}$$

is true and

$$\frac{(\underline{\mathbf{p}} - \underline{\mathbf{e}})^T (\underline{\mathbf{p}}'' - \underline{\mathbf{e}})}{(\underline{\mathbf{p}} - \underline{\mathbf{e}})^T (\underline{\mathbf{p}} - \underline{\mathbf{e}})} = \cos(\pi - \theta) \tag{12}$$

$$\mathbf{h}^T (\mathbf{p} + \mathbf{p}'') = 2\mathbf{h}^T \mathbf{e} \tag{13}$$

where for all reflected correspondences $(\mathbf{p}, \mathbf{p}'')$, $(\underline{\mathbf{p}}, \underline{\mathbf{p}}'')$ is the projection of $(\mathbf{p}, \mathbf{p}'')$ on the plane perpendicular to the fixed axis $\mathbf{h}$ and θ is a constant uniquely determined by the rigid transformation.

The point $\mathbf{e}$ in this theorem is called the essential point in 3D which encapsulates the rigid transformation information. Any line passing through the essential point $\mathbf{e}$ in 3D is called an essential line in 3D. Any essential line projected onto the plane perpendicular to the rotation axis $\mathbf{h}$ is called a projected essential line in 3D. The line with unit length parallel to the rotation axis $\mathbf{h}$ and passing through the essential point $\mathbf{e}$ is called the essential axis $\mathbf{h}_e$. Any cylinder with its centred axis $\mathbf{h}_e$ is called an essential cylinder. Any plane passing through the essential axis $\mathbf{h}_e$ is called an essential plane.

Theorem 2 has shown that any 3D reflected correspondence can be effected by a rotation around the essential axis $\mathbf{h}_e$ and a translation along the essential axis $\mathbf{h}_e$. In addition, the pure translational motion represents a singular case of the Chasles' screw motion theory. However, the above analysis is valid for pure translational motion. Thus this theorem has

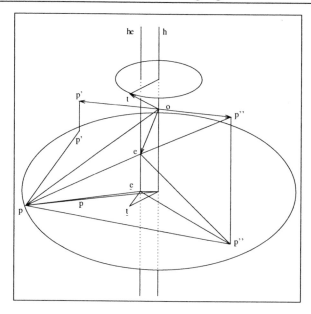

Figure 4: Any reflected correspondence $(\mathbf{p}, \mathbf{p}'')$ or its projections $(\underline{\mathbf{p}}, \underline{\mathbf{p}}'')$ must be equidistant to a fixed point $\mathbf{e}$ or the projections $(\underline{\mathbf{p}}, \underline{\mathbf{p}}'')$ of reflected correspondence $(\mathbf{p}, \mathbf{p}'')$ must be
equidistant to the projection $\underline{\mathbf{e}}$ of this fixed point $\mathbf{e}$ on the plane perpendicular to the rotation
axis $\mathbf{h}$, the including angle between $\underline{\mathbf{p}} - \underline{\mathbf{e}}$ and $\underline{\mathbf{p}}'' - \underline{\mathbf{e}}$ is equal to the supplement of the rotation angle θ of the rigid transformation, and the projection of any reflected correspondence
vector $\mathbf{p} + \mathbf{p}''$ along the rotation axis $\mathbf{h}$ is equal to a constant uniquely determined by the
rigid transformation.

extended the Chasles' screw motion theory [4] through investigating the properties of reflected image points rather than original image points and it is called the representation theorem of 3D rigid transformations based on reflected correspondence vectors with a reflected
correspondence vector defined as: $\mathbf{RCV} = \mathbf{p} - \mathbf{p}''$. Since Equations 9, 10 or 11 encapsulates
distance information, Equation 12 encapsulates angle information, and Equation 13 encap-
sulates projection information, thus the representation of 3D rigid transformation based on
reflected correspondence vectors also reflects exactly the three mutually independent rigid
constraints about distance, angle, and projection measures in 3D. The constraints represented by these equations also physically define the position of the reflected correspondent
$\mathbf{p}''$. Equations 9, 10 or 11 show that the reflected correspondent $\mathbf{p}''$ must lie on the essential
cylinder passing through the original point $\mathbf{p}$. Equation 12 shows that the reflected correspondent $\mathbf{p}''$ must lie on an essential plane. The including angle between this essential
plane and another passing through the original point $\mathbf{p}$ is equal to a constant. Equation 13
shows that the reflected correspondent $\mathbf{p}''$ must lie on a plane perpendicular to the essential
axis $\mathbf{h}_e$. The distance between this plane and another perpendicular to the essential axis
$\mathbf{h}_e$ and passing through the original point $\mathbf{p}$ is equal to a constant. As a result, any constraint represented by one of Equations 9, 10, 11, 12, and 13 confines the position of the

reflected correspondent $\mathbf{p}''$ from three to two dimensions, any constraint represented by one of Equations 9, 10, and 11, and another constraint represented by one of Equations 12 and 13 confine the position of the reflected correspondent $\mathbf{p}''$ from three to one dimension, any constraint represented by one of Equations 9, 10, and 11, and constraints represented by Equations 12 and 13 together uniquely define the position of the reflected correspondent $\mathbf{p}''$ to be the intersection point of two planes and a semicylinder. Such useful properties again provide an efficient method for the search of reflected correspondences in 3D space.

4   Motion Estimation from Motion Properties

Changing the viewpoint and looking again at 3D geometric properties as outlined in the last section, given a set of 3D correspondences $(\mathbf{p}_i, \mathbf{p}'_i)(i \in \{1, 2, \cdots, n\}, n \geq 3)$, closed form representations can all be developed for 3D rigid non-pure translational transformation parameters rotation axis $\mathbf{h}$, rotation angle θ, rotation matrix $\mathbf{R}$, and translation vector $\mathbf{t}$ which are outlined as follows.

4.1   Motion Estimation Based on Correspondence Vectors

In this section, we review methods used to calculate the motion parameters of interest based on correspondence vectors [76]:

Given three points $(\mathbf{p}_1, \mathbf{p}_2, \mathbf{p}_3)$ and their correspondents $(\mathbf{p}'_1, \mathbf{p}'_2, \mathbf{p}'_3)$, if their correspondence vector differences $\mathbf{CV_2} - \mathbf{CV_1}$ and $\mathbf{CV_3} - \mathbf{CV_1}$ are not parallel, then the rotation axis $\mathbf{h}$ can be estimated as:

$$\mathbf{h} = \frac{(\mathbf{CV_2} - \mathbf{CV_1}) \times (\mathbf{CV_3} - \mathbf{CV_1})}{||(\mathbf{CV_2} - \mathbf{CV_1}) \times (\mathbf{CV_3} - \mathbf{CV_1})||} \tag{14}$$

However, these is an ambiguity in deciding the direction of rotation axis. Thus, the following procedure is proposed to disambiguate the direction of the rotation axis. First, center all correspondence data $(\mathbf{p}_i, \mathbf{p}'_i)$ yielding $(\tilde{\mathbf{p}}_i, \tilde{\mathbf{p}}'_i)$, then, project all centred correspondences $(\tilde{\mathbf{p}}_i, \tilde{\mathbf{p}}'_i)$ on the plane perpendicular to this estimated rotation axis $\mathbf{h}$: $\underline{\tilde{\mathbf{p}}}_i = (\mathbf{I} - \mathbf{h}\mathbf{h}^T)\tilde{\mathbf{p}}_i$ and $\underline{\tilde{\mathbf{p}}}'_i = (\mathbf{I} - \mathbf{h}\mathbf{h}^T)\tilde{\mathbf{p}}'_i$. Since the rotation is from $\underline{\tilde{\mathbf{p}}}_i$ to $\underline{\tilde{\mathbf{p}}}'_i$, thus, the calibrated rotation axis $\mathbf{h}$ can be finally estimated as:

$$\mathbf{h} = \frac{\underline{\tilde{\mathbf{p}}}'_i \times \underline{\tilde{\mathbf{p}}}_i}{||\underline{\tilde{\mathbf{p}}}'_i \times \underline{\tilde{\mathbf{p}}}_i||} = \begin{pmatrix} h_x \\ h_y \\ h_z \end{pmatrix} \tag{15}$$

Once the rotation axis $\mathbf{h}$ has been estimated, any correspondence $(\mathbf{p}_i, \mathbf{p}'_i)$ $(i \in \{1, 2, \cdots n\}, n \geq 3)$ can be projected onto the plane perpendicular to the rotation axis $\mathbf{h}$:

$$\underline{\mathbf{p}}_i = (\mathbf{I} - \mathbf{h}\mathbf{h}^T)\mathbf{p}_i, \ \underline{\mathbf{p}}'_i = (\mathbf{I} - \mathbf{h}\mathbf{h}^T)\mathbf{p}'_i \tag{16}$$

Given two non-parallel projected correspondence vectors $\underline{\mathbf{CV}}_1$ and $\underline{\mathbf{CV}}_2$, then their corresponding perpendicular bisectors will intersect at the critical point $\mathbf{c}$ in 3D which can be uniquely estimated as:

$$\mathbf{c} = \frac{\mathbf{p}_2^T \mathbf{p}_2 - \mathbf{p}_2'^T \mathbf{p}_2'}{2\mathbf{CV}_2{}^T \mathbf{HCV}_1} \mathbf{HCV}_1 + \frac{\mathbf{p}_1^T \mathbf{p}_1 - \mathbf{p}_1'^T \mathbf{p}_1'}{2\mathbf{CV}_1{}^T \mathbf{HCV}_2} \mathbf{HCV}_2 \qquad (17)$$

where $\mathbf{H} = \begin{pmatrix} 0 & -h_z & h_y \\ h_z & 0 & -h_x \\ -h_y & h_x & 0 \end{pmatrix}$.

Once the critical point $\mathbf{c}$ in 3D has been estimated, for any projected correspondence $(\underline{\mathbf{p}}_i, \underline{\mathbf{p}}_i')$ ($i \in \{1, 2, \cdots, n\}$), the including angle measured at the critical point $\mathbf{c}$ in 3D between this projected correspondence must be equal to the rotation angle θ which can be uniquely estimated as:

$$\theta = \arccos \frac{(\underline{\mathbf{p}}_i - \mathbf{c})^T (\underline{\mathbf{p}}_i' - \mathbf{c})}{(\underline{\mathbf{p}}_i - \mathbf{c})^T (\underline{\mathbf{p}}_i - \mathbf{c})} \qquad (18)$$

Now that the rotation axis $\mathbf{h}$ and the rotation angle θ have been estimated, the rigid rotation matrix $\mathbf{R}$ can be estimated by the Rodrigues' formula as:

$$\mathbf{R} = \mathbf{I} - \mathbf{H}\sin\theta + (1 - \cos\theta)\mathbf{H}^2 \qquad (19)$$

From the above interesting phenomena regarding 3D rigid non-pure translational transformations, it is known that the projected translation vector $\underline{\mathbf{t}}$ on the plane perpendicular to the rotation axis $\mathbf{h}$ is equal to $(\mathbf{I} - \mathbf{R})\mathbf{c}$. On the other hand, once a correspondence $(\mathbf{p}_i, \mathbf{p}_i')$ ($i \in \{1, 2, \cdots, n\}, n \geq 3$) is known, then the projected translation vector along the rotation axis $\mathbf{h}$ can be estimated as: $-\mathbf{h}\mathbf{h}^T(\mathbf{p}_i - \mathbf{p}_i')$ leading to an estimation of the translation vector $\mathbf{t}$ as:

$$\mathbf{t} = (\mathbf{I} - \mathbf{R})\mathbf{c} - \mathbf{h}\mathbf{h}^T(\mathbf{p}_i - \mathbf{p}_i') \qquad (20)$$

It is worth pointing out that Equation 20 provides an alternative to the estimation of the translation vector $\mathbf{t}$ which is often estimated as the difference of centroids between two sets of image data [2, 53, 54, 65, 132] in the least squares sense.

4.2   Motion Estimation Based on Reflected Correspondence Vectors

In this section, we review the methods used to calculate the motion parameters of interest based on reflected correspondence vectors [107]:

Given three non-parallel $\mathbf{RCV}_i$ ($i \in \{1, 2, 3\}$), their perpendicular bisector planes intersect at the essential point $\mathbf{e}$ in 3D which can be uniquely estimated as:

$$\begin{pmatrix} \mathbf{RCV}_1^T \\ \mathbf{RCV}_2^T \\ \mathbf{RCV}_3^T \end{pmatrix} \mathbf{e} = \begin{pmatrix} (\mathbf{p}_1^T \mathbf{p}_1 - \mathbf{p}_1''^T \mathbf{p}_1'')/2 \\ (\mathbf{p}_2^T \mathbf{p}_2 - \mathbf{p}_2''^T \mathbf{p}_2'')/2 \\ (\mathbf{p}_3^T \mathbf{p}_3 - \mathbf{p}_3''^T \mathbf{p}_3'')/2 \end{pmatrix} = \mathbf{A}_{3\times3}\mathbf{b}_{3\times1} \qquad (21)$$

The rotation axis $\mathbf{h}$ can be estimated using the procedure as described in the last section. Once the rotation axis $\mathbf{h}$ has been estimated, any reflected correspondence $(\mathbf{p}_i, \mathbf{p}_i'')(i \in \{1, 2, \cdots n\}, n \geq 3)$ can be projected onto a plane perpendicular to the rotation axis $\mathbf{h}$:

$$\underline{\mathbf{p}}_i = (\mathbf{I} - \mathbf{h}\mathbf{h}^T)\mathbf{p}_i, \ \underline{\mathbf{p}}_i'' = (\mathbf{I} - \mathbf{h}\mathbf{h}^T)\mathbf{p}_i'' \tag{22}$$

Once the essential point $\mathbf{e}$ in 3D has been estimated, for any projected reflected correspondence $(\underline{\mathbf{p}}_i, \underline{\mathbf{p}}_i'')$ $(i \in \{1, 2, \cdots, n\}, n \geq 3)$, the including angle between the lines passing through the reflected correspondence $(\underline{\mathbf{p}}_i, \underline{\mathbf{p}}_i'')$ measured at the projection $\underline{\mathbf{e}}$ of the essential point $\mathbf{e}$ in 3D onto the plane perpendicular to the rotation axis $\mathbf{h}$ is equal to the supplement of the rotation angle θ of the transformation which can be uniquely estimated as:

$$\theta = \pi - \arccos \frac{(\underline{\mathbf{p}}_i - \underline{\mathbf{e}})^T (\underline{\mathbf{p}}_i'' - \underline{\mathbf{e}})}{(\underline{\mathbf{p}}_i - \underline{\mathbf{e}})^T (\underline{\mathbf{p}}_i - \underline{\mathbf{e}})} \tag{23}$$

Now that the rotation axis $\mathbf{h}$ and the rotation angle θ have been estimated, the rigid rotation matrix $\mathbf{R}$ can be estimated again by the Rodrigues' formula as:

$$\mathbf{R} = \mathbf{I} - \mathbf{H}\sin\theta + (1 - \cos\theta)\mathbf{H}^2 \tag{24}$$

As a result, the translation vector $\mathbf{t}$ can be perfectly estimated as:

$$\mathbf{t} = -(\mathbf{I} + \mathbf{R})\mathbf{e} \tag{25}$$

It is interesting to note that Equation 25 provides another alternative to the estimation of the translation vector $\mathbf{t}$ which is often estimated as the difference of centroids between two sets of image data [2, 53, 54, 65, 132] in the least squares sense.

4.3   Robust Motion Estimation

In principle, as long as three pairs of point correspondences that are not collinear are available, then a unique rigid motion can be estimated. However, in practice, due to noise presence in the point data, it is useful to adopt robust techniques to combat noise, leading to more accurate motion estimation results.

Based on the above geometric analysis of correspondence vectors and reflected correspondence vectors, it is known that given a set of either correspondences $(\mathbf{p}_i, \mathbf{p}_i')$ or reflected correspondences $(\mathbf{p}_i, \mathbf{p}_i'')$ $(i \in \{1, 2, \cdots, n\}, n \geq 3)$, in order to provide a closed form solution to all 3D rigid transformation parameters based on the explicit expressions of distance, angle, and projection information, it is necessary to calibrate the critical point $\mathbf{c}$ or the essential point $\mathbf{e}$ and the rotation axis $\mathbf{h}$. As a result, a number of robust linear methods, such as total least squares method [14, 45], can be used to accurately estimate either the critical point $\mathbf{c}$ or the essential point $\mathbf{e}$ and the rotation axis $\mathbf{h}$ leading to an accurate and robust estimation of 3D motion parameters [75, 71].

On the other hand, the Monte Carlo resampling technique can also be used for robust motion estimation. Assuming that a subsample contains m $(m \geq 3)$ correspondences, the whole set of image data are corrupted by false matches with probability ε, and the expected probability to obtain an accurate solution is P, then the number N $(N \geq 1)$ of required subsamples can be estimated as:

$$N = \frac{\log(1-P)}{\log(1-(1-\varepsilon)^m)} + 1$$

If $m = 4$, $\varepsilon = 0.7$, and $P = 0.99$, then $N = 567$. For each subsample, relative formulas described in the last section can be used to estimate candidates to the motion parameters of interest respectively where $(j = 1, 2, \cdots, N)$. The calibrated motion parameters of interest can be finally estimated by median filtering and normalising, if necessary, the corresponding components of these candidates.

5 Range Image Registration Based on Motion Properties

A range image is a rendering of a set of structured data points. Registration algorithms based on points is of general use, since all other free form shapes like triangular meshes, line segments, planar patches, or analytic forms can all be transformed into points [8]. Assume that the two overlapping range images to be registered are represented as two sets of either sparse or dense points $\mathbf{P} = \{\mathbf{p}_1, \mathbf{p}_2, \cdots, \mathbf{p}_{n_1}\}$ and $\mathbf{P}' = \{\mathbf{p}'_1, \mathbf{p}'_2, \cdots, \mathbf{p}'_{n_2}\}$. Due to occlusion and appearance and disappearance of points, n_1 here is not necessarily equal to n_2. The points with the same subscript do not mean that they are correspondences.

When the correspondences and the crude motion between the two images are not available, the automatic registration will be much more difficult, if not impossible. But under some special imaging conditions and when registration errors are allowable to some extent, a number of techniques for image registration can be feasible and practical, especially with the aid of human intervention. Since the ICP algorithm is a *de factor* standard method for registration, we first present a relatively detailed review in this section of the important variants of the ICP algorithm, then we review the advantages and disadvantages of the ICP algorithm, and finally we outline image registration using motion properties.

5.1 Iterative Closest Point(ICP) Algorithm

In this section, we review the ICP algorithm according to the main steps included (Figure 5). Doing so is justified since it is easy to follow and facilitates its implementation. The range images themselves must be complex enough so that the motion information can be extracted [7]. It was also pointed out in [109, 115] that for accurate image registration, the range images must have enough overlapping and texture. Laboureux and Hausler points out in [68] that the best results can be obtained using all object points. For computational cost reduction, the points from structure area should be preferentially considered over those in the smooth area.

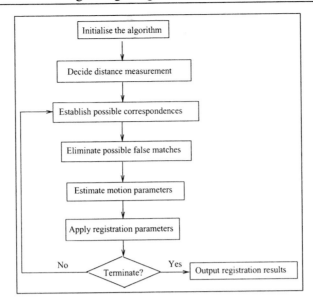

Figure 5: The main steps in the ICP algorithm.

5.1.1 Initialisation

The ICP algorithm requires a good initialisation which can be obtained by various ways:

- Manual selection of a number of point pairs from different images [5, 115];

- Feature extraction and matching: interpoint distance [38], point's fingerprint [127], bitangent curves [136], surface signatures [138], spin images [59], harmonic maps [143], line segments [1, 128], principal curvatures [37], bitangent points and distance [38], triangles [109], shape index [88], landmarks [125, 126]. In principle, all methods that extract and match features can be used to establish a set of possible correspondences between different images;

- Imaging device [30, 32, 40, 128, 66] like odometry or turn table;

- Pure translational motion derived from the two images to be registered [84, 107];

- Analytic transformation [47].

Gourdon and Ayache in 1993 [47] proposed the use of differential properties to register a curve on a surface. The paper proposes two algorithms: the first is based on the differential properties about Darboux and principal frames with respect to the arc length attached to each point and requires that the third order derivatives exist. The second is based on the differential invariants relative to each point pair and only requires that the first order derivatives exist. Both algorithms have to solve algebraic equations to find the registration parameters. In order to disambiguate multiple solutions, they examine the solutions through finding the corresponding points with the distances between the transformed points to the surface smaller than a threshold. If the number of such points is large enough, then the registration solution can be accepted.

5.1.2   Possible Correspondence Establishment

Once the registration parameters rotation matrix $\mathbf{R}$ and translation vector $\mathbf{t}$ have been initialised or estimated, the following criterion was proposed in 1992 [8, 141] for the establishment of possible correspondences $(\mathbf{p}_i, \mathbf{p}'_{c(i)})$ between the two images $\mathbf{P}$ and $\mathbf{P}'$ to be registered:

$$\mathbf{p}'_{c(i)} = argmin_{\mathbf{p}' \in \mathbf{P}'} ||\mathbf{p}' - \mathbf{R}\mathbf{p}_i - \mathbf{t}|| \tag{26}$$

which minimises the Euclidean distance between the transformed point $\mathbf{R}\mathbf{p}_i + \mathbf{t}$ and the points in the second image $\mathbf{P}'$. The search space is determined by the size of the second image $\mathbf{P}'$. This mapping means that a point $\mathbf{p}'_{c(i)}$ in the second image $\mathbf{P}'$ can be determined as the correspondent of a point $\mathbf{p}_i$ in the first image $\mathbf{P}$. As a consequence of this operation, a set of possible correspondences $(\mathbf{p}_i, \mathbf{p}'_i)(i = 1, 2, \cdots, n_1)$ between the two images to be registered have been established based on which the motion is estimated by the quaternion method.

Chen and Medioni in 1992 [16] also proposed a novel algorithm to register two-view and multi-view range images. It minimises the sum of distances between transformed points in the first frame and the tangent planes at the intersection points between the normals at the transformed points and the surfaces in the second image. The intersection point is computed by an iterative algorithm. The idea is similar to that proposed in [102] used to establish correspondences. Thus this algorithm does not need explicit point-to-point correspondences, but it does need to extract the tangent planes. This method was adopted later by [7, 30, 32] for pair-wise image registration. In [6], instead of minimising directly the point-to-plane distance, the point on the tangent plane that is closest to the transformed first frame point is explicitly solved, leading the registration parameters to be estimated with a close-form solution (quaternion method).

In 1992, Menq, Yau, and Lai [96] proposed minimising the sum of squared distances from transformed points in measurement space to a parametric surface in CAD design space. Since the objective function is highly non-linear involving time expensive optimisation, the paper proposed using the Euler angles to represent the relationship between measured and design points to be homogenous. As a result, the non-linear optimisation can be implemented by a least squares method without considering rigid constraints. But when the nearest point on the surface is calculated, the rigid constraint is imposed again. Experimental results have shown that the precision of the approximation is not lost while the algorithm could be 10 times faster. For error comparative analysis, a statistical model has been developed to determine the minimum number of required measurement points so that the sampling points can closely represent the entire population based on accuracy and tolerance specification. In [67], a similar idea was proposed, minimising the sum of distance between transformed points in regions in the first frame where all points can find correspondences in the second image and their projections on the parametric geometric models like planes, spheres, cylinders, cones, tori, obtained from the segmentation stage.

Brunnstrom and Stoddard in 1996 [12] proposed using genetic algorithm to establish point correspondences for the matching of free-form surfaces. The GA registration result is then refined by local gradient descent method like the ICP algorithm [8]. They first sample the scene and model surfaces to transform the continuous surfaces into discrete ones, then for each point, they define four invariants: one is about the distance between two points, three are about angles between the normals at these points and the line segment connecting these points. Based on these invariants, the matching quality can be defined assuming that the measurement and sampling error is modelled as simply probabilistic. Finally the cost function is defined as maximising the number of points between the scene and model points aligned by the rigid motion parameters.

5.1.3 Accurate Point Correspondence Establishment

The ICP criterion just applies a single distance constraint to establish point correspondences between two images and a single distance constraint cannot generally uniquely determine the exact position of the correspondent. On the other hand, inaccurate initialisation of camera pose, occlusion, appearance and disappearance of points, and noise distribution in image data render the ICP criterion to inevitably introduce false matches in almost every iteration of image registration. Thus, a number of techniques have been proposed to test whether the established correspondences are true or not. Specifically, the following techniques are used: orientation consistency test [141, 60, 66, 103, 109], discarding boundary points [131, 110], threshold [141, 66, 60, 62, 112], point-to-point distance consistency [29], the Mahalanobis distance based general χ^2 test [37, 38]. increasing dimensionality of points through incorporating colour [60, 114], laser reflectance strength [1, 99], normal vector [38, 30, 32, 114], invariants [116, 119], curvature [140], intensity [38], or brightness [98].

An overall analysis reveals that two kinds of information can be used to test possible point correspondences: one is the information extracted from a single image, the other is the information connecting point correspondences before and after a rigid motion. The former assumes that the same physical point should have the same local structural (geometric or optical) information that is invariant to the camera motion. The latter assumes that since the images are subject to a single camera motion, all the point correspondences from these images thus must satisfy that motion equation. Consequently, any point pair that does not possess similar local structural information or does not satisfy approximately the motion equation due to imaging noise cannot represent a real correspondence, leading to false matches to be eliminated.

5.1.4 Efficient Possible Point Correspondence Establishment

The search of the closest points dominates the ICP algorithm in the sense of computation. Thus, it is vital to accelerate the closest point search for computationally efficient image registration. There are a number of ways to achieve this: constraint nearest point [131], k-D tree [30, 32, 141, 123, 62, 64, 144, 98, 1], grid closest point [137], the cache of closest point found in previous iteration [123], z-buffer [5, 6], reverse calibration [10], and neighbour search [64]. Within these methods, the k-D tree is commonly used and it reduces the computational complexity of the ICP algorithm from $O(n_1n_2)$ to $O(n_1 \ln n_2)$. In [18], to reduce the computational cost, the data image is heavily sub-sampled randomly while the target image is untouched.

5.1.5 Robust Motion Estimation

Once the point correspondences have been established, the motion estimation is down to robust least squares estimation problem. Thus, all robust optimisation method can be applied in one way or another. In general, the robust motion estimation can be achieved by taking into consideration the following aspects:

- The reliability of the point correspondences [22, 28, 126, 131];

- The registration errors of the point correspondences [1, 98];

- Prediction of motion parameters [8, 123, 144];

- Sampling the motion parameter space [8];

- Sampling possible point correspondences [42, 125];

- Stochastic search for the motion parameters based on techniques like genetic algorithms [85, 122, 18], simulated annealing [91], or very fast simulated annealing [10];

- Heuristics [122].

In [95], Maurer et al. extended the classical ICP algorithm through weighting points and surfaces. For points not lying on a surface, they are weighted equally. For the points lying on a surface, they are weighted by the sum of the areas of all triangles with points as their vertices. Finally, the weights are normalised.

In [8], the motion parameters are predicted in 6D space: 3D for rotation, 3D for translation. In [123], it was proposed to de-couple the acceleration of rotation from that of translation. In practice, how to predict the new motion parameters have to be properly decided: a linear approximation or a parabolic interpolant? In addition, the prediction step is crucial, since if the step is too large, then an over-shooting problem will occur, leading the algorithm to oscillate or require intensive computation without a significant improvement to the existing registration results. If the step is too small, then the algorithm can hardly traverse local minimum, leading the algorithm not to be able to benefit from the motion parameter prediction.

It was proposed in [30, 32] that individual motion updates are computed using force-based optimisation. Thus in order to implement the proposed algorithm, several parameters need to be initialised: the relative weight of normal information, the damping matrices, and the time interval. Since the algorithm assumes that any point must be seen at least twice, thus all points in one image are guaranteed to be able to find correspondents from not necessarily a single, but the combined data sets from several images. As a result, no threshold is needed to reject false matches.

In reality, we only know that two images to be registered are overlapping in 3D space, the knowledge of the exact quantity of overlapping, the distribution of points, occlusion, appearance and disappearance of points, and noise level are not available. Thus, the final registration error is a complex function of a number of factors aforementioned. It is reasonable to believe that the objective function generally has a number of local minima. In order to increase the opportunity in finding the global minimum, it was proposed in [8] to sample the 6D motion space: 3D for rotation and 3D for translation. Each sampled parameters can then be used to initialise the ICP algorithm, leading different registration results, from which the best registration can be selected. A number of problems arise in doing this: (1) the number of sampling; (2) how to sample; (2) from where to sample. This means that we must have some rough ideas about the motion; (4) the resulting computation cost increase; (5) how to evaluate the quality of motion candidate: average registration error or the number of points in the overlapping area [37, 136]?

In order to increase the chance for the ICP algorithm to converge to the global minimum, Rusinkiewicz and Levoy proposed in 2001 [110] sampling image points based on an idea that the distribution of normals among selected points is as large as possible. However, the estimation of the surface normal is often sensitive to noise. For the improvement of the ICP algorithm proposed in [16], Gelfand etc. proposed in 2003 [41] sampling more points from informative area so that the stability of the ICP algorithm can be achieved.

In [42], it was proposed that a set of points are randomly selected from one image according to a probability function, which is estimated as the histogram of some attributes. For a point in this set, intensity-based attributes are first used to find all those in another with compatible attributes. Then from these compatible points, the closest point in the sense of Euclidean distance can be found. Finally the registration parameters is estimated through minimising a commutative objective function in the sense of image orders. To get a better result, the least median of squares method is used, which essentially deals with false matches.

To deal with false matches in the process of image registration, the trimmed closest point algorithm (TrICP) was proposed in [17]. Since the registration parameters are estimated in the least squares sense, it is natural to extend the ICP algorithm to the TrICP algorithm. This algorithm was later adopted in [85] to refine the existing registration results. The TrICP algorithm has a clear advantage that its convergence can be proved. However, it in practice can easily converge to local minima, leading to poor registration results.

A multi-resolution scheme was proposed in [64] for increasing the speed of the ICP algorithm and enabling the ICP algorithm to avoid local minimum. The experiments based on both synthetic data and real images have shown that the neighbour search based multi-resolution scheme is the most effective in accelerating ICP for accurate registration among kD tree, neighbour search, kD tree in conjunction with multi-resolution, and neighbour search with multi-resolution. However, the paper does not clarify which version of ICP was used for comparison. Since different versions of ICP have different behaviours in register-ing range images, the comparison is in general difficult. The multi-resolution scheme was also adopted in [126]. In [144], hierarchical point selection is employed for both efficient and accurate motion estimation.

In [98], it was proposed that the registration parameters were estimated through mini-mizing the weighted average of the point to plane error and normal flow constraint. In order to combat outliers, an M-estimator was employed for more accurate registration parameters estimation. The weight was computed using the sigmoid function with regard to the average distance of matched points during the closest point matching process. It was observed that while the ICP tracker can handle translation well and the normal flow constraint can handle rotation well, the hybrid method is able to track all the sequence reliably.

In [125, 126], the ICP algorithm was improved by bootstrapping. First, the point corre-spondences between two sets of blood vessel trace points are established by the traditional ICP criterion. The initial motion estimate is obtained by matching signatures computed from the widths and orientation of vessels meeting at landmarks. For a specific model, the relative parameters are then estimated from an objective function, minimising the point-to-line distances using an M-estimator optimised by the iteratively re-weighted least squares (IRLS) method. The point-to-line distance is defined to be the projection of point differ-ence along the normal to the trace point. Thirdly, they bootstrap whether the current simple lower-order model needs to be updated by a more complex high-order model based on the covariance matrix of the transformation parameters. Finally they bootstrap whether the re-gion needs to be expanded based on the uncertainty in the transformation estimate. As long as the overlap between the two sets of points is large enough, it was reported that virtually flawless performance has been obtained. But sometimes, an extremely large number of estimates is needed for success.

It was proposed in [122] that the genetic algorithm (global search) in conjunction with hill-climbing heuristics (local search) are used to search for the camera motion parameters in 6D space: 3D for rotation and 3D for translation. The fitness function is defined as the average registration error with the registration error of outliers capped by a threshold. To characterise the quality of registration, an interpenetration measure is defined. Experimental results based on real images have shown that a combination of mean square error (MSE) and surface interpenetration measure (SIM) can provide a good measure for the quality of registration algorithms.

5.1.6   Direction Motion Estimation without Involving Explicit Point Correspondences

This class of methods first constructs an objective function, then directly optimise this ob-jective function for motion estimation.

In [9], it was proposed to directly optimise the sum of the normal distributions from the second point cloud of all points in the first point cloud for the estimation of the motion parameters, using the Newton method. This method does not need explicit point correspon-dences.

While the point-to-point ICP algorithm [8] is sometimes slow for certain types of input data and initial positions, the point-to-plane ICP [16] tends to oscillate and fails to converge with the shapes starting far away from each other, or point clouds noisy, a novel distance is proposed in [97]. They use quadratic approximant to approximate the model cloud, leading to a second order accurate approximant to the squared distance objective function to be minimised for the motion parameters estimation. For the quadratic approximation, the computation of principal frame at each point in the model cloud is necessary.

5.1.7 Multi-view Registration

The methods that are used for the registration of multiple range images can be classified into two main categories: meta-view method and view graph balance method.

The main idea of the meta-view method [16] is to transform the multiple image registration to pair-wise image registration. Thus, all new images have to be registered onto the merged previous images. This method has an advantage of easy implementation and also a disadvantage that it may accumulate registration error so that the registration error of some images can be very large. This method is later adopted by [66, 67] for multi-view image registration.

In [89], the pair-wise registration is implemented through line segment matching. To avoid accumulating registration error, a network of pose relations is built on which two constraints are imposed: one is the pair-wise registration results, the other is the odometry measurement. Then through minimising the Mahalanobis distance about all combinations among different poses, the optimal globally consistent poses are obtained. To simplify the global optimisation, a linear measurement equation is adopted from which a closed form solution is derived to the global consistent poses.

It [117, 118], it was proposed that in order to avoiding accumulating registration error, the inconsistency of registration parameters in a cycle can be detected and the corresponding rotation error is then equally distributed among the views in that cycle and the translation vectors can then be optimised from a linear equation group.

The view graph balance approach proposed in [59] first uses pair-wise registration to estimate the motion information between neighbouring views, then construct an overall view graph. Since two views in the graph may have more than one path, then a number of constraints, visibility test, space occupancy test, can be carried out to evaluate the registration quality along the path. Eventually, the path connecting two views with the highest quality is found, leading to the best registration results with registration errors evenly distributed in different images. This method was also adopted by [1, 7] for multi-view image registration.

In [40, 7, 6], a star shape of network of views is adopted. A view is selected as the central of star. Then for any other view, it can be transformed into the coordinate frame of the central view, then the incremental matrix is estimated between this view and all other views. This process can be iterated until the incremental matrix is nearly equal to the identity matrix. As a result of this operation, the error of pair-wise registration can be equally distributed among different views. There are two strategies for the update of the network of views: one is immediately after any incremental matrix has been estimated [6] and the other is after all incremental matrixes corresponding to non-central views have been estimated [40, 7].

In [128], the robot odometry was employed to provide a good initialisation for the ICP algorithm. Then a simultaneous matching method was proposed for the construction of a consistent 3D model. This method is based on the method proposed in [103] and it iterates the registration of one scan with its neighbours until the registration is stable.

For the 3D modelling of historic sites, Allen et al. [1] first extract planar regions and line segments from range scans, then match the line segments to obtain a coarse estimation of registration parameters between pair-wise neighbouring scans. For more accurate registration, a global registration scheme is carried out where one scan is selected as anchor and all others register to it. Since two scans may have more then one path, the path with low confidence can be rejected, leading to high quality global registration results. Finally, the existing registration results are refined by a modified ICP algorithm. This modified ICP algorithm first searches for a set of points to a transformed point using K-D tree, then the laser reflectance strength value is used to get the closest point, and finally an M-estimator is applied to estimate the registration parameters, using a conjugate-gradient search to find the minimum.

A general global registration scheme was proposed in [13]. This scheme first extracts the regularised medial scaffolds which are then matched using the graduated assignment (GA) algorithm [43]. For the graph match of scaffolds, the GA algorithm was extended in the sense of creating a higher order objective function. The experimental results based on real images have shown that the algorithm sometimes does get stuck at local minimum, leading to poor registration results which may be enough to ICP for refinement.

5.2   Advantages of the ICP Algorithm

All the aforementioned publications show that the traditional ICP criterion has been popular and become a *de facto* standard for the establishment of possible correspondences between overlapping range images to be registered. This is due to the following widely acknowledged reasons: (1) Since it does not require image segmentation or feature extraction and matching and has a closest form solution to the registration parameters, it is easy to implement; (2) Since it is designed to be applicable to points and all geometric representation of 3D free form surface (such as triangular meshes, planar patches, line segments, or implicit or explicit representation) can be transformed to sets of points, it is of general use; and (3) Since elaborate data structures can be devised to accelerate the search for the closest points, it is computationally efficient. For example, if the k-D tree data structure [39] is used to accelerate the search for the closest points, then the ICP algorithm has a computational complexity of $O(n \log n)$.

5.3   Using Motion Properties to Eliminate False Matches

Due to inaccurate initial motion parameters and occlusion and appearance and disappearance of points, the point matches established by the traditional ICP criterion inevitably include false matches. Thus, the key to successfully applying the ICP criterion for image registration is to eliminate false matches. While the aforementioned approaches are mainly based on invariants described in a single coordinate frame, motion properties were employed in [84, 76, 107] to eliminate false matches.

Theorem 1 shows that once a set of image correspondences $(\mathbf{p}_i, \mathbf{p}'_{c(i)})$ $(i \in \{1, 2, \cdots, n\}, n \geq 3)$ satisfy the constraints represented by Equations 6, 7, and 8, then they must correspond to a 3D rigid non-pure translational transformation. If the 3D points $\mathbf{p}_i$ $(i \in \{1, 2, \cdots, n\}, n \geq 3)$ undergo a 3D rigid non-pure translational transformation, then the resulting image correspondences $(\mathbf{p}_i, \mathbf{p}'_{c(i)})$ $(i \in \{1, 2, \cdots, n\}, n \geq 3)$ must satisfy the constraints represented by Equations 6, 7, and 8 respectively. If a point pair does not satisfy any of the constraints represented by Equations 6, 7, and 8, then they by no means represent an image point correspondence.

Theorem 2 also shows that once a set of reflected correspondences $(\mathbf{p}_i, \mathbf{p}''_{c(i)})$ $(i \in \{1, 2, \cdots, n\}, n \geq 3)$ satisfy the constraints represented by Equations 9 or 10 or 11, and 12 and 13, then the original correspondences $(\mathbf{p}_i, \mathbf{p}'_{c(i)})$ $(i \in \{1, 2, \cdots, n\}, n \geq 3)$ must undergo a 3D rigid body transformation. If the 3D points $\mathbf{p}_i$ $(i \in \{1, 2, \cdots, n\}, n \geq 3)$ undergo a 3D rigid transformation, then the resulting reflected correspondences $(\mathbf{p}_i, \mathbf{p}''_{c(i)})$ $(i \in \{1, 2, \cdots, n\}, n \geq 3)$ must satisfy the constraints represented by Equations 9, 10, 11, 12, and 13 respectively. If a pair of points do not satisfy any of these equations, then they cannot possibly represent an image correspondence.

Thus, the relative difference g_i between the two sides of rigid motion constraints like Equations 6, 7, 8, 9, 10, 11, 12, and 13 was proposed in [76, 107, 81] for the measurement of the relative quality of possible point matches established by the traditional ICP criterion. The larger the difference, the worse the possible point matches. Then a stistical model is proposed to reject false matches: If $|g_i - \mu_g| > \kappa \, \sigma_g$ then the possible correspondence $(\mathbf{p}_i, \mathbf{p}'_{c(i)})$ is regarded as a false match where $\mu_g = \frac{1}{n_1} \sum_{i=1}^{n_1} g_i$ and $\sigma_g = \sqrt{\frac{1}{n_1} \sum_{i=1}^{n_1} (g_i - \mu_g)^2}$. Otherwise, it is regarded as a feasible one. All feasible correspondences are called the refined correspondences. The motion parameters are then updated using refined point correspondences. In general, if a point pair established by the traditional ICP criterion satisfies two rigid motion constraints, then they are very likely to represent a real correspondence. More constraints do not necessarily lead to better results except the fact that they often take more time for image registration. The algorithm using motion properties to eliminate false matches is called the Geometric ICP (GICP) algorithm.

In order to apply motion properties to reliably eliminate false matches, some motion parameters of interest like the critical point $\mathbf{c}$ or the essential point $\mathbf{e}$ have to be accurately estimated using, for example, the Monte Carlo resampling technique in conjunction with the median filter. Since the possible point matches are corrupted by false matches, the estimation of the motion parameters of interest is not easy and this is also a "Chicken and egg" issue. To address this issue, the collinear ICP (CICP) algorithm was proposed in [82, 83]. The CICP algorithm defines novel point correspondence quality measures without involving additional parameter estimation from false matches corrupted data. For this purpose, the

CICP algorithm defines two point correspondence quality measures based on registration error and collinearity constraint.

Basically, given accurate motion parameters rotation matrix $\mathbf{R}$ and translation vector $\mathbf{t}$, the relationship between a real correspondence $(\mathbf{p}_i, \mathbf{p}'_{c(i)})$ can be represented as:

$$\mathbf{p}'_{c(i)} \approx \mathbf{R}\mathbf{p}_i + \mathbf{t} \tag{27}$$

where $\approx$ denotes "approximately equals" due to imaging noise. Equation 27 essentially describes a rigid motion constraint bridging the points $(\mathbf{p}_i, \mathbf{p}'_{c(i)})$ described in two different coordinate frames before and after a rigid motion $(\mathbf{R}, \mathbf{t})$. This constraint implies the following three constraints: (1) $\|\mathbf{p}'_{c(i)}\| \approx \|\mathbf{R}\mathbf{p}_i + \mathbf{t}\|$; (2) points $\mathbf{p}'_{c(i)}$, $\mathbf{R}\mathbf{p}_i + \mathbf{t}$, and the optical centre $\mathbf{O}$ of the range camera are as collinear as possible; and (3) the including angle between vectors $\mathbf{p}'_{c(i)}$ and $\mathbf{R}\mathbf{p}_i + \mathbf{t}$ is smaller than $180°$. These constraints represent necessary conditions for a pair of points to represent a real correspondence. If a pair of points cannot satisfy any of these constraints, then they cannot represent a real correspondence. Thus these constraints can also be used to evaluate whether a pair of points represents a real correspondence.

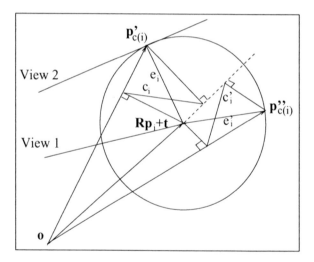

Figure 6: Collinearity constraint is used to reject false matches. With the same registration error e_i, since $c'_i < c_i$, point $\mathbf{p}''_{c(i)}$ is a preferred correspondent over point $\mathbf{p}'_{c(i)}$ of point $\mathbf{p}_i$.

Since both $\mathbf{p}'_{c(i)}$ and $\mathbf{R}\mathbf{p}_i + \mathbf{t}$ are in front of the range camera, the third constraint is automatically satisfied. This means that the third constraint cannot provide useful information for the evaluation of possible point matches $(\mathbf{p}_i, \mathbf{p}'_{c(i)})$. The former two constraints imply that the transformed point $\mathbf{R}\mathbf{p}_i + \mathbf{t}$ should be as close as possible to point $\mathbf{p}'_{c(i)}$ and to the ray passing through $\mathbf{p}'_{c(i)}$ (Figure 6). The former essentially says that the registration error e_i between the point correspondence $(\mathbf{p}_i, \mathbf{p}'_{c(i)})$ should be minimised:

$$e_i = \|\mathbf{p}'_{c(i)} - \mathbf{R}\mathbf{p}_i - \mathbf{t}\| \tag{28}$$

The latter says that the collinearity error between transformed point $\mathbf{Rp}_i + \mathbf{t}$ and the ray passing through $\mathbf{p}'_{c(i)}$ should be minimised. The collinearity error is defined as:

$$c_i = |\mathbf{p}'^{T}_{c(i)}(\mathbf{Rp}_i + \mathbf{t})| \; \Big|\Big| \frac{\mathbf{p}'_{c(i)}}{\mathbf{p}'^{T}_{c(i)}\mathbf{p}'_{c(i)}} - \frac{\mathbf{Rp}_i + \mathbf{t}}{(\mathbf{Rp}_i + \mathbf{t})^T(\mathbf{Rp}_i + \mathbf{t})} \Big|\Big| \tag{29}$$

Note that the collinearity error defined in [83] is symmetric about the transformed point $\mathbf{Rp}_i + \mathbf{t}$ and $\mathbf{p}'_{c(i)}$. Thus, it eliminates the bias of the distances defined either from $\mathbf{Rp}_i + \mathbf{t}$ to the ray passing $\mathbf{p}'_{c(i)}$ or from $\mathbf{p}'_{c(i)}$ to the ray passing through $\mathbf{Rp}_i + \mathbf{t}$ in the process of rejecting false matches. Consequently, it is expected that this definition of collinearity error is more robust in dealing with false matches for the automatic registration of overlapping 3D range images with various qualities. The collinearity constraint is justified by the assumption that the scanning error mainly occurs along the ray shot from the range camera [112].

The difference between the GICP and CICP algorithms lies in that the latter does not need to estimate additional motion parameters of interest from false matches corrupted correspondence data and does not need to employ the Monte Carlo resampling technique to estimate any additional parameters of interest. Consequently, the CICP algorithm is expected to be more accurate and robust than the GICP algorithm.

6   Range Image Registration Using Both Motion and Structural Properties

Rigid motion constraints were proposed to bridge the points described in two different coordinate frames before and after a rigid motion. These constraints were successfully used to deal with false point matches, thus leading to more accurate image registration and motion estimation results. However, the Geometric ICP (GICP) algorithm was designed for the automatic registration of sets of unorganised points and did not make use of any geometric information existing between neighbouring points. As a result, the GICP algorithm sometimes does converge to a local minimum leading to inaccurate image registration and motion estimation results. It is proposed to combine the advantage of GICP with those of using structured data, since it is acknowledged that, in general, algorithms based on structured data yield better registration results [22]. To extract useful structural information for the evaluation of possible matches established by the traditional ICP criterion, the proximity and closeness constraints were proposed in [77, 78, 80] which are outlined as follows.

6.1   Proximity Constraint

Independent of the criterion used to establish correspondences, the correspondents of neighbouring points in the first image $\mathbf{P}$ should also be neighbouring points in the second image $\mathbf{P}'$. This is illustrated in Figure 7, where $\mathbf{NN}_1$, $\mathbf{NN}_2$, $\mathbf{NN}_3$, $\mathbf{NN}_4$, $\mathbf{DN}_1$, $\mathbf{DN}_2$, $\mathbf{DN}_3$, $\mathbf{DN}_4$

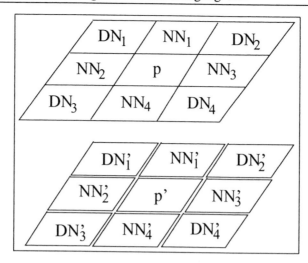

Figure 7: The neighbouring points should be neighbouring before and after a rigid motion with no occlusion.

and $\mathbf{p}$ are neighbouring points in the first image. For the sake of brevity, $\mathbf{NN}_1$, $\mathbf{NN}_2$, $\mathbf{NN}_3$, and $\mathbf{NN}_4$ are referred to as the non-diagonal neighbouring points of $\mathbf{p}$: $NN(\mathbf{p}) = \{\mathbf{NN}_1, \mathbf{NN}_2, \mathbf{NN}_3, \mathbf{NN}_4\}$ and $\mathbf{DN}_1$, $\mathbf{DN}_2$, $\mathbf{DN}_3$, and $\mathbf{DN}_4$ are referred to as the diagonal neighbouring points of $\mathbf{p}$: $DN(\mathbf{p}) = \{\mathbf{DN}_1, \mathbf{DN}_2, \mathbf{DN}_3, \mathbf{DN}_4\}$. In principle, their possible correspondents $\mathbf{NN}_1'$, $\mathbf{NN}_2'$, $\mathbf{NN}_3'$, $\mathbf{NN}_4'$, $\mathbf{DN}_1'$, $\mathbf{DN}_2'$, $\mathbf{DN}_3'$, $\mathbf{DN}_4'$, and $\mathbf{p}'$ in the second image should also be neighbouring points. Such topology among points should not be changed with different viewpoints without occlusion and appearance and disappearance of points. This structural constraint is implemented for the elimination of false point matches as follows.

For each possible correspondence $(\mathbf{p}_i, \mathbf{p}'_{c(i)})$, it was proposed to first check whether or not the non-diagonal neighbouring points $\mathbf{NN}_1$, $\mathbf{NN}_2$, $\mathbf{NN}_3$ and $\mathbf{NN}_4$ of $\mathbf{p}_i$ are all valid. If so, then the differences between the rows and columns of their correspondents $g(\mathbf{p}_n)$ $(\mathbf{p}_n \in NN(\mathbf{p}_i))$ and $g(\mathbf{p}_i)$ in the second raster image file is computed as:

$$\{|u'(g(\mathbf{p}_n)) - u'(g(\mathbf{p}_i))|, |v'(g(\mathbf{p}_n)) - v'(g(\mathbf{p}_i))|$$
$$|\mathbf{p}_n \in NN(\mathbf{p}_i)\} = \{RC_j | j = 1, 2, \cdots, 8\} \tag{30}$$

Otherwise, it was proposed to further check whether or not the diagonal neighbouring points $\mathbf{DN}_1$, $\mathbf{DN}_2$, $\mathbf{DN}_3$ and $\mathbf{DN}_4$ of $\mathbf{p}_i$ are all valid. If so, then the differences between the rows and columns of their correspondents $g(\mathbf{p}_d)$ $(\mathbf{p}_d \in DN(\mathbf{p}_i))$ and $g(\mathbf{p}_i)$ in the second raster image file can be similarly computed as:

$$\{|u'(g(\mathbf{p}_d)) - u'(g(\mathbf{p}_i))|, |v'(g(\mathbf{p}_d)) - v'(g(\mathbf{p}_i))|$$
$$|\mathbf{p}_d \in DN(\mathbf{p}_i)\} = \{RC_j | j = 1, 2, \cdots, 8\} \tag{31}$$

Sort RC_j in ascending order and let $a_i = RC_4$. Since the row and column difference between any neighbouring point $\mathbf{p}_b$ $(\mathbf{p}_b \in NN(\mathbf{p}_i) \bigcup DN(\mathbf{p}_i))$ of $\mathbf{p}_i$ and $\mathbf{p}_i$ in the first raster

image file is either one or zero, the row and column difference between their possible correspondents $g(\mathbf{p}_b)$ and $g(\mathbf{p}_i)$ in the second raster image file must also be either one or zero.
Based on this observation, the following rule was proposed to label the correspondence: if
$a_i < 2$, then the correspondence $(\mathbf{p}_i, \mathbf{p}'_{c(i)})$ is labelled as $d_i = 1$, otherwise it is labelled as
$d_i = -1$. This constraint is called the *proximity* constraint which measures the similarity
between the local structures around a possible correspondence $(\mathbf{p}_i, \mathbf{p}'_{c(i)})$ in the two images
to be registered.

In [78, 80, 81], the rows and columns of four neighbours were considered. It is certainly
the case that one could consider more neighbours. The proximity constraint describes the
neighbouring relationship among points in a single image. In order to aid the computation
of the proximity constraint, it was assumed that the images to be registered are of the same
resolution, which is normally the case when the same range camera is used to capture
the two images. If the images are not of the same resolution, then an interpolation or a
resampling scheme is required to pre-process the images.

6.2   Closeness Constraint

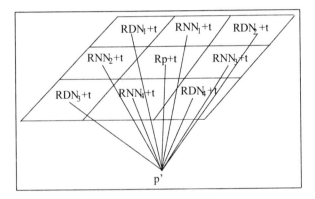

Figure 8: The distance between $\mathbf{R}\mathbf{p} + \mathbf{t}$ and $\mathbf{p}'$ is smaller than that between any other transformed neighbouring point of $\mathbf{p}$ and $\mathbf{p}'$ with a smooth surface.

Given the motion parameters rotation matrix $\mathbf{R}$ and translation vector $\mathbf{t}$, the traditional
ICP criterion (Equation 26) assumes that the correspondent of a point $\mathbf{p}$ in the first image is
the closest point $\mathbf{p}'$ in the second image to the transformed point $\mathbf{R}\mathbf{p} + \mathbf{t}$. If we change our
viewpoint to look carefully at this assumption from the second image to the first image, then
we can conclude that, for any point $\mathbf{p}'$ in the second image, the closest point in the first image
to the inversely transformed point $\mathbf{R}^T(\mathbf{p}' - \mathbf{t})$ should also represent its correspondent. This
is illustrated in Figure 8. In general, for a real correspondence $(\mathbf{p}, \mathbf{p}')$, the distance between
$\mathbf{R}\mathbf{p} + \mathbf{t}$ and $\mathbf{p}'$ should be smaller than that between any other transformed point in the first
image and $\mathbf{p}'$. This implies that in Figure 8 if points $(\mathbf{p}, \mathbf{p}')$ represent a real correspondence,
then the distance between points $\mathbf{RNN}_1 + \mathbf{t}$ and $\mathbf{p}'$, $\mathbf{RNN}_2 + \mathbf{t}$ and $\mathbf{p}'$, $\mathbf{RNN}_3 + \mathbf{t}$ and $\mathbf{p}'$,
$\mathbf{RNN}_4 + \mathbf{t}$ and $\mathbf{p}'$, $\mathbf{RDN}_1 + \mathbf{t}$ and $\mathbf{p}'$, $\mathbf{RDN}_2 + \mathbf{t}$ and $\mathbf{p}'$, $\mathbf{RDN}_3 + \mathbf{t}$ and $\mathbf{p}'$, and $\mathbf{RDN}_4 + \mathbf{t}$ and

$\mathbf{p}'$ should be larger than that between $\mathbf{Rp} + \mathbf{t}$ and $\mathbf{p}'$. Based on this assumption, a structural constraint to deal with false matches can be implemented as follows.

For a possible correspondence $(\mathbf{p}_i, \mathbf{p}'_{c(i)})$, if the non-diagonal neighbouring points $\mathbf{NN}_1$, $\mathbf{NN}_2$, $\mathbf{NN}_3$ and $\mathbf{NN}_4$ of $\mathbf{p}_i$ are all valid and the object surface is smooth, then the following distances between these transformed neighbouring points and $\mathbf{p}'_{c(i)}$ can be computed as:

$$dis_1 = ||\mathbf{Rp}_i + \mathbf{t} - \mathbf{p}'_{c(i)}||$$
$$dis_2 = ||\mathbf{RNN}_1 + \mathbf{t} - \mathbf{p}'_{c(i)}||, \ dis_3 = ||\mathbf{RNN}_2 + \mathbf{t} - \mathbf{p}'_{c(i)}||$$
$$dis_4 = ||\mathbf{RNN}_3 + \mathbf{t} - \mathbf{p}'_{c(i)}||, \ dis_5 = ||\mathbf{RNN}_4 + \mathbf{t} - \mathbf{p}'_{c(i)}||$$

Otherwise, it was proposed to check whether or not the diagonal neighbouring points $\mathbf{DN}_1$, $\mathbf{DN}_2$, $\mathbf{DN}_3$ and $\mathbf{DN}_4$ of $\mathbf{p}_i$ are all valid. If so, then the following distances between these transformed neighbouring points and $\mathbf{p}'_{c(i)}$ can be similarly computed as:

$$dis_1 = ||\mathbf{Rp}_i + \mathbf{t} - \mathbf{p}'_{c(i)}||$$
$$dis_2 = ||\mathbf{RDN}_1 + \mathbf{t} - \mathbf{p}'_{c(i)}||, \ dis_3 = ||\mathbf{RDN}_2 + \mathbf{t} - \mathbf{p}'_{c(i)}||$$
$$dis_4 = ||\mathbf{RDN}_3 + \mathbf{t} - \mathbf{p}'_{c(i)}||, \ dis_5 = ||\mathbf{RDN}_4 + \mathbf{t} - \mathbf{p}'_{c(i)}||$$

Then the following rule is used to label the correspondence: if $dis_1 < dis_2$, then the correspondence $(\mathbf{p}_i, \mathbf{p}'_{c(i)})$ is labelled as $c_i = 1$, otherwise it is labelled as $c_i = -1$ where dis_2, dis_3, dis_4 and dis_5 have been sorted in ascending order. This constraint is called the *closeness* constraint. It describes the relationship among points in different images.

6.3   Image Registration Based on Both Motion and Structural Properties

Research has shown that false matches can accelerate the image registration process. Thus, a novel strategy was proposed in [78, 80] that the structural constraints like proximity and closeness constraints outlined above are used in the middle of the process of image registration. The algorithm using both the motion and structural constraints to eliminate false matches is called the Structured ICP (SICP) algorithm. In principle, all invariants to either rotation or translation or both such as spin image [57, 58], colour [60, 61, 62], laser reflectance strength value [99], normal vectors [37, 38] can all be used to evaluate the false matches. But of course, different structural constraints have different sensitivities to noise in extraction and different efficiencies in computation.

Clearly, the SICP algorithm represents a compromise between algorithms based on pure motion properties and pure structural feature matching. If the structural constraints are used at the beginning of image registration, then the SICP algorithm degenerates to feature matching based ones and false matches are eliminated at an early stage in the image registration process, yielding a slow progress of image registration. If the structural constraints are used too late, then the SICP algorithm degenerates to the GICP algorithm without using structural information to eliminate false matches, leading to poor registration results. In practice, the firing point when the structural constraints are used can be experimentally determined.

7   Range Image Registration Based on the Two-way Constraint

The false matches introduced by the ICP criterion are, to some degree, caused by the ignorance of a two-way constraint: if point $\mathbf{p}'$ is the correspondent of point $\mathbf{p}$, then point $\mathbf{p}$ should also be the correspondent of point $\mathbf{p}'$. In [21, 49, 27], only a one-way constraint was enforced. Obviously, the one-way constraint is a general case of the two-way constraint. While outliers were modelled in [27], they were not addressed in [49].

However, in practice, once this two-way constraint is enforced, the algorithm often becomes much more complex not only in establishing possible point matches, but the optimisation of camera motion parameters as well, given the possible point matches between the images to be registered.

In [44], a graduated assignment (softassign) algorithm is proposed for the 3D free form shape matching. In [24], the softassign algorithm was extended for the alignment of a 3D model and a projective image. The advantages of the softassign algorithm lie in that while it explicitly models outliers in the process of 3D free form shape matching, it adopts the Sinkhorn iterative alternate row and column normalization procedure [124] to gradually impose the two-way constraint, greatly simplifying the establishment of point matches between the images to be registered. Moreover, the softassign algorithm can be treated as an EM algorithm [26]: in the expectation step, assume that the camera motion parameters are fixed, the match matrix $\mathbf{M}$ is then estimated; in the maximization step, assume that the match matrix $\mathbf{M}$ is fixed, the camera motion parameters rotation matrix $\mathbf{R}$ and translation vector $\mathbf{t}$ are then estimated in the weighted least squares sense. As a result, the softassign algorithm is theoretically elegant since it enforces the two-way constraint and can be justified in the framework of the EM algorithm. Practically, once the algorithm correctly matches 3D free form shapes with small motions, it can match 3D free form shapes undergoing a motion with a rotation angle up to $90°$ around a certain rotation axis. This characteristic implies that, in practice, the number of images to be captured, and thus time required for processing, can be greatly reduced.

7.1   Objective Function

Intuitively, any point $\mathbf{p}'_j$ in the second image can be a candidate correspondent for a point $\mathbf{p}_i$ in the first image. However, the point matches $(\mathbf{p}_i, \mathbf{p}'_j)$ should satisfy the two-way constraint aforementioned: while point $\mathbf{p}'_j$ is the correspondent of point $\mathbf{p}_i$, point $\mathbf{p}_i$ should also be the correspondent of point $\mathbf{p}'_j$. However, due to inaccurate motion parameters at the beginning of registration and occlusion and appearance and disappearance of points, the two-way constraint is often violated. For characterising the two-way constraint violation, a parameter m_{ij} in the unit interval $[0, 1]$ has to be used to represent the probability for point $\mathbf{p}'_j$ to be the correspondent of point $\mathbf{p}_i$. The two-way constraint can then be used to optimise m_{ij}.

Since occlusion and appearance and disappearance of points almost always occur, two slack variables $\mathbf{p}_{n_1+1}$ and $\mathbf{p}'_{n_2+1}$ have to be introduced so that any point $\mathbf{p}_i$ in $\mathbf{P}$ can correspond to $\mathbf{p}'_{n_2+1}$ and any point $\mathbf{p}'_i$ in $\mathbf{P}'$ can correspond to $\mathbf{p}_{n_1+1}$. As a result, the two-way constraint can be formulated as: $\sum_{j=1}^{n_2+1} m_{ij} = 1$ and $\sum_{i=1}^{n_1+1} m_{ij} = 1$. For the optimisation of the match matrix $\mathbf{M} = \{m_{ij}\}$ and the camera motion parameters rigid rotation matrix $\mathbf{R}$ and translation vector $\mathbf{t}$, given the initial camera motion parameters rotation matrix $\mathbf{R}$ and translation vector $\mathbf{t}$, the following objective function can then be constructed [43, 106] as:

$$E_{3D}(\mathbf{M}, \mathbf{R}, \mathbf{t}) = arg \min_{\mathbf{M}, \mathbf{R}, \mathbf{t}} \sum_{i=1}^{n_1} \sum_{j=1}^{n_2} m_{ij} \|\mathbf{p}'_j - \mathbf{R}\mathbf{p}_i - \mathbf{t}\|^2$$

$$- \alpha \sum_{i=1}^{n_1} \sum_{j=1}^{n_2} m_{ij} + \frac{1}{\beta} \sum_{i=1}^{n_1} \sum_{j=1}^{n_2} m_{ij}(\ln m_{ij} - 1)$$

$$+ \sum_{i=1}^{n_1} u_i \left(\sum_{j=1}^{n_2+1} m_{ij} - 1 \right) + \sum_{j=1}^{n_2} v_j \left(\sum_{i=1}^{n_1+1} m_{ij} - 1 \right). \tag{32}$$

In this objective function, the first term is the matching error for the point match $(\mathbf{p}_i, \mathbf{p}'_j)$, the second term is to maximise the overlapping area between the images to be matched, the third term is a barrier function making sure that m_{ij} is positive, the fourth and fifth terms are Lagrange parameters that enforce the two-way constraint, α is a parameter indicating when a point pair $(\mathbf{p}_i, \mathbf{p}'_j)$ should be regarded as a feasible point match, and β is the temperature control parameter indicating how chaotic the assignment of possible point matches is.

It was proved in [44] that the optimisation of Lagrange parameters can be implemented by the Sinkhorn iterative alternate row and column normalisation [124]. The application of the Sinkhorn iterative alternate row and column normalisation thus removes the two-way constraint from the above objective function, leading to:

$$E_{3D}(\mathbf{M}, \mathbf{R}, \mathbf{t}) = arg \min_{\mathbf{M}, \mathbf{R}, \mathbf{t}} \sum_{i=1}^{n_1} \sum_{j=1}^{n_2} m_{ij} \|\mathbf{p}'_j - \mathbf{R}\mathbf{p}_i - \mathbf{t}\|^2$$

$$- \alpha \sum_{i=1}^{n_1} \sum_{j=1}^{n_2} m_{ij} + \frac{1}{\beta} \sum_{i=1}^{n_1} \sum_{j=1}^{n_2} m_{ij}(\ln m_{ij} - 1) \tag{33}$$

For the optimisation of m_{ij}, we differentiate this new objective function about it and set the result equal to zero:

$$\frac{\partial E_{3D}(\mathbf{M}, \mathbf{R}, \mathbf{t})}{\partial m_{ij}} = \|\mathbf{p}'_j - \mathbf{R}\mathbf{p}_i - \mathbf{t}\|^2 - \alpha$$

$$+ \frac{1}{\beta}\left(\ln m_{ij} - 1 + m_{ij} \times \frac{1}{m_{ij}}\right)$$

$$= \|\mathbf{p}'_j - \mathbf{R}\mathbf{p}_i - \mathbf{t}\|^2 - \alpha + \frac{1}{\beta} \ln m_{ij} = 0. \tag{34}$$

Then we get:

$$m_{ij} = exp(-\beta(\|\mathbf{p}'_j - \mathbf{R}\mathbf{p}_i - \mathbf{t}\|^2 - \alpha)). \tag{35}$$

This shows that the match quality of point pair $(\mathbf{p}_i, \mathbf{p}'_j)$ is an exponential function of the corresponding point match error $\|\mathbf{p}'_j - \mathbf{R}\mathbf{p}_i - \mathbf{t}\|$. This definition is justified by the universal law of generalisation [120]. That law says that the probability for a stimulus to confuse another decreases exponentially with the distance between these two stimuli. Thus, the points farther from the closest points are unlikely to ambiguate the range image registration and thus, the closest points alone may be enough for accurate range image registration. Moreover, when human brain responds to a stimulus, it inhibits the response to other stimuli [129]. This shows that the response from the human brain to the closest points may inhibit the response to the points farther away from the closest points.

For a robust estimation of m_{ij}, the deterministic annealing scheme is adopted. When the temperature control parameter β is increased for actual image points, it is proposed not to do so for slack variables in order to guarantee that they can cover a large area and correspond to all occluded and appearing and disappearing points.

Once the point match probabilities have been estimated through Sinkhorn procedure, the camera motion parameters can then be estimated by the quaternion method [8] from the following objective function:

$$E_{3D}(\mathbf{R}, \mathbf{t}) = arg\min_{\mathbf{R}, \mathbf{t}} \sum_{i=1}^{n_1} \sum_{j=1}^{n_2} m_{ij} \|\mathbf{p}'_j - \mathbf{R}\mathbf{p}_i - \mathbf{t}\|^2 \tag{36}$$

in the weighted least squares sense.

7.2 Matching Algorithm

The following algorithm was proposed in [44] to optimise the above objective function (Equation 32), taking the two-way constraint into account:

Initialize R to the identity matrix, **t**, β to β_0, $\hat{m}_{ij}$ to $(1 + \varepsilon)$

Begin A: Do A until $(\beta \geq \beta_f)$

Begin B: Do B until **M** converges or # of iterations $> I_0$

Begin C (update correspondence parameters by softassign):

$Q_{ij} \leftarrow \frac{\partial E_{3D}}{\partial m_{ij}}$

$m_{ij}^0 \leftarrow exp(-\beta(Q_{ij} - \alpha))$

Begin D: Do D until $\hat{\mathbf{M}}$ converges or # of iterations $> I_1$

Update $\hat{\mathbf{M}}$ by nomalising across all rows: $\hat{m}_{ij}^1 \leftarrow \frac{\hat{m}_{ij}^0}{\sum_{j=1}^{j=n_2+1} \hat{m}_{ij}^0}$

Update $\hat{\mathbf{M}}$ by nomalising across all columns:

$\hat{m}_{ij}^0 \leftarrow \frac{\hat{m}_{ij}^1}{\sum_{i=1}^{i=n_1+1} \hat{m}_{ij}^1}$

End D

End C

Begin E (update pose parameters using Walker et al.'s method):

Update **R, t**

End E

End B

$\beta \leftarrow \beta_r \beta$

End A

where β_0 denotes the initial temperature control parameter for deterministic annealing, β_r the temperature control parameter increasing rate, β_f the final temperature control parameter, I_0 and I_1 are the maximum iteration numbers, α determines when two points should be regarded as a plausible point match.

7.3   Match Matrix Property

The continuous match matrix $\mathbf{M}$ in the softassign algorithm outlined above converges toward a discrete matrix due to the following two mechanisms that are used concurrently:

1. First, a technique due to Sinkhorn [124] is applied. When each row and column of $\mathbf{M}$ is normalised (several times, alternately) by the sum of the elements of that row and column respectively, the resulting matrix has positive elements with all rows and columns summing to one (doubly stochastic matrix);

2. The term β is increased as the iteration proceeds. As β increases and each row or column of $\mathbf{M}$ is renormalized, the terms m_{ij} corresponding to the smallest Q_{ij} in the overlapping area tend to converge to non-zero and all others except those corresponding to slack variables tend to converge to zero. This is a deterministic annealing process known as softmax. This is desirable behaviour, since it leads to an assignment of point correspondences that satisfy the matching constraints and the minimization of the objective function.

An example of the final match matrix $\mathbf{M}$ estimated by the softassign algorithm is presented in Table 1 where data points were generated using computer simulation.

Table 1: The final match matrix $\mathbf{M}$ estimated by the softassign algorithm using synthetic points data (n=6) with 20% appearing and disappearing points.

	$\mathbf{p}'_1$	$\mathbf{p}'_2$	$\mathbf{p}'_3$	$\mathbf{p}'_4$	$\mathbf{p}'_5$	$\mathbf{p}'_6$
$\mathbf{p}_1$	0.000	0.000	0.000	0.000	0.000	1.000
$\mathbf{p}_2$	0.407	0.000	0.000	0.000	0.000	0.593
$\mathbf{p}_3$	0.000	0.410	0.000	0.000	0.000	0.590
$\mathbf{p}_4$	0.000	0.000	0.413	0.000	0.000	0.587
$\mathbf{p}_5$	0.000	0.000	0.000	0.412	0.000	0.588
$\mathbf{p}_6$	0.593	0.590	0.587	0.588	1.000	

From Table 1, it can be seen that the match matrix $\mathbf{M}$ here satisfies the property described above. $\mathbf{p}_6$ and $\mathbf{p}'_6$ are slack variables, $(\mathbf{p}_2, \mathbf{p}'_1)$, $(\mathbf{p}_3, \mathbf{p}'_2)$, $(\mathbf{p}_4, \mathbf{p}'_3)$, and $(\mathbf{p}_5, \mathbf{p}'_4)$ are correspondences, $\mathbf{p}_1$ is a disappearing point, and $\mathbf{p}'_5$ is an appearing point. The probabilities for these correspondences are not necessarily 1, effected by a number of factors: α, β, their actual match errors, and slack variables. Even though they can be processed so that some of them or all of them are then set to be 1, the motion estimation results are similar, if not the same.

7.4   Extended EM-ICP Algorithm

The EM-ICP algorithm is originally proposed in [49], enforcing the one-way constraint for image registration. This algorithm is extended in [27]. The reason why this extended algorithm is reviewed here is that even though its computational complexity is still $O(n_1 n_2)$, however, its space complexity has been cleverly reduced from $O(n_1 n_2)$ to $O(n_1 + n_2)$. The extended EM-ICP algorithm is implemented as follows.

- The probability of a match between two points has the form:

$$\alpha_{ij} = \frac{1}{C_i} exp(-d_{ij}^2/\sigma_p^2)$$

  where $d_{ij} = ||\mathbf{p}_j' - \mathbf{R}\mathbf{p}_i - \mathbf{t}||$, σ_p is a characteristic distance which decreases over ICP iterations (p stands for points) and C_i is computed so that the sum of each row of the correspondence matrix α_{ij} is 1. Since a virtual point is added to the second point set which is at distance d_0 of all points. This virtual point is here to allow the outliers in $\mathbf{P}$ to be matched to no point within $\mathbf{P}'$. Thus, C_i can be computed as:

$$C_i = exp(-d_0^2/\sigma_p^2) + \sum_{j=1}^{n_2} exp(-d_{ij}^2/\sigma_p^2)$$

- Then the objective function is constructed as:

$$J(\mathbf{R}, \mathbf{t}) = \min_{\mathbf{R}, \mathbf{t}} \sum_{i=1}^{n_1} \lambda_i^2 ||\mathbf{Z}_i - \mathbf{R}\mathbf{p}_i - \mathbf{t}||^2$$

  where

$$\lambda_i = \sum_{j=1}^{n_2} \sqrt{\alpha_{ij}},$$

$$\mathbf{Z}_i = \frac{1}{\lambda_i} \sum_{j=1}^{n_2} \sqrt{\alpha_{ij}} \mathbf{p}_j'$$

which can be easily optimised using the traditional quaternion method [8]. The decrease of parameter σ_p here can be implemented by deterministic annealing [44].

7.5 Strong and Weak Points of the Softassign Algorithm

The softassign algorithm is theoretically elegant, of general use, easy to implement. It only assumes that the free form shapes to be registered are represented as sets of discrete points. This is a reasonable assumption, since other representation of free form shapes like triangular meshes, planar patches, line segments, or analytic forms can all be transformed into sets of points [8].

However, the softassign algorithm has a fatal shortcoming in that its time and space complexities, $O(n_1 n_2)$, are too high [63] for the matching of 3D free form shapes with thousands of points, which will hinder its application. In this case, scene point merging [49], resampling [79], feature point extraction and resampling [19], and feature point extraction and fusion [20] have to be used to reduce the number of points for feasible matching.

8 Conclusion

In this chapter, we have reviewed the range image registration techniques from five aspects: motion representation, motion estimation, image registration based on motion properties, image registration using both motion and structure properties, and the two way constraints. Our review not only emphasizes the evaluation of possible point matches established, but the developability of registration algorithms as well. As a result, this review is hopefully of both theoretic and practical use in robotics research and applications.

Even though a large number of techniques have been proposed so far and they often succeed in various situations, the automatic range image registration is still an open problem. Under general imaging conditions, no algorithms have been proven to converge to the global minimum. Thus, in principle, all algorithms are likely to produce poor registration results. However, with human intervention at different stages in the process of image registration, good results can often be obtained and thus satisfy the requirement of some applications. Therefore, the problem left is whether we are happy to accept the laborious human intervention involved in the process of image registration. If not, we then still have to develop novel algorithms to automate the image registration process.

References

[1] P.K. Allen, I. Stamos, A. Troccoli, B. Smith, M. Leordeanu, Y.C. Hsu. 3D modelling of historic sites using range and image data. *Proc. ICRA*, 2003.

[2] K. S. Arun, T.S. Huang, and S.D. Blostein. Least-squares fitting of two 3-D point sets. *IEEE Trans. PAMI* 1987, 9, 698-700.

[3] A. P. Ashbrook, R. B. Fisher, et al. Finding surface correspondences for object recognition and registration using pair-wise geometric histogram. *Proc. 5th ECCV*, 1998.

[4] R. S. Ball. *A treatise on the theory of screws*. Cambridge University Press, 1900.

[5] R. Benjemaa, F. Schmitt. Registering range views of complex objects. *Proc. 4th European Conference on Rapid Prototyping*, 1995.

[6] R. Benjemaa, F. Schmitt. Fast global registration of 3D sampled surfaces Using a multi-z-buffer technique. *Image and Vision Computing*, 1999, 17, 113-123.

[7] R. Bergevin, M. Soucy, H. Gagnon, and D. Laurendeau. Towards a general multi-view registration technique. *IEEE Trans. PAMI*, 1996, 18, 540-547.

[8] P. J. Besl, N. D. McKay. A method for registration of 3D shapes. *IEEE Trans. PAMI*, 1992, 14, 239-256.

[9] P. Biber. The normal distribution transform: a new approach to laser scan matching. *Proc. IROS*, 2003.

[10] G. Blais, M.D. Levine. Registering Multiview Range Data to Create 3D Computer Objects. *IEEE Trans. PAMI*, 1995, 17, 820-824.

[11] L.G. Brown. A survey of image registration techniques. *ACM Computing Surveys (CSUR)*, 1992, 24, 325 - 376

[12] K. Brunnstrom and A.J. Stoddart, Genetic algorithms for free-form surface matching. *Proc. ICPR*, 1996, pp. 689-693.

[13] M. Chang, F.F. Leymarie, B.B. Kimia. 3D shape registration using regularized medial scaffolds. *Proc. 3DPVT*, 2004.

[14] S. Chaudhuri and S. Chatterjee. Performance analysis of total least squares methods in three dimensional motion estimation. *IEEE Transactions on Robotics and Automation*, 1991, 7, 707-714.

[15] S. Chaudhuri, S. Sharma and S. Chatterjee. Recursive estimation of motion parameters. *Computer Vision and Image Understanding* 1996, 64, 434-442.

[16] Y. Chen and G. Medioni. Object modelling by registration of multiple range images. *IVC*, 1992, 10, 145-155.

[17] D. Chetverikov, D. Svirko, D. Stepanov, P. Krsek. The trimmed iterative closest point algorithm. *Proc. ICPR*, 2002, pp. 545-548.

[18] C.K. Chow, H. T. Tsui, T. Lee. Surface registration using a dynamic genetic algorithm. *Pattern Recognition*, 2004, 37, 105-17.

[19] H. Chui and A. Rangarajan. A new point matching algorithm for non-rigid registration. *Computer Vision and Image Understanding*, 2003, 89, 114-141.

[20] H. Chui and A. Rangarajan. A unified non-rigid feature registration method For brain mapping. *Medical Image Analysis*, 2003, 7, 113-130.

[21] H. Chui, A. Rangarajan, J. Zhang, and C. M. Leonard. Unsupervised learning Of at atlas from unlabeled point-sets. *IEEE Trans. PAMI*, 2004, 26, 160-172.

[22] B. Curless and M. Levoy, Volumetric method for building complex models from range images, in: *Proc. SIGGRAPH*, 1996, pp. 303-312.

[23] P. David, D.DeMenthon, R. Duraiswami, and H. Samet. SoftPOSIT: Simultaneous pose and correspondence determination. *Proc. ECCV*, 2002, pp. 698-714.

[24] P. David, D.DeMenthon, R. Duraiswami, and H. Samet. Simultaneous pose and correspondence determination using line features. *Proc. CVPR*, 2003.

[25] E. R. Davis. *Machine vision: theory, algorithms, practicalities*. Academic Press, 1990.

[26] A.P. Dempster, N.M. Laird, and D.B. Rubin. Maximum likelihood from incomplete data via the EM algorithm. *J. Royal Statistical Society Series B*, 1977, 39, 1-38.

[27] G. Dewaele, F. Devernay, and H. Horaud. Hand motion from 3D point trajectories and a smooth surface model. *Proc. ECCV*, 2004

[28] C. Dorai, J. Weng, and A. K. Jain. Optimal registration of multiple range views. *Proc. ICPR*, 1994, pp. 569-571.

[29] C. Dorai, G. Wang, A.K. Jain, and C. Mercer. From images to models: automatic model construction from multiple views. *Proc. ICPR*, 1996, pp. 770-774.

[30] D. W. Eggert, A. W. Fitzgibbon, and R. B. Fisher. Simultaneous registration of multiple range images satisfying global consistency constraints for use in reverse engineering. *Proceedings of the IAPR International Conference on Pattern Recognition*, Vienna, August 1996, pp. 243-247.

[31] D. W. Eggert, A. Lorusso, R. B. Fisher. Estimating 3-D rigid body transformations: a comparison of four major algorithms. *Machine Vision and Applications*, 1997, 10, 272-290.

[32] D. W. Eggert, A. W. Fitzgibbon, and R. B. Fisher. Simultaneous registration of multiple range images for use in reverse engineering of CAD models. *Computer Vision and Image Understanding*, 1998, 69, 253-272.

[33] P. A. Van den Elsen, E. J. D. Pol, and M. A. Viergever. Medical image matching: a review with classification. *IEEE Transactions on Medical Imaging*, 1993, 12, 26-39.

[34] O. D. Faugeras and M. Hebert. A 3D recognition and positioning algorithm using geometric matching between primitive surfaces. *Proceedings of the Eighth International Joint Conference on Artificial Intelligence*, Karlsruhe, West Germany, August 8-12, 1983, pp. 996-1002.

[35] O. D. Faugeras, M. Hebert, E. Pauchon, and J. Ponce. Object representation, identification, and positioning from range data. In M. Brady and R. Paul (editors): *Robotics Research*, MIT Press, 1984.

[36] O. Faugeras and M. Hebert. The representation, recognition and locating of 3-D objects. *International Journal of Robotics Research* 1986, 5, 27-52.

[37] J. Feldmar and N. Ayache. Rigid, affine, and locally affine registration of free-form surface. *Technical report*, no. 2220, INRIA, 1994. Also appeared in *Proceedings of European Conference on Computer Vision* (rigid and affine, 1994) and *Proceedings of IEEE International Conference on Computer Vision and Pattern Recognition* (locally affine, 1994), *International Journal of Computer Vision* 1996, 18, 99-119.

[38] J. Feldmar, N. Ayache, and F. Betting. 3D-2D projective registration of free-form curves and surfaces. *CVIU*, 1997, 65, 403-424.

[39] J. H.Friedman, J. L. Bently, and P. A. Finkel. An algorithm for finding best matches in logarithmic expected time. *ACM Trans. Math. Soft.*, 1977, 3, 209-226.

[40] G. Gagnon, M. Soucy, R. Bergevin, and D. Laurendeau. Registration of multiple range views for automatic 3-D model building. *Proc. CVPR*, 1994, pp. 581-586.

[41] N. Gelfand, L. Ikemoto, S. Rusinhiewicz, and M. Levoy. Geometrically stable sampling for the ICP algorithm. *Proc. 3DIM*, 2003.

[42] G. Godin, D. Laurendeau, R. Bergevin. A method for the registration of attributed range images. *Proc. 3DIM*, 2001, pp. 179-186.

[43] S. Gold and A. Rangarajan. A graduated assignment algorithm for graph matching. *IEEE PAMI*, 1996, 18, 377-388.

[44] S. Gold, A. Rangarajan, et al. New algorithms for 2-D and 3-D point matching: pose estimation and correspondence. *Pattern Recognition* 1998, 31, 1019-1031.

[45] G. H. Golub and C. F. Van Loan. *Matrix Computations*. Third edition, The John Hopkins University Press, 1996.

[46] D. Goryn. On the estimation of rigid body rotations from noisy data. *IEEE Trans. PAMI*, 1995, 17, 1219-1220.

[47] A. Gourdon and N. Ayache. Registration of a curve on a surface using differential properties. *Technical report*, no. 2145, INRIA, France, 1993.

[48] J. C. Gower. Generalised procrustes analysis. *Psychometrika*, 1975, 40, 33-51.

[49] S. Granger,and X. Pennec. Multi-scale EM-ICP: a fast and robust approach for surface registration. *Proc. ECCV*, 2002, 418-432.

[50] B. F. Green. The orthonormal approximation of an oblique structure in factor analysis. *Psychometrika*, 1952, 17, 429-440.

[51] D. Hahnel, S. Thrun, and W. Burgard. An extension of the ICP algorithm for modelling non-rigid objects with mobile robots. *Proceedings of International Joint Conference on Artificial Intelligence*, 2003.

[52] R. M. Haralick, H. Joo, C. Lee, X. Zhuang, V. G. Vaidya, and M. B. Kim. Pose estimation from corresponding point data. *IEEE Transactions on Systems, Man, and Cybernetics*, 1989, 19, 1426-1446.

[53] B. K. P. Horn. Closed-form solution of absolute orientation using unit quaternions. *Journal of Optical Society of America (A)*, 1987, 4, 629-642.

[54] B. K. P. Horn. Closed-form solution of absolute orientation using orthonormal matrices. *Journal on Optical Society of America (A)*, 1988, 5, 1127-1135.

[55] B. K. P. Horn. Relative orientation revisited. *Journal of Optical Society of America (A)*, 1991, 8, 1630-1638.

[56] T. S. Huang and A. N. Netravali. Motion and structure from feature correspondence: a review. *Proceedings of the IEEE*, 1994, 82, 252-268.

[57] D. Huber, M. Hebert. Fully automatic registration of multiple 3D data sets, *IVC*, 2003, 21, 637-650.

[58] D.F. Huber. Automatic 3D modeling using range images obtained from unknown viewpoints. *Proc. Int. Conf. 3D Digital Imaging and Modelling*, pp. 153-160, 2001.

[59] D. Huber and M. Hebert 3D Modeling Using a Statistical Sensor Model and Stochastic Search *Proc. CVPR*, 2003, pp. 858-865.

[60] A. Johnson and S. B. Kang. Registration and integration of textured 3-D data. *Technical report*, CRL 96/4, Cambridge research Lab, Digital Equipment Corporation, 1996.

[61] A.E. Johnson and M. Hebert. Surface registration by matching oriented points. *Proc. Int. Conf. 3D Digital Imaging and Modelling*, pp. 234-241, 1997.

[62] A.E. Johnson and S.B. Kang. Registration and integration of textured 3-D data. *IVC*, 1999, 17, 135-147.

[63] H. Jonsson and B. Soderberg. Deterministic annealing and nonlinear assignment. *Technical report 01-16*, Department of Theoretical Physics, Lund University, 2001.

[64] T. Jost and H. Hugli. A multi-resolution scheme ICP algorithm for fast shape registration. *Proc. 3DPVT*, 2002.

[65] K. Kanatani. Analysis of 3-D rotation fitting. *IEEE Trans. PAMI*, 1994, 16, 543-549.

[66] V. Koivunen, J. M. Vezien. Multiple representation approach to geometric model construction from range data. *Proceedings of the Second CAD-based Vision Workshop*, Champion, Pennsylvania, February 1994.

[67] B. Kverh and A. Leonardis. Registration of range images based on segmented data. *Proceedings of 8th International Conference on Computer Analysis Of Images and Patterns*, Ljubljana, Slovenia, September 1-3, 1999, pp. 347-356.

[68] X. Laboureux and G. Hausler. Localization and registration of three-dimensional objects in space-where are the limits? *Applied optics*, 2001, 40, 5206-5216.

[69] S. P. Langron and A. J. Collins. Perturbation theory of generalised procrustes analysis. *Journal of the Royal Statistical Society(B)*, 1985, 47, 277-284.

[70] Z. Lin, H. Lee, and T. Huang, Finding 3D point correspondences in Motion Estimation. *Proc. 8th ICPR*, 1986, pp. 303-305.

[71] Y. Liu and M. A. Rodrigues. Using rigid constraints to analyse motion parameters from two sets of 3D corresponding point pattern. *Proceedings of 8th International Conference on Computer Analysis of Images and Patterns (CAIP'99)*, Lecture Notes on Computer Science, vol. 1689, Springer Verlag Press, Ljubljana, Slovenia, 1-3 September, 1999, pp. 321-328.

[72] Y. Liu and M. A Rodrigues. Geometric understanding of rigid body transformations. *Proceedings of 1999 IEEE International Conference on Robotics and Automation(ICRA'99)*, Detroit, USA, May 10-15, 1999, pp. 1275-1280.

[73] Y. Liu and M. A. Rodrigues. Geometrical analysis of motion parameters from two sets of range image data. *Proceedings of 1999 IEEE International conference on Systems, Man, and Cybernetics (SMC'99)*, IEEE Computer Science Press, Tokyo, Japan, Oct. 12-15, 1999.

[74] Y. Liu and M. A. Rodrigues. Analysing the geometric properties of reflected correspondence vectors for the registration of free form shapes. *Proceedings of 2000 IEEE International Conference on Systems, Man, and Cybernetics (SMC'00)*, Nashville, Tennessee, 8-11 October 2000.

[75] Y. Liu and M. A Rodrigues. Essential representation and calibration of rigid body transformations. *Machine Graphics and Vision*, 2000, 9, 123-138.

[76] Y. Liu and M.A. Rodrigues. Geometrical analysis of two sets of 3D correspondence data patterns for the registration of free-form shapes. *Journal of Intelligent and Robotic Systems*, 2002, 33, 409-436.

[77] Y. Liu and F. Labrosse. Inverse Validation for Accurate Range Image Registration with Structured Data. *Proceedings of 16th International Conference on Pattern Recognition*, August 11-15, 2002, Qubec, Canada, vol. 3, pp. 537-540.

[78] Y. Liu, M.A. Rodrigues. Accurate registration of structured data using two overlapping range images. *Proc. IEEE ICRA*, 2002, vol. 3, pp. 2519-2524.

[79] Y. Liu, L. Li, and Y. Wang. Free form shape matching using deterministic annealing and softassign. *Proc. ICPR*, 2004.

[80] Y. Liu and B. Wei. Developing structural constraints for accurate registration of overlapping range images. *Robotics and Autonomous Systems*, 2004, 47, 11-30.

[81] Y. Liu and B. Wei. Evaluating structural constraints for accurate range image registration. *Proceedings of The 4th International Conference on 3-D Digital Imaging and Modeling(3DIM'03)*, October 6-10, 2003, Banff, Alberta, Canada, pp. 187-194.

[82] Y. Liu. Improving ICP with easy implementation for free form surface matching. *Pattern Recognition*, 2004, 37, 211-226.

[83] Y. Liu, L. Li, and B. Wei, 3D shape matching using collinearity constraint. *Proceedings of IEEE 2004 International Conference on Robotics and Automation*, April 26-1 May, 2004, New Orleans, LA, USA.

[84] Y. Liu, M.A. Rodrigues, and Y. Wang, Developing rigid motion constraints for the registration of free-form shapes. *Proc. IEEE/RSJ IROS*, 2000, pp. 2280-2285.

[85] E. Lomonosov, D. Chetverikov, and A. Ekart. Fully automatic, robust and precise alignment of measured 3D surfaces for arbitrary orientations. *Proc. 28th OAGM/AAPR workshop*, 2004, pp. 39-46.

[86] H. C. Longuet-Higgins. A computer program for reconstructing a scene from two projections. *Nature*, 1981, 293, 133-135.

[87] A. Lorusso, D. W. Eggert and R. B. Fisher. A comparison of four algorithm for estimating 3-D rigid transformations. *Proceedings of the Sixth British Machine Vision Conference*, September 1995, pp. 239-246.

[88] X. Lu, D. Colbry, and A.K. Jain. Three-dimensional model based face recognition. *Proc. ICPR*, 2004.

[89] F. Lu, E. Milios. Globally consistent range scan alignment for environment mapping. *Autonomous Robots*, 1997, 4, 333-349.

[90] L. Lucchese, G. Doretto, and G.M. Cortelazzo. A frequency domain technique for range image registration. *IEEE Trans. PAMI*, 2002, 24, 1468-1484.

[91] J. Luck, C. Little, and W. Hoff. Registration of range data using a hybrid simulated annealing and iterative closest point algorithm. *Proc. ICRA*, 2000, pp. 3739-3744.

[92] J.B. Antoine Maintz and M.A. Viergever. A survey of medical image registration. *Medical Image Analysis*, 1998, 2, 1-36.

[93] B. Matei, P. Meer, and D. Tyler. Performance assessment by resampling: rigid motion estimators. *Proceedings of IEEE Computer Society Workshop on Empirical Evaluation of Computer Vision Algorithms*, Santa Barbara, California, June 21-22, 1998, pp. 72-79.

[94] C. R. Maurer and J. M. Fitzpatrick. A review of medical image registration. In R. J. Maciunas, editor. *Interactive Image-Guided Neurosurgery*, Park Ridge, IL: American Association of Neurological Surgeons, 1993, pp. 17-44.

[95] C. R. Maurer, G. B. Aboutanos, B. M. Dawant, R. J. Maciunas, J. M. Fitzpatrick. Registration of 3-D images using weighted geometric features. *IEEE Transactions on Medical Imaging*, 1996, 15, 836-849.

[96] C. Menq, H. Yau, and G. Lai. Automated precision measurement of surface profile in CAD-directed inspection. *IEEE Transactions on Robotics and Automation*, 1992, 8, 268-278.

[97] N.J. Mitra, N. Gelfand, H. Pottmann, and L. Guibas. Registration of point cloud data from a geometric optimisation perspective. *Pro. Eurographics Symposium on Geometry Processing*, 2004, pp. 23-32.

[98] L.-P. Morency, T. Darrell. Stereo tracking using ICP and normal flow constraint. *Proc. ICPR*, 2002.

[99] K. Nishino and K. Ikeuchi. Robust simultaneous registration of multiple range images. *Proc. 5th ACCV*, 2002, pp. 454-461.

[100] D. Page, A. Koschan, Y. Sun and M. Abidi. Laser-based imaging for reverse engineering. *Sensor review*, 2003, 23, 223-229.

[101] T. Pajdla, L. Van Gool. Registration of 3-D curves using semi-differential invariants. *Proc. ICCV*, 1995, pp. 390-395.

[102] M. Potmesil. Generating models of solid objects by matching 3D surface segments. *Proceedings of 8th International Joint Conference on Artificial Intelligence*, 1983.

[103] K. Pulli. Multiview registration for large data sets. *Proc. Int. Conf. 3DIM*, 1999, pp. 160-168.

[104] L. Quan and Z. Lan. linear 4-point camera pose determination. *Proceedings of Sixth IEEE International Conference on Computer Vision*, Bombay, India, 1998.

[105] L. Quan and Z. Lan. linear N-point camera pose determination. *IEEE Trans. PAMI*, 1999, 21, 774-780.

[106] A. Rangarajan. Self annealing and self annihilation: unifying deterministic annealing and relaxation labelling. *Pattern Recognition*, 2000, 33, 635-649.

[107] M.A. Rodrigues and Y. Liu. On the representation of rigid body transformations for accurate registration of free form shapes. *Robotics and Autonomous Systems*, 2002, 39, 37-52.

[108] M. Rioux. Laser range finder based on synchronized scanners. *Applied Optics*, 1984, 23, 3837-3844.

[109] G. Roth. An automatic registration algorithm for two overlapping range images. *Proceedings of the 8th International Conference on Computer Analysis of Images and Patterns*, Lecture Notes in Computer Science, vol. 1689, Ljubljana, Slovenia, September 1999, pp. 329-338.

[110] S. Rusinkiewicz and M. Levoy. Efficient variants of the ICP algorithm. *Proc. Int. Conf. 3DIM*, 2001, pp. 145-152.

[111] B. Sabata and J. K. Aggarwal. Estimation of motion from a pair of range images: a review. *CVGIP: Image Understanding*, 1991, 54, 309-324.

[112] R. Sagawa, T. Osihi, A. Nakazawa, R. Kurazume, K. Ikeuhi. Iterative refinement of range images with anisotropic error distribution. *Proc. IROS*, 2002, pp. 79-85.

[113] P. H. Schonemann. A generalised solution of the orthonormal procrustes problem. *Psychometrika*, 1966, 31, 1-10.

[114] C. Schtz, T. Jost and H. Hgli. Multi-feature matching algorithm for free-form 3D surface registration. *Proc. ICPR*, 1998, pp. 982-984.

[115] C. Schutz, T. Jost, and H. Hugli. Semi-automatic 3D Object Digitizing System Using Range Images. *Proc. ACCV*, 1998, pp. 490-497.

[116] G. C. Sharp, S. W. Lee, and W. K. Wehe. Invariant features and the registration of rigid bodies. *Proceedings of the 1999 IEEE International Conference on Robotics and Automation*, Detroit, Michigan, May 1999, pp. 932-937.

[117] G.C. Sharp, S.W. Lee, and D.K. Wehe. Toward multiview registration in frame space. *Proc. ICRA*, 2001.

[118] G.C. Sharp, S.W. Lee, and D.K. Wehe. Multiview Registration of 3D Scenes by Minimizing Error Between Coordinate Frames *Proc. ECCV*, 2002.

[119] G.C. Sharp, S.W. Lee, and W.K. Wehe. ICP registration using invariant features. *IEEE Trans. PAMI*, 2002, 24, 90-112.

[120] R.N. Shepard. Toward a universal law of generalisation for psychological science. *Science*, 1987, 237, 1317-1323.

[121] R. Sibson. Studies in the robustness of multidimensional scaling: procrustes. *Journal of Royal Statistical Society(B)*, 1978, 24, 234-238.

[122] L. Silva, Olga R.P. Bellon, K. L. Boyer, Paulo F.G. Gotardo. Low-overlap range image registration for archaeological applications. *Proc. CVPR Workshop*, 2003.

[123] D. A. Simon, M. Hebert, and T. Kanade. Real-time 3D pose estimation using a high speed range sensor. *Technical report*, CMU-RI-TR-93-24, Robotics Institute, Carnegie Mellon University, 1993.

[124] R. Sinkhorn. A relationship between arbitrary positive matrices and doubly stochastic matrices. *Ann. Math. Statist.*, 1964, 35, 876-879.

[125] C.V. Stewart, C.-L. Tsai, and B. Roysam. The dual-bootstrap iterative closest point algorithm with application to retinal image registration. *IEEE Trans. Medical Imaging*, **22**(2003) 1379-1394.

[126] C. V. Stewart, C.-L. Tsai, and A. Perera. A view-based approach to registration: theory and application to vascular image registration. *Proc. IPMI*, 2003.

[127] Y. Sun and M.A. Abidi. Surface matching by 3D point's fingerprint. *Proc. ICCV*, 2001, pp. 263-269.

[128] H. Surmann, A. Nuchter, and J. Hertzberg. An autonomous mobile robot with a 3D laser range finder for 3D exploration and digitalisation of indoor environment. *Robotics and Autonomous Systems*, 2003, 45, 181-198.

[129] L. Tamm, V. Menon, and A.L. Reiss, Maturation of brain function associated with response inhibition. *J Am Acad Child Adolesc Psychiatry*, 2002, 41, 1231-1238.

[130] R. Y. Tsai and T. S. Huang. Uniqueness and estimation of three-dimensional motion parameters of rigid objects with curved surfaces. *IEEE Trans. PAMI*, 1984, 6, 13-27.

[131] G. Turk and M. Levoy. Zippered polygon meshes from range images. *Proc. SIGGRAPH*, 1994, pp. 311-318.

[132] S. Umeyama. Least-squares estimation of transformation parameters between two point pattern. *IEEE Trans. PAMI*, 1991, 13, 376-380.

[133] M. W. Walker, L. Shao, and R. A. Volz. Estimating 3-D location parameters using dual number quaternions. *CVGIP: Image Understanding*, 1991, 54, 358-367.

[134] M.D. Wheeler and K. Ikeuchi. Sensor modelling, probabilistic hypothesis generation, and robust localization for object recognition. *IEEE Trans. PAMI*, 1995, 17, 252-265.

[135] R. T. Whitaker, et al. Indoor scene reconstruction from sets of noisy range images. *Proc. 3DIM*, 1999, pp. 348-357.

[136] J.V. Wyngaerd, L.V. Gool, B. Koch, and M. Proesmans. Invariant-based registration of surface patches. *Proc. ICCV*, 1999, pp. 301-306.

[137] S.M. Yamany, M.N. Ahmed, E.E. Hemayed, and A.A. Farag, Novel surface registration using
the grid closest point transform. *Proc. ICIP*, 1998, pp. 809-813.

[138] S. M. Yamany, A. A. Farag, and A. El-Bialy. Free-form object recognition and registration using surface signature. *Proceedings of 1999 IEEE International Conference on Image
Processing*, Kobe, Japan, 24-28 October 1999.

[139] S.M. Yamany and A.A. Farag. Surface signatures: an orientation independent free-form surface representation scheme for the purpose of objects registration and matching. *IEEE Trans.
PAMI*, 2002, 24, 1105-1120.

[140] R. Yang and P.K. Allen. Registering, integrating, and building CAD models from range data.
Proc. ICRA, 1998, pp. 3115-3120.

[141] Z. Zhang. Iterative point matching for registration of free-form curves. *Technical report*, no.
1658, INRIA, France, May 1992.

[142] G. Zhang, W. Jiang, and D. Li. A new algorithm of automatic relative orientation. *Proc. 22nd
Asian Conf. on Remote Sensing*, 2001.

[143] D. Zhang and M. Hebert. Harmonic maps and their applications in surface matching. *Proc.
CVPR*, 1999.

[144] T. Zinber, J. Schmidt, H. Niemann. A refined ICP algorithm for robust 3-D correspondence
estimation. *Proc. ICIP*, 2003.

In: Computer Vision and Robotics
Editor: John X. Liu, pp. 99-134

ISBN 1-59454-357-7
© 2006 Nova Science Publishers, Inc.

Chapter 4

NOT ALL MOTIONS ARE EQUIVALENT IN TERMS OF DEPTH RECOVERY

Loong-Fah Cheong [a], Tao Xiang [b],
Valerie Cornilleau-Pérès [c,d], and Ling Chiat Tai [a]
[a]Department of Electrical and Computer Engineering
National University of Singapore, Singapore 119260 [*]
[b]Department of Computer Science
Queen Mary, University of London, London E1 4NS, UK [†]
[c]Lab. Neurosciences Fonctionnelles and Pathologies
CNRS–Univ. Lille2, France[‡]
[d]Singapore Eye Research Institute, Singapore

Abstract

Given that errors in the estimates for the intrinsic and extrinsic camera parameters are inevitable, it is important to understand the behaviour of the resultant distortion in depth recovered under different motion-scene configurations. The main goal in this study is to look for a generic motion type that can render depth recovery more robust and reliable. To this end, lateral and forward motions are compared both under cali- brated and uncalibrated scenarios. For lateral motion, we find that although Euclidean reconstruction is difficult, ordinal depth information is obtainable; while for forward motion, depth information (even partial one) is difficult to recover. We obtain the same conclusion in the uncalibrated case when the intrinsic camera parameters are fixed. However, when these parameters are not fixed, then lateral motion allows only a local recovery of depth order. In general, the depth distortion transformation is a Cremona transformation, and becomes a simple projective one in the case of lateral motion. We applied the above analysis to the scenario of recovering curvature of a quadric surface

---

[*]E-mail address: eleclf@nus.edu.sg
[†]E-mail address: txiang@dcs.qmul.ac.uk
[‡]E-mail address: c.peres2@wanadoo.fr

under lateral motion and showed that the shape estimates are recovered with varying degrees of uncertainty depending on the motion-scene configuration. Specifically, the reconstructed second order shape tends to be more distorted in the direction parallel to the translational motion than that in the orthogonal direction. We present the result of a psychophysical experiment, which confirms that in human vision, curvature estimates tend to be more erroneous and variable along the direction of lateral motion, than along its orthogonal direction.

Keywords: Structure from motion, Uncalibrated motion analysis, Depth distortion, Shape representation, Surface curvature, Depth perception.

1 Introduction

The estimation of 3-D motion and structure from optic flow is notorious for its noise sensitivity; a small amount of error in the image measurements can lead to very different solutions. Previous structure from motion (SFM) algorithms faced this problem to various extent. This has led to many error analyses (Adiv, 1989; Daniilidis and Spetsakis, 1996; Dutta and Snyder, 1990; Thomas et al., 1993; Weng and Huang, 1991; Young, 1992) in the past. Recently, a number of papers further investigated the error behaviour of SFM algorithms in terms of their local minima and ambiguities (Oliensis, 2000b; Soatto and Brockett, 1998). Furthermore, in view of the emergence of active vision systems, where the calibration parameters (also known as intrinsic parameters) are no longer known or fixed, the ambiguities associated with uncalibrated motions have been analyzed (Kahl, 1999; Sturm, 1997, 2002). In a recent critique (Oliensis, 2000a), Oliensis expressed the view that since current SFM algorithms perform well only in restricted domains, and different types of algorithms do well on quite different types of sequences, it was important to evaluate the limits of applicability of each algorithm. If such understanding could be achieved, an optimal strategy could then consist in fusing the outputs of different algorithms.

The main concern of these previous approaches is usually on the reliability of the motion estimates. Fewer studies have addressed the reliability of the depth estimates. Some of them (Weng and Huang, 1991; Szeliski and Kang, 1997; Grossmann and Victor, 2000) quantified the influence of image noise on the error in the depth estimates. The case where the depth errors are due to an erroneous estimation of the 3D motion has been addressed only for critical surface pairs (Horn, 1987; Negahdaripour, 1989). In the case of uncalibrated motion where both the extrinsic and the intrinsic parameters are unknown, the projective transformation is used to characterize the effect of these unknown parameters on the recovered depth. Again, the depth distortions arising from errors in the camera parameters have not been described, except for special motions termed as critical motions (Kahl, 1999; Sturm, 1997, 2002). The need to characterize such depth distortion arising from errors in the motion estimates prompted our previous work (Cheong et al., 1998), which gives an account of the systematic nature of the errors in the depth estimates via the so-called iso-distortion framework. Indeed, in the general case, the transformation from physical to perceptual space is more complicated than a projective transformation, and belongs to the family of Cremona transformations.

In that previous work, we observe that most properties of the visual space are not preserved under the Cremona transformation. Thus, not only recovering metrical depth information is generally difficult—errors could be very large near the fundamental elements of the Cremona transformation, but even the recovery of partial depth such as ordinal depth might not be possible in some situations. The latter result lends support to Oliensis' view (Oliensis, 2000a) that SFM algorithms perform well only in restricted domains and that it is important to evaluate the limits of applicability of these algorithms. Adopting this viewpoint, this chapter attempts to characterize the reliability of depth recovery under different motion-scene configurations. In particular, we use the iso-distortion framework (Cheong et al., 1998) to investigate motion types that allow robust recovery of depth information.

The work detailed here differs from previous approaches (Oliensis, 2000b; Soatto and Brockett, 1998) since it deals with the reliability of the depth estimates rather than that of the motion estimates. In general, the reliability of a reconstructed scene might have quite a different behaviour from that of the motion estimates. For instance, if the motion contains a dominant lateral translation, it might be very difficult to lift the ambiguity between translation and rotation. However, in spite of such ambiguity, some depth information can be robustly recovered. On the other hand, psychophysical experiments (Ullman, 1979) reported that under pure forward translation, human subjects were unable to recover structure unless under favorable conditions (for instance a large field of view). This suggests that all motions are not equivalent in terms of robust depth recovery and that there exists a certain dichotomy between forward and lateral translation. Similarly, in the case of an uncalibrated motion, the reliability of a reconstructed scene might have quite a different behaviour from that of the calibration process. For instance, the inclusion of the camera zoom operation complicates the estimation procedure for both the extrinsic and the intrinsic parameters; however, despite an ambiguity between the zoom flow and the focal length, it was observed that certain qualitative aspects of the recovered depth such as parallelism remain unaffected under specific conditions (Bougnoux, 1998; Cheong and Peh, 2004). The preceding arguments stress the need to treat the question of the reliability of depth reconstruction in its own right, apart from that of motion estimation. As a corollary, by analyzing how errors in the intrinsic parameters affect different aspects of depth recovery, we can address the fundamental question of the need for camera calibration to recover those aspects of depth robustly.

If we understand the reliability of the depth estimates under different motion-scene configurations, we can design an optimal motion strategy to reveal reliable depth information. The idea of executing intelligent controlled movements so as to accomplish tasks robustly is of course the central tenet of the active vision paradigm. While there have been many motion-based works under this paradigm, we find that most of them dealt with problems whose purpose is to perform robust navigation. For instance, Santos-Victor et al. (Santos-Victor et al., 1993) and Coombs and Roberts (Coombs and Roberts, 1993) present methods to steer a camera between two walls, and to veer around obstacles, both methods being based on simple analysis of the optical flow without going through depth recovery. Fewer analyses have been conducted on how to execute movements so as recover interesting structure information (besides that used for avoiding obstacles in navigation). If self motions can be controlled perfectly or if there are no other constraints, then of course any pure translation would be a good motion strategy to recover depth. However rotation often accompanies translation, either involuntarily as in the case of the human during locomotion, or voluntarily as in the case of wasps performing object-directed zig-zag flights (Voss, 1998). In such cases, the rotation inevitably confounds the recovery of the translation. The question becomes: What *kind* of translation is the best strategy, given the mechanical constraints of the visual systems or the ecological constraints of the environment? Chaumette et al. (Chaumette et al., 1994) dealt with the optimal estimation of 3-D structures using visual servoing, but the errors they examined concerned only those of discretization and measurement, and the analysis was applied only to specific shape primitives such as spheres and cylinders.

In this chapter, we address depth recovery under the case where the 3-D motion parameters themselves are estimated with some errors and where the scene in view is arbitrary in shape. In view of the various results that seem to imply that not all motions are equivalent in terms of depth recovery, we use the iso-distortion framework to analyze the following scenarios:

- We investigate the sensitivity of the depths recovered under lateral motion and forward motion.

- For each motion type, we compare the cases where the intrinsic parameters are known, unknown but fixed, and, unknown and varying respectively.

- We also briefly deal with an alternative scheme of recovering depth (the least square procedure) in the appendix.

Our investigation has wider implication in the field of bio-robotics or biologically inspired vision. Indeed, the prevalence of lateral motion (the so-called 'peering' motion) used by different animals to appreciate distances (locusts (Collett, 1978; Sobel, 1990); grasshopper (Eriksson, 1980); honeybee (Srinivasan et al., 1990); praying mantis (Poteser and Kral, 1995); wasp (Voss, 1998); Barn Owl (Wagner, 1989)) underlines the ecological relevance of studying the perception of depth in that case. Given the limitation of mechanisms extracting egomotion parameters in biological and artificial systems, our approach allows one to relate depth distortions to the errors in the 3D motion estimates.

A final application of the iso-distortion framework is the prediction of how the geometry of the 3D space is actually perceived by human observers from optic flow. Apparent distortions of the visual scene have been reported in the field of stereoscopic vision (Foley, 1980) (note that human stereopsis is mathematically equivalent to a lateral monocular translation along the inter-ocular distance, followed by an eye rotation equal to the convergence angle). In particular, the experiments by Ogle (Ogle, 1964) showed that the apparent frontoparallel plane (AFPP) tends to be concave along the inter-ocular axis for near distances, and convex for far distances (i.e. a near plane tends to be perceived as convex, while a far plane tends to be perceived as concave). As a consequence of our theoretical work, we point to a directional anisotropy in the recovery of depth from lateral motion. More precisely, we predict that the optimal image direction for the recovery of surface curvature is in the direction orthogonal to the translation. The fact that discriminating between different surfaces is easier along that direction than along the translation direction has been demonstrated for a small field of view (Cornilleau-Pèrés and Droulez, 1989; Norman, 1992) and explained computationally via the spin variation model (Droulez and Cornilleau-Pèrés, 1990). Here we explore in more details how surfaces are apparently distorted when motion parallax is the only depth cue, under a large field of view, and we relate the psychophysical results to the iso-distortion model.

The organization of this chapter is as follows. After reviewing the iso-distortion framework in Section 2, we consider various aspects of depth recovery under generic types of calibrated motion (Section 3). The case of uncalibrated motion is analyzed in Section 4. In particular, we first look at the perceived space obtained with inaccurate estimates of the fixed focal length f and the fixed principal point (O_x, O_y). Then we examine the case of a dynamically changing focal length which results in a zoom field and a changing principal point in the motion recovery process. In Section 5, we conduct experiments on images to verify our various theoretical predictions. Section 6 describes our psychophysical investigation, aimed at demonstrating that one such prediction—namely the anisotropy in motion direction with respect to depth reconstruction—is supported by human perceptive responses.

2   The Iso-distortion Framework

In (Cheong et al., 1998), the geometric laws under which the recovered scene is distorted due to some errors in the estimated motion parameters is represented by a distortion transformation. The distortion in the perceived space can then be visualized by looking at the locus of constant distortion. This approach was termed the iso-distortion framework.

We adopt the standard perspective image formation model. A camera is moving rigidly with respect to a coordinate system $OXYZ$ fixed to its nodal point O with a translation (U, V, W) and a rotation (α, β, γ); the image plane is located at a focal length f pixels from O along the Z-axis; a point P at (X, Y, Z) in the world produces an image point p at (x, y) on the image plane where (x, y) is given by $(\frac{fX}{Z}, \frac{fY}{Z})$. The resulting optical flow (u, v) at an image location (x, y) can then be expressed with the following well-known equations (Longuet-Higgins, 1981):

$$
\begin{aligned}
u &= u_{trans} + u_{rot} \\
&= (x - x_0)\frac{W}{Z} + \frac{\alpha xy}{f} - \beta\left(\frac{x^2}{f} + f\right) + \gamma y \\
v &= v_{trans} + v_{rot}
\end{aligned}
$$

$$= (y-y_0)\frac{W}{Z} + \alpha\left(\frac{y^2}{f} + f\right) - \frac{\beta xy}{f} - \gamma x$$

(1)

where $(x_0, y_0) = (f\frac{U}{W}, f\frac{V}{W})$ is the focus of expansion (FOE), Z is the depth of a scene point, u_{trans}, v_{trans} are the horizontal and vertical components of the flow due to translation, and u_{rot}, v_{rot} the horizontal and vertical components of the flow due to rotation, respectively.

Since the depth can only be derived up to a scale factor, we can set $W = 1$ without loss of generality. Then the scaled depth of a scene point recovered can be written as

$$Z = \frac{(x-x_0, y-y_0).(n_x, n_y)}{(u-u_{rot}, v-v_{rot}).(n_x, n_y)}$$

(2)

where (n_x, n_y) is a unit vector which specifies a direction.

If there are some errors in the estimation of the extrinsic parameters, this will in turn cause errors in the estimation of the scaled depth, and thus a distorted version of the space will be computed. Denoting the estimated parameters with the hat symbol ($\hat{\ }$) and errors in the estimated parameters with the subscript e (where error of any estimate r is defined as $r_e = r - \hat{r}$), the estimated depth $\hat{Z}$ can be readily shown to be related to the actual depth Z as follows:

$$\hat{Z} = Z\left(\frac{(x-\hat{x}_0, y-\hat{y}_0).(n_x, n_y)}{(x-x_0, y-y_0).(n_x, n_y) + (u_{rot_e}, v_{rot_e}).(n_x, n_y)Z + (u_e, v_e).(n_x, n_y)Z}\right)$$

(3)

where (u_e, v_e) is a noise term representing error in the estimate for the optical flow. In the forthcoming analysis we do not attempt to model the statistics of the noise and we will therefore ignore the noise term, that is, $(\hat{u}, \hat{v}) = (u, v)$.

In (3) $\hat{Z}$ is obtained from Z through multiplication by a factor given by the terms inside the bracket, which we call the distortion factor D. The expression for D contains the term (n_x, n_y), which here does not necessarily refer to the image intensity gradient direction. Its value depends on the scheme we use to recover depth. For instance, the normal flow approach (Fermüller, 1995) recovers depth along the normal direction in which case (n_x, n_y) is the gradient direction. In the full optical flow approach, however, a possible scheme is to recover depth along the estimated epipolar direction, based on the intuition that the epipolar direction contains the strongest translational flow. It means that we first project optical flow along the direction emanating from the estimated FOE and then recover depth along that direction, i.e. $(n_x, n_y) = \frac{(x-\hat{x}_0, y-\hat{y}_0)}{\sqrt{(x-\hat{x}_0)^2 + (y-\hat{y}_0)^2}}$, or in the case of $\hat{W} = 0$ where the estimated FOE is at infinity, $(n_x, n_y) = -\frac{(\hat{U}, \hat{V})}{\sqrt{\hat{U}^2 + \hat{V}^2}}$. In the forthcoming analysis, we will study the properties of the recovered depth based on the epipolar reconstruction approach. Another important alternative of recovering depth, which we do no more than performing a brief analysis in this chapter, is the linear least square reconstruction approach where $(n_x, n_y) = \frac{(u-\hat{u}_{rot}, v-\hat{v}_{rot})}{\sqrt{(u-\hat{u}_{rot})^2 + (v-\hat{v}_{rot})^2}}$. See the appendix for the derivation of this expression, as well as the statistical and geometrical reasons for not choosing this scheme of reconstructing depth as the main focus of the present study.

Upon substituting the corresponding value of (n_x, n_y) for the case of epipolar reconstruction approach, we obtain the following expression for the distortion factor:

$$D = \frac{(x - \hat{x}_0)^2 + (y - \hat{y}_0)^2}{(x - x_0, y - y_0).(x - \hat{x}_0, y - \hat{y}_0) + (u_{rot_e}, v_{rot_e}).(x - \hat{x}_0, y - \hat{y}_0)Z} \tag{4}$$

For specific values of the parameters $x_0, y_0, \hat{x}_0, \hat{y}_0, \alpha_e, \beta_e$, and γ_e, and for any fixed distortion factor D, equation (4) describes a surface $f(x, y, Z) = 0$ in the xyZ-space, which we call an iso-distortion surface. This iso-distortion surface has the obvious property that its points are distorted in depth by the same multiplicative factor D. The systematic nature of the distortion can then be studied by looking at the organization of these iso-distortion surfaces. In order to facilitate the pictorial description of these surfaces, we slice them with planes parallel to either the xZ-plane or the xy-plane. We call the curves thus obtained on the planar slice the iso-distortion contours. The examples in Figures 1(a) and 1(b) illustrate that the distortion is usually complex.

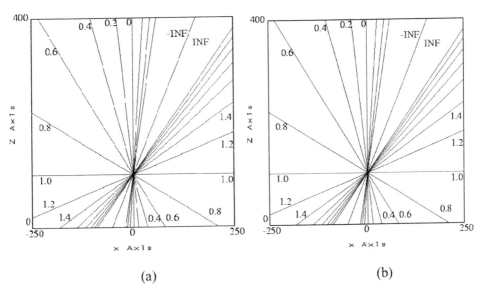

(a)   (b)

Figure 1:   Families of iso-distortion contours in the xZ-plane, parameterized by $x_0, y_0, \hat{x}_0, \hat{y}_0, y, \alpha_e, \beta_e$ and γ_e. The number beside each contour denotes the distortion factor D of that contour. INF denotes ∞. 1(b) illustrates the effects of including second order terms on the iso-distortion contours of 1(a), with FOV=50°.

Algebraically, the transformation from physical to perceptual space belongs to the family of Cremona transformations (Cheong and Ng, 1999). We write the homogeneous coordinates of a point P^3 as $[X, \mathcal{Y}, Z, \mathcal{W}]$, which is related to the non-homogeneous coordinates (X, Y, Z) by $(X, Y, Z) = \left[\frac{X}{\mathcal{W}}, \frac{\mathcal{Y}}{\mathcal{W}}, \frac{Z}{\mathcal{W}}, 1\right]$. Denoting the homogeneous co-ordinates of the estimated position $\hat{P}^3$ by $\left[\hat{X}, \hat{Y}, \hat{Z}, \hat{W}\right]$, we look for a distortion transformation $\phi : P^3 \rightarrow \hat{P}^3$. Note that to obtain the estimated $\hat{X}$, we use the back-projection given by $\hat{X} = \frac{x\hat{Z}}{f} = D\frac{xZ}{f} = DX$; similarly, $\hat{Y} = DY$. The image $\left[\hat{X}, \hat{Y}, \hat{Z}, \hat{W}\right]$ of a point $[X, Y, Z, W]$ can then be expressed as follows:

$$\left[\hat{X}, \hat{Y}, \hat{Z}, \hat{W}\right] = [\phi_1, \phi_2, \phi_3, \phi_4]$$

Similarly, the inverse transformation $\phi^{-1} : \hat{P}^3 \rightarrow P^3$ can be expressed as:

$$[X, Y, Z, W] = \left[\phi_1^{-1}, \phi_2^{-1}, \phi_3^{-1}, \phi_4^{-1}\right]$$

where the quantities ϕ_i are homogeneous polynomials in $[X, Y, Z, W]$ and ϕ_i^{-1} are homogeneous polynomials in $\left[\hat{X}, \hat{Y}, \hat{Z}, \hat{W}\right]$. In general, these homogeneous polynomials are of degree greater than one. The resulting transformation ϕ is a Cremona transformation; such transformation is bijective almost everywhere except on the set of fundamental elements where all the ϕ_i's vanish, under which the correspondence is one-to-many. However, under some special cases, the transformation may reduce to that of a projective transformation, in which case the homogeneous polynomials ϕ_i and ϕ_i^{-1} are of degree one.

Even from such brief geometric and algebraic analyses, it is clear that given the complex nature of a Cremona transformation, it is in general very difficult to recover metric depth accurately. What is less clear is the feasibility of recovering some of the less metrical depth representations. For instance, the ordinal representation of depth constitutes one such reduced representation of depth argued by researchers. In many cases, knowing that ordinal depth is preserved is enough for us to carry out some visual tasks. Unfortunately, the distortion equation in the most general case (as illustrated in Figure 1) shows that it may not be possible to recover even ordinal relationships under all situations. Nevertheless, we show below that some generic motions allow a robust recovery of partial depth information. In particular, when translation is coupled with rotation, with known or unknown intrinsic parameters, a lateral motion yields a better reconstruction of ordinal depth than a forward motion.

Before embarking on such analysis, a few reasonable assumptions have to be made. Since these generic types of motions are likely to be purposely executed for depth recovery, we expect that the agent executing such motion is at least aware that such generic type of motion is being executed. That is,

- When lateral motion is executed, $\hat{W} = W = 0$.

- When forward motion is executed, $\hat{U} = U = \hat{V} = V = 0$

Furthermore, the following assumption allows us to better grasp the geometrical organization of the iso-distortion surfaces: within a limited field of view (FOV), quadratic terms in the image co-ordinates are small relative to linear and constant terms. This is typically the case when the field of view is small or when the visual system focuses its attention on the foveal region. Finally, we assume that the contribution of γ_e is small, so that $(u_{\text{rot}_e}, v_{\text{rot}_e})$ becomes $(-\beta_e f, \alpha_e f)$. Indeed, in most visual systems, the rotation about the optical axis is usually not executed unless as a result of perturbation. In any case, given their typical magnitudes, these terms do not qualitatively affect the organization of the iso-distortion surfaces (see Figure 1(b)).

3 Depth Recovery under Calibrated Motion

3.1 Lateral Motion

We derive the distorted depth under lateral motion following the same procedure as in section 2, except that we express the translational parameters in equations (3) in terms of U and V to handle the case of FOE at infinity:

$$\hat{Z} = Z\left(\frac{(\hat{U},\hat{V}) \cdot (n_x, n_y)}{(U,V) \cdot (n_x, n_y) + Z(\beta_e, -\alpha_e) \cdot (n_x, n_y)}\right) \tag{5}$$

For the epipolar reconstruction scheme of recovering depth, since the estimated FOE lies in the infinity, all (n_x, n_y) will be in the same direction given by $-\frac{(\hat{U},\hat{V})}{\sqrt{\hat{U}^2+\hat{V}^2}}$. For notational convenience, we can set this (n_x, n_y) to be $(1,0)$ via a rotation of the x- and y-axes without loss of generality (it can be easily shown that even without such a change in the coordinate system, the distortion expression obtained has identical form). After this simplification, we obtain the distortion factor as follows:

$$D = \frac{\hat{U}}{U + Z\beta_e} \tag{6}$$

where $U, \hat{U}$ and β_e are understood to be the corresponding quantities in the rotated coordinate system. Thus the equation of the iso-distortion surface is:

$$Z = \frac{1}{D}\frac{\hat{U}}{\beta_e} - \frac{U}{\beta_e}$$

which represents plane parallel to the image plane.

Figure 2 depicts how the perceived space is distorted. It shows that there exists a $D = 1$ iso-distortion surface which divides the whole space into two parts: one in which the space is expanded $(D > 1)$ and the other in which the space is compressed $(D < 1)$. Its equation is given by $Z = \frac{\hat{U}-U}{\beta_e}$. Whether the D values increase or decrease with Z depends on the sign of β_e. However, we shall show later that in both cases, we are able to recover the ordinal depth, provided that we take proper care of the sign. An estimated depth will be negative if it falls between the region bounded by the $D = 0$ and the $D = -\infty$ surfaces. In this case, the $D = 0$ surface is always located at infinity as its equation is given by $Z = \pm\infty$, and the $D = \pm\infty$ surface is located at $Z = -\frac{U}{\beta_e}$.

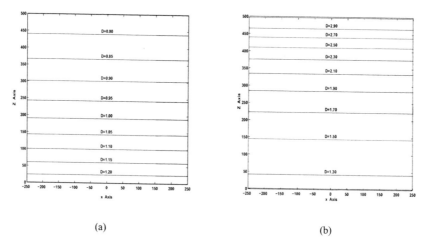

(a)　　　　　　　　　　　　　　　(b)

Figure 2: Families of iso-distortion contours for lateral motion in calibrated case. The parameters are: $U = 0.81$, $\hat{U} = 1.0$, and $\beta_e = 0.001$ for (a) and $\beta_e = -0.001$ for (b).

3.2　Forward Motion

For the case of forward motion, we again make use of the assumptions stated at the end of last section. Conducting an "epipolar reconstruction", the direction (n_x, n_y) can be expressed as $\frac{(x,y)}{\sqrt{x^2+y^2}}$. Substituting into equation (3), we obtain:

$$\hat{Z} = Z\left(\frac{x^2+y^2}{x^2+y^2+Z(-\beta_e f, \alpha_e f)\cdot(x,y)}\right) \tag{7}$$

or expressing $\hat{Z}$ in terms of DZ, the above equation can be expressed as:

$$x^2+y^2+\left(\frac{DZf}{D-1}\right)(-\beta_e x+\alpha_e y) = 0 \tag{8}$$

For a particular value of D, the corresponding iso-distortion surface is a cone. The $D = \pm\infty$ surface is of special interest as all other region in space where D is negative is encompassed by the cone formed by this $D = \pm\infty$ surface. This negative volume is illustrated schematically in Figure 3(a). If we slice these cones with planes parallel to the image plane, we obtain a family of circles, each with center at $\left(\frac{DZf\beta_e}{2(D-1)}, -\frac{DZf\alpha_e}{2(D-1)}\right)$ and radius as $\frac{1}{2}\left|\frac{D}{D-1}\right|Zf\sqrt{\beta_e^2+\alpha_e^2}$. It can also be shown readily that all D surfaces intersect on a common line, which is the Z-axis (see Figure 3(a)). In other word, on this line, the distortion factor is undefined, or equivalently, the Z-axis is the fundamental element of the Cremona transformation describing the distortion.

If we further intersect these cones with planes parallel to the xZ-plane, we obtain the iso-distortion contours as shown in Figures 3(b), (c) and (d), respectively for the cases of $y = 0, y = 50$, and $y = -50$. Several salient features can be identified from the plot:

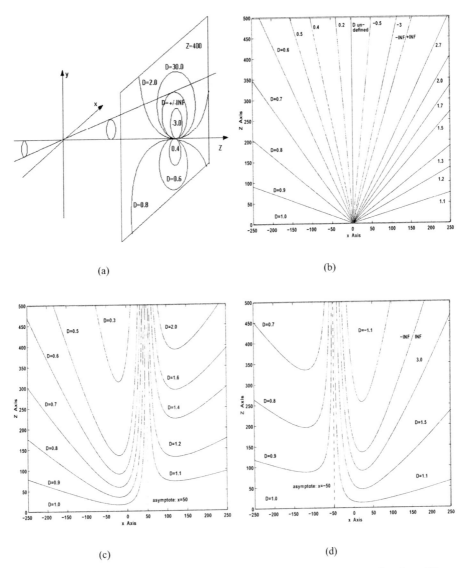

Figure 3: Families of iso-distortion surfaces and contours for forward motion in calibrated case. The distortion surfaces are all cones in the xyZ-space. In Figure (a), the shaded region depicts the volume formed by the $D = \pm\infty$ surface. It contains all other cones with negative D values. We also slice the iso-distortion surfaces with a frontal parallel plane. As can be seen, all cones intersect on the Z-axis. Figures (b), (c), and (d) show the results of slicing the iso-distortion surfaces in Figure (a) with the $y = 0$, $y = 50$, and $y = -50$ planes respectively. Parameters for these plots are: $f = 309.0$, $\alpha_e = 0.001$ and $\beta_e = 0.001$.

1) As far as depth reconstruction is concerned, the key observation here is that, unlike the case of the lateral motion, the value of the distortion factor depends on the image co-ordinate position, which renders depth reconstruction more difficult.

2) In terms of depth recovery, the fundamental element Z-axis indicates that around this axis, the distortion factor changes value rapidly in a small neighborhood (Figure 3(b)), resulting in poor depth estimates that are not even locally smooth. As we move away from the fundamental element (e.g. Figures 3(c) and (d)), the distortion contours no longer intersect together. Nevertheless, as $Z \to \infty$, the contours approach asymptotically towards the line $x = \frac{\alpha_e}{\beta_e} y$, again resulting in rapidly changing distortion values (on the line itself, D happens to have the value of one in this case). The size of the region where distortion changes rapidly depends on the magnitude of α_e, β_e. In the limiting case where α_e, β_e approach zero, this region shrinks to the asymptotic line itself.

3) For forward motion, the $D = 1$ surface always coincides with the plane $Z = 0$. This means that metrical values of depths in the near ground can be judged relatively accurately, which is not necessarily the case for lateral motion.

3.3   Ordinal Depth

Looking at the specific case of the lateral motion, the distortion factor expressed in equation (6) has the form $\frac{1}{a+bZ}$, where $a = \frac{U}{\hat{U}}$ and $b = \frac{\beta_e}{\hat{U}}$ are constants for all the scene points. Such distortion has the effect of generating a relief transformation which has some nice properties (Koenderink and Van Doorn, 1995). In particular, consider two points in space with depths $Z_1 > Z_2$. It can be shown that, given the following conditions:

$$(a+bZ_1)(a+bZ_2) > 0 \text{ if } a > 0$$
$$(a+bZ_1)(a+bZ_2) < 0 \text{ if } a < 0$$

the transformation preserves the depth order of the two points, that is, $\hat{Z}_1 > \hat{Z}_2$. Since $a = \frac{U}{\hat{U}}$ here, the condition $a > 0$ means that U and $\hat{U}$ have the same sign. This condition can easily be met by most visual systems; thus we can just focus on the first condition. The requirement $(a+bZ_1)(a+bZ_2) > 0$ simply means that the two estimated depths have the same sign. However, even if the two estimated depths have different signs, we know that if $a > 0$, it is the greater depth whose estimate will have a negative sign, which means that we can always restore the correct depth order under all conditions.

If we take into account the full rotational error flow (γ_e and the second order terms), then the b term in the transformation $\frac{1}{a+bZ}$ is no longer constant. What this means is that global ordinality is no longer preserved; we can only obtain ordinality within a neighborhood where b can be approximately treated as constant (the size of this neighborhood depends on the size of the motion errors, the respective image co-ordinates, and the depth differences). This means that even with lateral motion, global ordinal depth information may not be obtainable over a large field of view, or if motion perturbation results in unaccounted-for rotation about the Z-axis. Indeed, the presence of the second order terms can be used to explain various psychophysical phenomena in stereoscopic perception, such as:

- Apparent Fronto Parallel Plane (AFPP)

- Apparent Distance Bisection (ADB)

- Distance Judgment from Motion and Stereo

For details, see (Fermüller et al. 1997).

For the case of forward motion, even local ordinal depth information is difficult to obtain, as the distortion factor changes value significantly in a local region. Regions near the fundamental element of the distortion transformation (Figures 3(a) and (b)) or near the asymptotic lines illustrated in Figures 3(c) and (d) are particularly susceptible to depth reversal. The size of the neighborhood in which we can determine ordinal relationship is in general small and depends on several factors. Again, if we consider two points in space with depths $Z_1 > Z_2$, we found that given the following condition, the depth order will be preserved:

$$\frac{Z_1 - Z_2 + (b_2 - b_1)Z_1Z_2}{(1 + b_1Z_1)(1 + b_2Z_2)} > 0$$

where $b_1 = \left(\frac{(-\beta_e f. \alpha_e f).(x_1, y_1)}{x_1^2 + y_1^2}\right)$ and $b_2 = \left(\frac{(-\beta_e f. \alpha_e f).(x_2, y_2)}{x_2^2 + y_2^2}\right)$. From this expression, we can only say that, in general, as we approach the image periphery, the size of this ordinal neighborhood increases (this is also reflected in Figure 3, where the iso-distortion contours become more parallel to the image plane near the periphery, i.e. less dependent on the image co-ordinates).

3.4   Distortion Transformation under Small FOV

Under lateral motion with small field of view, we obtain from equation (6) linear expressions in ϕ_i's; thus the distortion transformation can be expressed with the following matrix:

$$\begin{bmatrix} \hat{X} \\ \hat{Y} \\ \hat{Z} \\ \hat{W} \end{bmatrix} = \begin{bmatrix} \phi_1 \\ \phi_2 \\ \phi_3 \\ \phi_4 \end{bmatrix} = \begin{bmatrix} \hat{U} & 0 & 0 & 0 \\ 0 & \hat{U} & 0 & 0 \\ 0 & 0 & \hat{U} & 0 \\ 0 & 0 & \beta_e & U \end{bmatrix} \begin{bmatrix} X \\ Y \\ Z \\ W \end{bmatrix} \qquad (9)$$

It is obvious that the inverse transformation can be expressed as a matrix with similar form.

As can be seen, the original complex Cremona transformation has now reduced to an invertible projective transformation. Furthermore most of the elements of the matrix representing the transformation are zero, which means that it is really a "well-behaved" kind of projective transformation. In particular, the tilt of a surface, which represents the ordinal aspect of depth information, is preserved although the slant is not. Furthermore, looking at the matrix, if the term β_e in the last row approaches zero, the transformation will tend to preserve the plane at infinity (i.e. it is an affine transformation), in which case all the first order and second order shapes are preserved. In general, the nice properties of approaching such an affine transformation close enough may indeed be sufficient for most vision systems.

For the case of forward motion, we obtain from equation (7) homogeneous polynomials ϕ_i of degree three, given by:

$$\hat{X} = \phi_1 = \left(X^2 + Y^2\right) X$$
$$\hat{Y} = \phi_2 = \left(X^2 + Y^2\right) Y$$
$$\hat{Z} = \phi_3 = \left(X^2 + Y^2\right) Z$$
$$\hat{W} = \phi_4 = \left(X^2 + Y^2\right) W + \left(-\beta_e X + \alpha_e Y\right) Z^2$$

Under this transformation, we can only say that a general element is distorted into an element of the same nature: a point remains as a point, a surface remains as a surface, and a curve remains as a curve. By general element, we mean that the element does not contain any fundamental elements (in this case the Z-axis). If an element is not general, then a point may blow up into a plane, or a plane may reduce to a line under such a transformation.

4 Depth Recovery under Uncalibrated Motion

We use three parameters to describe the intrinsic parameters, namely, the focal length of the optical sensor and the principal point position.

4.1 Intrinsic Parameters Unknown but Fixed

The true optical flow can be expressed in the following form to take into account intrinsic parameters:

$$u = \frac{1}{Z}((x_s - O_x)W - fU) - \beta f + \gamma(y_s - O_y) + O_u^2(x_s, y_s)$$

$$v = \frac{1}{Z}((y_s - O_y)W - fV) + \alpha f - \gamma(x_s - O_x) + O_v^2(x_s, y_s)$$

where (x_s, y_s) represents the image pixel location in a new co-ordinate system with origin located at the lower left corner of the image, (O_x, O_y) is the location of the principal point in the new co-ordinate system, and $O_u^2(x_s, y_s), O_v^2(x_s, y_s)$ represent second order terms in (x_s, y_s). Note that (x_s, y_s) is related to (x, y) by $(x, y) = (x_s - O_x, y_s - O_y)$.

In the uncalibrated case, both the focal length and the location of the principal point are unknown or estimated with error. We denote the estimated focal length and the estimated principal point as $\hat{f}$ and $(\hat{O}_x, \hat{O}_y)$ respectively.

If the influence of the second order terms and γ_e is ignored, we can rewrite the iso-distortion factors for the case of lateral motion as:

$$D = \frac{\hat{f}\hat{U}}{fU + \hat{\beta}_f Z} \tag{10}$$

where $U, \hat{U}$ and β_e are again quantities in the rotated coordinate system. For the case of forward motion, we have:

$$D = \frac{x^2 + y^2}{(x'x + y'y) + \left(-\widehat{\beta_f} x + \widehat{\alpha_f} y\right)Z}$$

where $\left(\widehat{\beta_f}, \widehat{\alpha_f}\right) = \left(\beta f - \hat{\beta}\hat{f}, \alpha f - \hat{\alpha}\hat{f}\right)$, and $(x', y') = (x - O_{xe}, y - O_{ye})$.

It can be seen from equation (10) that the error in the principal point estimate has no impact on the depth reconstruction for the case of lateral motion at all, while error in the focal length estimate alters some constant parameters in the expression for the iso-distortion factor without changing its form. The upshot is that all the previous results regarding ordinal depth is still applicable and that the distortion transformation is still a projective transformation with equation (9) revised as:

$$\begin{bmatrix} \hat{X} \\ \hat{Y} \\ \hat{Z} \\ \hat{W} \end{bmatrix} = \begin{bmatrix} \hat{f}\hat{U} & 0 & 0 & 0 \\ 0 & \hat{f}\hat{U} & 0 & 0 \\ 0 & 0 & \hat{f}\hat{U} & 0 \\ 0 & 0 & \widehat{\beta_f} & fU \end{bmatrix} \begin{bmatrix} X \\ Y \\ Z \\ W \end{bmatrix}$$

For the case of forward motion, we again resort to the iso-distortion plot to visualize the distortion. Figure 4 shows that not much difference from Figure 3 can be found. Each iso-distortion surface is still a cone:

$$x^2 + y^2 + \frac{D}{D-1}\left((-Z\widehat{\beta_f} - O_{xe})x + (Z\widehat{\alpha_f} - O_{ye})y\right) = 0 \tag{11}$$

If we slice these cones with planes parallel to the image plane, we obtain a family of circles, each with center at $\left(\frac{D(O_{xe} + Z\widehat{\beta_f})}{2(D-1)}, \frac{D(O_{ye} - Z\widehat{\alpha_f})}{2(D-1)}\right)$ and radius equal to $\frac{1}{2}\left|\frac{D}{D-1}\right|$
$\sqrt{(O_{xe} + Z\widehat{\beta_f})^2 + (O_{ye} - Z\widehat{\alpha_f})^2}$. Slicing these cones with planes parallel to the xZ-plane yields the iso-distortion contours as illustrated in Figure 4(b). The distortion transformation remains a Cremona one; we will not show the expressions for its homogeneous polynomials ϕ_i and ϕ_i^{-1} here.

As a whole, we can say that fixed uncalibrated intrinsic parameters may change the distortion factor equations, but they do not alter the essential properties of the distortion for both the cases of lateral motion and forward motion. For the case of lateral motion, depth order is still preserved, given the same conditions stated in the calibrated case. For the case of forward motion, as can be seen from Figure 4, there are regions which exhibit

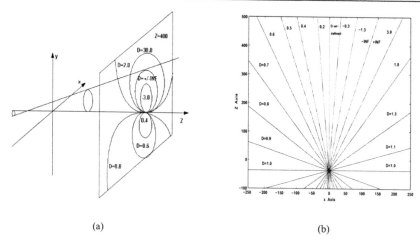

(a)   (b)

Figure 4: Families of iso-distortion contours for forward motion in the uncalibrated case with fixed intrinsic parameters. (a) Schematic representation of the iso-distortion surfaces in the xyZ-space, as in Figure 3(a). (b) Iso-distortion contours obtained by slicing the iso-distortion surfaces with the $y = 0$ plane. $(O_{x_e}, O_{y_e}) = (10, -10), \beta = 0, \hat{\beta} = -0.001, \alpha = 0, \hat{\alpha} = -0.001, f = 309.0,$ and $\hat{f} = 280.0$.

large distortion variation as before and are therefore likely to undergo depth reversal. The $D = 1$ surface has also shifted away from the $Z = 0$ plane due to the presence of the term (O_{x_e}, O_{y_e}).

4.2   Intrinsic Parameters Unknown and Varying

Varying the focal length will result in a zoom field (considering infinitesimal motion) which is hard to separate from a field of translation along the optical axis. Usually a changing focal length will also be accompanied by a change in the principal point. We will see that in such cases, depth ordinality in the global sense is lost even for the case of lateral motion.

With focal length and principal point variation, the resulting optical flow (u, v) can be expressed as:

$$u = \frac{1}{Z}((x_s - O_x)W - fU) - \beta f + \gamma(y_s - O_y) + \dot{O}_x + \frac{\dot{f}}{f}(x_s - O_x) + O_u{}^2(x_s, y_s)$$

$$v = \frac{1}{Z}((y_s - O_y)W - fV) + \alpha f - \gamma(x_s - O_x) + \dot{O}_y + \frac{\dot{f}}{f}(y_s - O_y) + O_v{}^2(x_s, y_s)$$

where $(\dot{O}_x, \dot{O}_y)$ is the rate of change of the principal point and $\dot{f}$ is the rate of change of the focal length.

We assume that the principal point and focal length and their corresponding change rates are all estimated with errors. We make use of the following notations: $(\zeta_{u_e}, \zeta_{v_e}) = (\dot{O}_x - \hat{\dot{O}}_x, \dot{O}_y - \hat{\dot{O}}_y)$ and $\sigma_e = \frac{\dot{f}}{f} - \left(\frac{\hat{\dot{f}}}{\hat{f}}\right)$.

4.2.1 Lateral Motion

Again, by an appropriate rotation of the xy-coordinate system, we can express the iso-distortion factor as follows:

$$D = \frac{\hat{f}\hat{U}}{fU + \left(\widehat{\beta_f} - \dot{\zeta}_{u_e} + \frac{\dot{f}}{f}O_{xe}\right)Z - \sigma_e x Z} \qquad (12)$$

from which we derive:

$$Z = \frac{\hat{f}\hat{U} - DfU}{D\left(\widehat{\beta_f} - \dot{\zeta}_{u_e} + \frac{\dot{f}}{f}O_{xe} - \sigma_e x\right)}$$

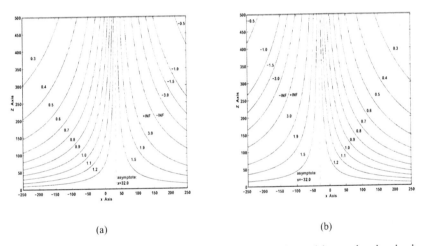

(a)   (b)

Figure 5: Families of iso-distortion contours for lateral motion with varying intrinsic parameters. The parameters are: $U = 0.81$, $\hat{U} = 1.0$, $\beta = 0$, $\hat{\beta} = -0.001$, $f = 309.0$, $\hat{f} = 330.0$, $O_{xe} = 10.0$, $\zeta_{u_e} = 0.01$, and $\frac{\dot{f}}{f} = 0$. The two figures illustrate the effects of the sign of σ_e: $\sigma_e = 0.01$ for (a) and $\sigma_e = -0.01$ for (b). On the asymptote, the value of D is $\frac{\hat{U}}{U} = 1.23$.

The presence of an error in the zoom estimate σ_e could significantly change the topological distribution of the iso-distortion surfaces. As illustrated in Figure 5, the iso-distortion contours become reciprocal curves with a common vertical asymptote given by $x = \frac{\beta_f - \dot{\zeta}_{u_e} + \frac{\dot{f}}{f}O_{xe}}{\sigma_e}$. As can be seen, the position of this asymptote depends very much on the zoom error term σ_e. This means that any flow due to zoom motion must be estimated accurately for meaningful depth information to be derived from lateral motion. When σ_e approaches zero, the vertical asymptote approaches infinity, and the contours will tend to flatness again.

Algebraically, the distortion transformation can be written as:

$$\begin{bmatrix} \hat{x} \\ \hat{y} \\ \hat{z} \\ \hat{w} \end{bmatrix} = \begin{bmatrix} f\hat{U} & 0 & 0 & 0 \\ 0 & f\hat{U} & 0 & 0 \\ 0 & 0 & f\hat{U} & 0 \\ -f\sigma_e & 0 & \widehat{\beta_f} - \zeta_{u_e} + \frac{\dot{f}}{f}O_{xe} & fU \end{bmatrix} \begin{bmatrix} x \\ y \\ z \\ w \end{bmatrix}$$

It is still a projective transformation with a simple form. However the presence of the σ_e term in the last row of the matrix means that the distortion transformation is no longer a relief transformation (see also equation (12)). Ordinality is only preserved if the image distance between two points satisfies certain inequality, which we derive as follows. Consider two points with true depth Z_1 and Z_2 respectively and that $Z_1 > Z_2$. To determine the geometry of this local neighborhood within which ordinal depth is preserved, we need to determine the sign of $\hat{Z}_1 - \hat{Z}_2$. However, directly deciding the sign of $\hat{Z}_1 - \hat{Z}_2$ is difficult; instead we consider $\frac{1}{\hat{Z}_1} - \frac{1}{\hat{Z}_2}$. If $\hat{Z}_1$ and $\hat{Z}_2$ have the same sign, then if $\frac{1}{\hat{Z}_1} - \frac{1}{\hat{Z}_2} < 0$, we can say that the depth order in the perceived space is preserved.

$$\frac{1}{\hat{Z}_1} - \frac{1}{\hat{Z}_2} = \frac{f\frac{U}{\hat{U}}(Z_2 - Z_1) + \frac{\sigma_e}{\hat{U}}(x_2 - x_1)Z_1 Z_2}{fZ_1 Z_2}$$

Since the denominator on the right hand side of the above equation is positive, the condition needed for preservation of depth ordinality can be expressed as:

$$\frac{\sigma_e}{\hat{U}}(x_2 - x_1) < -f\frac{U}{\hat{U}}\frac{(Z_2 - Z_1)}{Z_1 Z_2}$$

If we further make the reasonable assumption that $\hat{U}$ and U have same sign, we have:

$$x_2 - x_1 < -f\frac{U}{\sigma_e}\frac{(Z_2 - Z_1)}{Z_1 Z_2} \quad \text{if } \frac{\sigma_e}{U} > 0 \tag{13}$$

$$x_2 - x_1 > -f\frac{U}{\sigma_e}\frac{(Z_2 - Z_1)}{Z_1 Z_2} \quad \text{if } \frac{\sigma_e}{U} < 0 \tag{14}$$

In either cases, given two fixed depths whose estimates have same sign, the geometry of the local neighborhood on the image plane is that of a half-plane. Furthermore, it is noted that errors in the principal point and the rotational parameter estimates do not influence the properties of the neighborhood. The simple geometry of this local neighborhood means that if bounds can be given to the various terms found in the inequalities, the region of the neighborhood can be approximated.

4.2.2 Forward Motion

The expression for the iso-distortion factor is complex:

$$D = \frac{x^2 + y^2}{(x'x + y'y) + \left(-\widehat{\beta_f} + \zeta_{u_e} - \frac{\dot{t}}{f}O_{xe}, \widehat{\alpha_f} + \zeta_{v_e} - \frac{\dot{t}}{f}O_{ye} \right) \cdot (x,y)Z + \sigma_e(x^2 + y^2)Z} \tag{15}$$

Each iso-distortion surface has the following expression:

$$x^2 + y^2 + \frac{D}{D + D\sigma_e Z - 1}\left((C_1 Z - O_{xe})x + (C_2 Z - O_{ye})y \right) = 0 \tag{16}$$

where $(C_1, C_2) = \left(-\widehat{\beta_f} + \zeta_{u_e} - \frac{\dot{t}}{f}O_{xe}, \widehat{\alpha_f} + \zeta_{v_e} - \frac{\dot{t}}{f}O_{ye} \right)$.

In the xyZ-space, each iso-distortion surface is no longer a cone, but a third-order surface, due to the σ_e term. Slicing these iso-distortion surfaces with a frontal-parallel plane would still yield circles, all with one end anchored at the Z-axis. Each circle has its center at $\left(-\frac{1}{2}\frac{D(C_1 Z - O_{xe})}{D + D\sigma_e Z - 1}, -\frac{1}{2}\frac{D(C_2 Z - O_{ye})}{D + D\sigma_e Z - 1} \right)$ and radius as $\frac{1}{2}\left| \frac{D}{D + D\sigma_e Z - 1} \right| \sqrt{(C_1 Z - O_{xe})^2 + (C_2 Z - O_{ye})^2}$. As $Z \to \infty$, the circle radius becomes constant, that is, in the xyZ-space, the iso-distortion surface forms a cylinder. Slicing these iso-distortion surfaces with the xZ-plane yields the iso-distortion contour plot shown in Figure 6(b)

In contrast to the previous cases, each iso-distortion surface now also approaches asymptotically towards the frontal parallel plane given by $Z = \frac{1-D}{D\sigma_e}$ (see the iso-distortion contour plots in Figure 6). The asymptotic plane of the $D = \pm\infty$ surface is determined by σ_e only as the plane is given by $Z = \frac{1}{\sigma_e}$. Its position has particular significance as it determines the distribution of the positive and negative distortion regions. If the sign of σ_e is positive (Figure 6(b)), then the space in front of the image plane mostly experiences a positive distortion factor, except for the small negative distortion region enclosed by the $D = \pm\infty$ surface. Conversely, if the sign of σ_e is negative (Figure 6(c)), the distortion configuration flips about the $Z = 0$ plane, and most of the region in front of the image plane would have negative distortion factor. In other words, if we impose the "depth is positive" constraint in our motion estimation algorithm, the sign of the error for the zoom estimate is more likely to be positive than negative.

Figure 6(d) shows that the larger σ_e is, the flatter the iso-distortion contours will be (with a smaller dependence of the distortion factor on the image coordinates). Indeed by letting the σ_e term in equation (15) approach infinity, we obtain $D = \frac{1}{1 + \sigma_e Z}$ which is a relief transformation. This might seem to suggest that ironically, large error in estimating zoom field parameter will result in more "well-behaved" recovered depths, in the sense of preserving its ordinality. However these recovered depths are very much compressed and 3-D information is to a great extent lost.

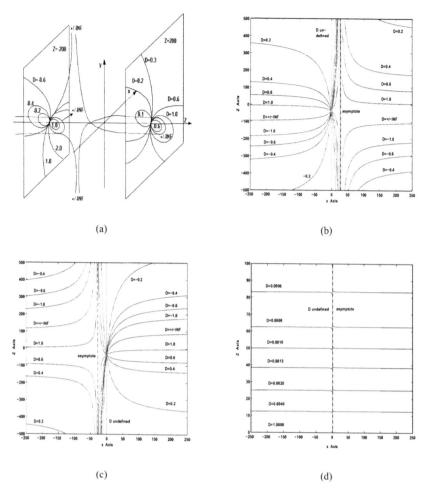

(a)   (b)

(c)   (d)

Figure 6: Distribution of iso-distortion surfaces and contours for forward motion with varying intrinsic parameters. (a) Schematic representation of the iso-distortion surfaces in the xyZ-space. The region where D is negative is not shaded due to the complexity of the region. Figure (b) depicts the iso-distortion contours obtained by slicing the iso-distortion surfaces with the $y = 0$ plane. Figures (c) and (d) illustrate the effects of various values of σ_e on the iso-distortion contours. In (a) and (b), $\left(\frac{\hat{f}}{f}\right) = -0.01$. (c) Reverse the sign of σ_e: $\left(\frac{\hat{f}}{f}\right) = 0.01$. (d) Large σ_e: $\left(\frac{\hat{f}}{f}\right) = -0.01$. Other parameters are the same for all the plot: $\beta = 0$, $\alpha = 0$, $\hat{\alpha} = -0.001, \hat{\beta} = -0.001$, $\frac{i}{f} = 0$, $f = 309.0$, $\hat{f} = 280.0$, $\left(O_{xe}, O_{ye}\right) = (10.0, -10.0)$ and $(\zeta_{u_e}, \zeta_{v_e}) = (0.01, -0.01)$ for (a) and (b); $\left(\frac{\hat{i}}{f}\right) = -20.0$ for (c) and (d); $y = 0$ for (a), (b) and (c) and $y = 50$ for (d).

5 Experiments

In order to support the above theoretical findings, we performed computational experiments on SOFA[1] image sequences. SOFA is a package of 9 synthetic sequences designed for testing research on motion analysis. It includes full ground truth on motion and camera parameters. Sequence 1 and 5 (henceforth abbreviated as SOFA1 and SOFA5) were chosen for our experiments, the former depicting a lateral motion and the latter a forward motion. Both of them have an image dimension of 256×256 pixels, a focal length of 309 pixels and a field of view of approximately $45°$. The camera focal length and principal point were fixed for the whole sequence. The optical flow was obtained using Lucas and Kanade's method (Lucas, 1984), with a temporal window of 15 frames. Depth was recovered for frame 9.

The 3-D scene for SOFA1 consisted of a cube resting on a cylinder (Figure 7(a)). The camera trajectory was a circular route on a plane perpendicular to the world Y-axis, with a constant translational $(U, V, W) = (0.8137, 0.5812, 0)$ and a constant rotational $(\alpha, \beta, \gamma) = (-0.0203, 0.0284, 0)$. If the observer or the system is aware that a lateral motion is being executed, then $\hat{W} = 0$, and the following equation will be used to recover depth:

$$\hat{Z} = \frac{-\hat{f}(\hat{U}, \hat{V}) \cdot (n_x, n_y)}{(u, v) \cdot (n_x, n_y) - (\hat{u_{rot}}, \hat{v_{rot}}) \cdot (n_x, n_y) - \left(\frac{\hat{f}}{f}x, \frac{\hat{f}}{f}y\right) \cdot (n_x, n_y)}$$

ignoring terms involving the principal point (since they have little effect on depth reconstruction, we ignore them in the experiments). The erroneous motion estimates were arbitrarily fixed at $(\hat{U}, \hat{V}) = (1, 0)$ and $(\hat{\alpha}, \hat{\beta}, \hat{\gamma}) = (-0.0213, 0.0274, 0.001)$. We further chose the scheme where the depth was recovered along the estimated epipolar directions. Since we estimated $(\hat{U}, \hat{V})$ to be $(1, 0)$, it means that (n_x, n_y) would be fixed in the horizontal direction $(1, 0)$.

We first simulated the case of no errors in the intrinsic parameters. Using the erroneous motion estimates, we performed depth reconstruction. In Figure 7(b), the recovered depths are depicted with a color coding scheme; cool colors such as deep blue correspond to points close to the observer, while points in warm colors such as red are far away from the observer. The mapping between the colors and the depth ranges was performed individually for each experiment so as to render the plots readable. In Figure 7(c), the reconstructed 3-D depth is displayed using a 3-D plot viewed from the side. Despite significant errors in the motion parameters, these figures indicate a fairly good depth reconstruction in that depth order is preserved for most of the feature points, except for those which are probably affected by noise. Figure 7(c) also shows that the recovered depths tend to be under-estimated, this effect being stronger as the physical depth increases. This can be explained by the contours in Figure 2, where the value of the iso-distortion factor decreases with depth.

Next we simulated the case where the intrinsic parameters are fixed and estimated with errors. Figure 7(d) was obtained by adding a 10% error on the focal length ($\hat{f} = 330.0$). No significant difference is observed with Figure 7(b), which corroborates our theoretical

---

[1]courtesy of the Computer Vision Group, Heriot-Watt University (http://www.cee.hw.ac.uk/ mtc/sofa)

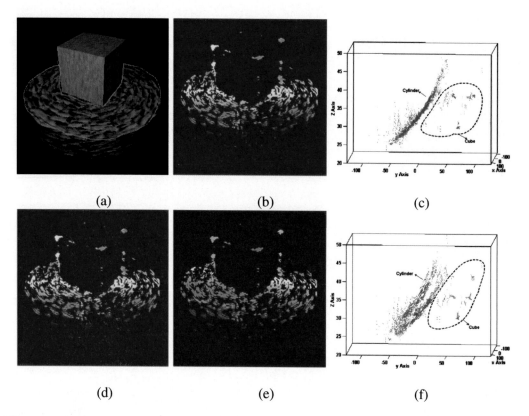

(a)   (b)   (c)

(d)   (e)   (f)

Figure 7: Lateral motion sequence and depth reconstructions. (a) SOFA1 frame 9 with the top face of the cylinder delineated. $(b), (c)$ Reconstruction with true focal length and zoom parameter (i.e. no zoom). (d) Reconstruction with erroneous focal length $\hat{f} = 330.0$ and true zoom parameter. $(e), (f)$ Reconstruction with true focal length and an erroneous zoom field of $\frac{\hat{f}}{f} = -0.01$. The errors used in the other parameters were the same throughout: $(\hat{U}, \hat{V}, \hat{W}) = (1, 0, 0)$ and $(\hat{\alpha}, \hat{\beta}, \hat{\gamma}) = (-0.0213, 0.0274, 0.001)$.

prediction that focal length errors have a weak influence on the distortions in the case of lateral motion. Finally, we simulated the case of varying intrinsic parameters estimated with errors. Figure 7(e) and (f) were obtained for an erroneous zoom field with $\frac{\hat{f}}{f} = -0.01$ (or $\sigma_e = 0.01$). In this situation (Section 4.2.1), we concluded that if $\frac{\sigma_e}{U} > 0$, ordinal depth is preserved only if the image distance between the two feature points satisfies the inequality (13). Figure 7(e) and (f) indicate that ordinal depth is preserved locally, but not for points located far from each other (see for instance the depth reversals between points located on the top face of the cylinder, and on the cube sides). Also, the fact that the recovered depths on the left side of the image are under-estimated more than those on the right side is consistent with the iso-distortion contours depicted in Figure 5.

Since no ground truth for the depth orders is available for the SOFA sequences, it is difficult to quantify the validity of the recovered depth orders. Nevertheless, we observed that for the points on the top face of the cylinder in SOFA1 (which are delineated in Figure 7(a) and account for 4328 of the 5092 feature points), the true depth orders are known since Z increases with y. For this particular region, the rates of correct ordinal depth recovery are tabulated in Table 1.

Table 1: Rates of correct ordinal depth recovery for points on the cylinder's top face (SOFA1)

Figures 7(b) and (c)	Figures 7(d)	Figures 7(e) and (f)
92.42%	92.50%	82.28%

The SOFA5 sequence was used to test our predictions in the case of forward motion. The 3-D scene for SOFA5 comprises a pile of 4 cylinders in front of a frontal-parallel background (Figure 8(a)). The camera trajectory for SOFA5 is parallel to the world Z-axis and the corresponding translation and rotation are $(U,V,W) = (0,0,1)$ and $(\alpha,\beta,\gamma) = (0,0,0)$ respectively. We assume that the observer or the system is aware that a forward motion is being executed, i.e. $\hat{U} = \hat{V} = 0$. Performing an epipolar reconstruction, the equation for calculating the depth of each feature point would be (again ignoring terms involving the principal point):

$$\hat{Z} = \frac{x^2 + y^2}{(u,v).(x,y) - (\hat{u_{rot}}, \hat{v_{rot}}).(x,y) - \frac{\hat{f}}{f}(x^2 + y^2)}$$

Figures 8 (b),(c) and (d) use the aforementioned color coding scheme. Figure 8 (b) depicts the case of error-free extrinsic and intrinsic parameters. Here, the ordinality of the depths recovered is error-prone, possibly as a result of the noise in optical flow computation. In the case of erroneous extrinsic parameters (Figure 8 (c)), the recovered depth is expanded in the bottom right image part, and compressed in the top left area. Such a distortion is consistent with the iso-distortion contours depicted in Figure 3. Figure 8 (d) was obtained for an arbitrarily large estimated zoom field (the true zoom field being zero).

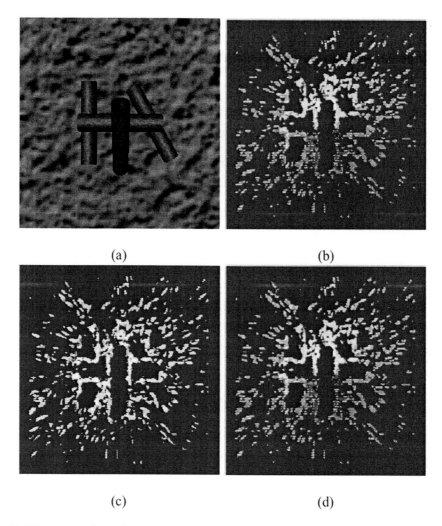

(a)   (b)

(c)   (d)

Figure 8: The forward motion sequence and the depth reconstructions. (a) SOFA5 frame 9. (b) Reconstruction with no errors in both the extrinsic and intrinsic parameters. (c) Reconstruction with errors only in the extrinsic parameters. (d) Reconstruction with an erroneous focal length $\hat{f} = 280.0$ and a large erroneous zoom field of $\frac{\hat{f}}{f} = -20.0$. ($c$) and ($d$) have the same translational and rotational parameter estimates: $(\hat{U}, \hat{V}, \hat{W}) = (0, 0, 1)$ and $(\hat{\alpha}, \hat{\beta}, \hat{\gamma}) = (-0.001, -0.001, 0.01)$.

This corresponds to the case where some intrinsic parameters' variations are not estimated correctly. The resultant iso-distortion contours should approximate that of Figure 6 (d), although we did not introduce any error in the principal point estimate in our experiments. As predicted, the depths recovered seem "better" than that of Figure 8 (c), in the sense that there is no systematic bias in the global arrangement of the depths. However, the recovered depths have values compressed within a very small range, representing a loss of most depth details. Thus it still cannot be deemed a good reconstruction.

Since no ground depth values are available, we conducted the following check on the recovered depth orders. We singled out those points on the cylinders as foreground points and the points on the fronto-parallel plane as background points. We then compared the recovered depth of each foreground point with that of each background point. The former should be smaller than the latter if the depth order is preserved. Although such a test is not a comprehensive one, it gives a fair indication on the rates of correct ordinal depth recovery under various error scenarios. Using 1509 foreground points and 5925 background points, we obtained the following results. With true 3D motion (Figure 8 (b)), 75.14% of such foreground-background depth orders were preserved. Most of these errors were randomly distributed in the image, as could be seen from Figure 8 (b), and could be attributed to image noise. This corroborates our prediction that forward motion is not well suited for depth recovery because of a high sensitivity to noise. For comparison, in the case of SOFA1 (lateral motion configuration) under no errors in the motion parameters, the rate of correct ordinal depth recovery for points on the cylinder's top face was a much better figure of 92.89%[2]. For the case of forward motion with errors in the extrinsic parameters (Figure 8 (c)), the rate of correct ordinal depth recovery dropped to an almost chance level of 56.60%. Adding an error to the focal length estimate ($\hat{f} = 280$) yielded a similar figure of 56.91%. With a small zoom error of 0.01, we obtained 56.67%. Finally, for the case of a large zoom error of 20 (Figure 8(d)), we obtained a poor result of 50.15%, confirming our earlier prediction that under this scenario, ordinal depth recovery is in practice very difficult.

6   Visual Psychophysics

In this section, we aim at investigating the kind of shape distortion experienced by human observers under lateral head translation. In particular, we shall show that our theory points to a directional anisotropy in recovered shape, which suggests that the optimal orientation for detection or recovery of curvature is in the direction orthogonal to the motion. Conversely, the recovery of curvature along the direction of motion is more sensitive to noise and error prone.

---

[2]No ground truth was available for comparing the relative accuracy of the optical flow fields computed for SOFA1 and SOFA5, but from a visual inspection of the flow fields computed, they appeared to be of similar quality and thus should not be a significant factor contributing to the different rates of correct ordinal depth recovery.

6.1 Methods

Four naive subjects participated, all of them had normal or corrected to normal (with contact lenses) vision. Computer generated images were projected onto a large translucent screen (dimension $2.5 \times 2m^2$, refresh rate: 75 Hz.). They showed the sinusoidal oscillations of a surface about a vertical or horizontal fronto-parallel axis, tangent to the surface in a point located at the center of the screen (Figure 9(b)) (rotation frequency 0.5 Hz, amplitude 10 cm). Such rotations in depth, when viewed in the reference frame of the observer, contain a component of frontal translation which is orthogonal to the rotation axis. Hence, a vertical motion refers to a rotation about a horizontal axis, and vice versa. The duration of each stimulus was 4 s (2 oscillations). The simulated distance to the center of the screen was 1.2 m, and the true distance to the screen was 0.87 m. The luminance of a dot and the background were 0.4 cd/m^2 and 0.001 cd/m^2 respectively.

The stimuli were sets of green dots randomly spread over quadrics (Figure 9(b)), which were seen under perspective projection. The dots were displayed only if located within the field of view (60° diameter, or 384 pixels). The surfaces had their principal curvatures aligned with the horizontal and vertical directions. Their values were $\pm 0.27, \pm 0.18, \pm 0.09$ or 0/m, with a positive curvature indicating convexity. We presented 24 horizontal/vertical curvature pairs (Figure 9(a)), corresponding to 8 parabolic cylinders, 8 ellipsoids and 8 hyperboloids.

With head maintained in a chin-rest, and the non-dominant eye covered with an eye-patch, the subject fixated a cross at the center of the stimulus, and reported the apparent curvature of the stimulus within a psychophysical scale. Each trial started with the display of a pair of lines (horizontal or vertical) indicating the direction of judgment (JD), for a duration of 0.4 s. Then the stimulus was displayed for a duration of 4 s, after which the panel shown in Figure 9(c) appeared, carrying 11 lines of different curvatures, and a representation of the eye's position. With the help of a joystick, the subject could select one of the lines in the scale. He/she then pressed a key to confirm his/her answer and proceed to the next trial.

In each session of 48 trials, the 24 surfaces were seen under horizontal or vertical motion in random order. Sessions for the horizontal and vertical JD alternated randomly. For each JD, the first 4 sessions were considered as training (discarded data). Then 10 sessions served in the result analysis, making a total of 10 curvature responses for each stimulus.

The subjects' responses, ranging between -5 and +5, were averaged over 10 trials for each stimulus and JD to obtain the mean perceived curvature PC, and its standard deviation SD. These measures were compared across conditions using the Wilcoxon Matched Pair test (WMPT). The correlation coefficients R are non parametric (Spearman rank).

6.2 Results

- **Curvature reversals**. For the largest curvatures ($\pm 0.27, \pm 0.18$/m), there were only 0.7% of curvature reversals, which confirms that the recovery of the depth sign is essentially unambiguous in large field motion, in this case 60°. Within these "reversal"

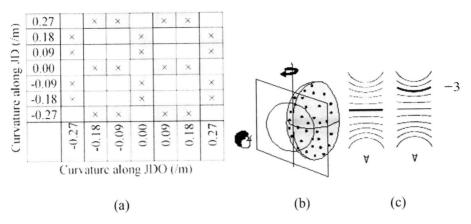

(a)   (b)   (c)

Figure 9: (a). The shaded boxes show the selected combinations of curvatures along JD (judgment direction) and JDO (its orthogonal) for the quadratic surfaces. (b). Dotted surfaces rotating about a vertical or horizontal axis were presented under perspective projection on a large translucent screen. (c). The line pattern presented to the subject after each trial (left). The subject indicates the perceived curvature by shifting the position of the bold line across a series of curved lines (right).

trials, we discarded the 6 trials (0.27%) for which the reversed response could affect the mean of the perceived curvature.

- **Global effect of the variables.** We call PARA and ORTHO the cases where the curvature to be judged (Curv_JD) is parallel (respectively orthogonal) to the motion direction (MD). Curv_JDO is the curvature in the direction orthogonal to the curvature judgment. We performed an ANCOVA (analysis of covariance) on the subjects' responses using Curv_JD and Curv_JDO as covariants, and MD (horizontal or vertical) and the configuration (PARA or ORTHO) as categorical factors. Table 2 indicates that the main influential factor is Curv_JD (as would be expected). The responses are also significantly affected by Curv_JDO and the configuration (PARA or ORTHO), but do not depend on the direction of the motion in space. Therefore, in what follows, we group the results for the horizontal and vertical motion directions. In order to examine the sensitivity of the responses to the curvatures along JD and JDO (Curv_JD and Curv_JDO), we performed a linear fitting of the responses with these curvatures (Figure 10 and Table 3) according to the equation:

  $$\text{Response}(\text{Curv\_JD}) = a_{JD} \, \text{Curv\_JD} + a_{JDO} \, \text{Curv\_JDO} + \text{offset}$$

- **The convexity trend.** In both the PARA and ORTHO cases, the offset of the fitting is significantly positive (Figure 10 and Table 3). Hence, there is general trend to perceive a plane as curved in the convex direction. In addition, the direction of null curvature of the cylinders is also perceived as convex, and this effect increases with the convexity of the orthogonal direction (JDO) for PARA.

Table 2: Results of the ANCOVA on the mean reported curvature and its standard deviation. Continuous factors: curvatures along JD and JDO (Curv_JD and Curv_JDO). Categorical factors: motion direction MD (horizontal/ vertical), and configuration PARA/ORTHO (see text). The right column gives the interaction between the categorical factors.

Factor		Curv_JD	Curv_JDO	MD	PARA/ORTHO	MD*PARA/ORTHO
Mean perceived curvature	F(1,378) (p)	3550 (1E-4)	32 (1E-3)	1 (0.28)	22 (1E-3)	2 (0.13)
Standard deviation	F(1,378) (p)	.18 (.67)	2.88 (.09)	.07 (.79)	23.25 (1E-3)	.26 (.61)

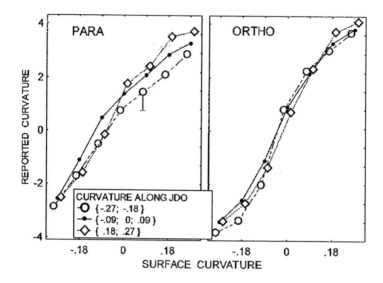

Figure 10: Mean perceived curvature as a function of the curvature along the judgment direction JD. The curvature along JDO (orthogonal to JD) is classified in 3 categories (circles: concave, dot: low and null curvatures, losanges: convex), for the PARA (left graph) and ORTHO (right graph) configurations.

- **Analysis of the precision and variability of the responses**. The coefficient a_{JD} is significantly higher for ORTHO than for PARA, which indicates a lower discriminability of surface curvature in the latter case. Also, the average absolute perceived curvature is significantly higher for ORTHO than for PARA (Table 4). Hence, the curvature is reported as higher in magnitude for ORTHO than for PARA. The coefficient a_{JDO} is significantly positive for both PARA and ORTHO (Table 3), which shows that there is an overall positive effect of the curvature in the orthogonal direction, upon the curvature to be judged (i.e. curvature tends to be perceived as more convex when the orthogonal curvature is convex, rather than concave). For instance the curvatures of hyperboloid look flatter than the curvatures of the corresponding ellipsoid (obtained by reversing the sign of one of the curvatures). This effect is significantly stronger for PARA than for ORTHO. Hence, the curvature responses are

Table 3: Coefficients of the linear fitting of the mean responses with Curv_JD and Curv_JDO. The standard deviations of the coefficients are indicated in the parenthesis. All coefficients are significantly positive at p<1E-3.

	Offset	Curv_JD	Curv_JDO
PARA	.74(.05)	.11(.003)	.016(.003)
ORTHO	.35(.05)	.15(.003)	.009(.003)

Table 4: Paired Student t-test between the configurations PARA and ORTHO, for the mean perceived curvature, its absolute value, and its standard deviation. All differences significant at p<1E-6

N=192	PARA	ORTHO	t
PC	0.74	0.35	5.54
|PC|	2.07	2.62	-9.9
SD	0.908	0.759	5.4

more vulnerable to the overall shape of the surface in the PARA case. However, even then, the overall effect of JDO remains small quantitatively (Figure 10). As noted above, the standard deviations of the curvature estimates are significantly larger for PARA than for ORTHO (Table 2).

In conclusion, the responses along MD show (1) a larger convexity trend, (2) a lower discriminability between different surface curvatures, (3) a larger variability in curvature estimates, and (4) a larger sensitivity of the responses to the overall surface shape, as compared to the responses along MDO. Hence, the motion direction corresponds to a direction with lower reliability of the directional curvature estimates and larger apparent space distortion, as compared to the orthogonal direction.

6.3 Explanation and Discussion

Consider the case where the stimulus was rotating about a vertical axis through its centroid with a rotational velocity β_0 (the corresponding case for rotation about a horizontal axis can be developed analogously and shall be omitted here). Under such motion, the equivalent egomotion for an observer is given by $(\alpha, \beta, \gamma) = (0, -\beta_0, 0)$ and $(U, V, W) = (\beta_0 Z_0, 0, 0)$ where Z_0 is the distance between the optical center and the centroid of the surface. Using the expression for distortion factor given by equation (6), the distorted normal curvatures along the horizontal and vertical directions can be readily obtained as:

$$\hat{Z}_{\hat{X}\hat{X}} = \frac{U}{\hat{U}}Z_{XX} - \frac{2\beta_e}{\hat{U}}$$

$$\hat{Z}_{\hat{Y}\hat{Y}} = \frac{U}{\hat{U}}Z_{YY} \tag{17}$$

Equation (17) shows that normal curvatures along different directions have different distortion properties. The perceived curvature along the horizontal direction (which is the motion direction in this case) suffers from an additional biasing factor caused by the inaccurate β estimates. Thus, even in the absence of the curvature in the X direction (i.e. $Z_{XX} = 0$), some curvature $\hat{Z}_{\hat{X}\hat{X}}$ might still be perceived. The curvature orthogonal to the motion direction, on the other hand, does not suffer from the biasing effect. Assuming that the observer knows the direction of the lateral motion, U and $\hat{U}$ would have the same sign. Without loss of generality, we can let both U and $\hat{U}$ be positive. Then the systematic distortion towards convex shape (at least at this viewing distance) suggests an overestimation of the rotational velocity β, i.e. $\beta_e < 0$.

We might also expect a greater variability in the perception of curvatures along direction of motion. This is because if we assume a random distribution for each of the estimated motion parameters $\hat{U}$ and $\hat{\beta}$ (say Gaussian), we would expect the variance to be larger for $\hat{Z}_{\hat{X}\hat{X}}$ than for $\hat{Z}_{\hat{Y}\hat{Y}}$.

Our psychophysical results clearly demonstrate an apparent convexity of straight lines during the surface motion. To our knowledge, the iso-distortion model is the only approach that explains this phenomenon, at least along the motion direction. In particular, the spin variation model (Droulez and Cornilleau-Pérés, 1990) does not predict any offset in the perceived surface curvature. However, the iso-distortion model only accounts for convexity trend in the direction parallel to the translation, but not along its orthogonal direction. The results obtained in this section suggest the existence of a spatial integration process which averages curvatures through several visual directions during curvature perception. This process is not explained by any directional model so far.

7   Conclusion

Assuming errors in the estimates for the intrinsic and extrinsic parameters, we sought for generic motion types that render depth recovery more robust and reliable. To this end, we compare lateral and forward motions under calibrated and uncalibrated scenarios. For lateral movement, although Euclidean space reconstruction is difficult, the resulting geometrical distortion possesses many nice properties. For the case of calibrated motion, the distortion preserves the depth relief, i.e. ordinal depth. In the uncalibrated case, if the intrinsic parameters are fixed, the situation remains largely the same with regards to ordinal depth recovery. If focal length variation is allowed and the resulting zoom field is not estimated correctly, the ordinal depth can be preserved only locally. In all these three cases, the transformations from the physical space to the perceived space are projective transformations of simple forms. For forward movement, whether calibrated or uncalibrated, depth information is hard to recover—even partially—except for those points close to the observer. Again, if the intrinsic parameters are fixed but estimated with some errors, things change very little from that of the calibrated case. If intrinsic parameters variations are allowed, a large error in the zoom estimate has in theory the potential of improving ordinal depth recovery, although the resultant drastic loss of 3-D depth information means that this recovery is in practice very suspect. In all these three cases, the transformation relating the physical space to the perceived space is a Cremona transformation of degree three. Experiments conducted on images supported the preceding theoretical predictions of a dichotomy between lateral and forward motions.

The conclusion is that under lateral movement, while it might be very difficult to resolve the ambiguity between translation and rotation, ordinal depth can be recovered with robustness. Conversely, under forward motion, depth recovery is difficult unless favorable conditions such as large field of view exist. In the case of uncalibrated motion, in spite of uncertainty in the focal length, the properties of the recovered depth is largely the same as those recovered from calibrated motion, be it a lateral or a forward motion. Thus, as far as depth recovery is concerned, if the intrinsic parameters are fixed, then the calibration of these parameters is not the determining factors for accurate ordinal depth recovery. However, if the intrinsic parameters are allowed to vary, then it is important to estimate the zoom field correctly, although it is safe to forgo the variation of the principal point for depth reconstruction purpose. Otherwise, even in the case of lateral translation, global ordinality of depth will be lost.

Within the hypothesis that observers perform erroneous estimates of the 3D motion in our psychophysical stimuli, the results of the psychophysical experiments are consistent with the iso-distortion prediction of a directional anisotropy in curvature perception. Specifically, the reconstructed space in the direction parallel to the translation is more likely to be distorted than its orthogonal component. More investigations have to be performed to determine if our conclusions are also valid for active human vision (i.e. when motion parallax is generated through head movement). In any case, the iso-distortion approach remains, to our knowledge, the only theory explaining the convexity effect (i.e. a straight line in space appears as curved) along the motion direction.

One major implication of our work is that the orientation of the surface to be captured and reconstructed in 3D is important both in artificial and biological vision systems. For instance, to detect surface curvatures in camera and conveyor belt systems, it is optimal to orientate the surface such that the main curvature to be detected is orthogonal to the motion. As far as bio-robotic works are concerned, understanding the nature of the space distortion under lateral motion has important epistemological implications for systems attempting to mimic such behaviour for tasks such as navigation and predation. Lastly, our psychophysical experiments point to the need for further modeling of the human visual processes. For instance, the theoretical approach does not account for the fact that curvature judgments in one direction seem to be affected by curvatures in the neighbouring directions.

Appendix: Using the Linear Least Square Scheme to Recover Depth

For any SFM algorithm based on optical flow as input, after obtaining the 3-D motion parameters, the next task would be to recover depth for each scene point according to equation (2). For notational convenience, we re-write them as follows:

$$\hat{Z} = \frac{\hat{\mathbf{e}} \cdot \mathbf{n}}{\hat{\mathbf{r}} \cdot \mathbf{n}} \tag{18}$$

where $\hat{\mathbf{e}} = (x - \hat{x_0}, y - \hat{y_0})$, $\hat{\mathbf{r}} = (u - \hat{u_{rot}}, v - \hat{v_{rot}})$ and $\mathbf{n} = (n_x, n_y)$. If there is no error in the optical flow and the 3-D motion estimates, the choice of $\mathbf{n}$ is immaterial, for any direction can give rise to correct depth recovery. However, when the 3-D motion estimates contain errors, $\hat{\mathbf{e}}$ and $\hat{\mathbf{r}}$ would not be parallel, and choosing different $\mathbf{n}$ will yield different depth recovery, which leads to different depth recovery schemes. In particular, the standard linear least square estimate $\hat{Z}_{LLSR}$ is given by $\hat{Z}$ which minimizes the "estimated measurement error" $\| \hat{\mathbf{e}} - \hat{Z}\hat{\mathbf{r}} \|$, from which the following is obtained:

$$\hat{Z}_{LLSR} = \frac{\hat{\mathbf{e}} \cdot \hat{\mathbf{r}}}{\hat{\mathbf{r}} \cdot \hat{\mathbf{r}}}$$

In other words, $\mathbf{n}$ is given by $\frac{\hat{\mathbf{r}}}{\|\hat{\mathbf{r}}\|}$, instead of $\frac{\hat{\mathbf{e}}}{\|\hat{\mathbf{e}}\|}$ in the case of epipolar reconstruction.

Statistically, while the least square scheme is optimal given the necessary conditions, it must be noted that this approach has a serious qualification. In this problem, not only the observation term $\hat{\mathbf{e}}$ contains error, the measurement matrix also contains errors since the entries of the matrix are themselves estimates. Such errors, depending on their magnitudes, could be malign and affect the validity of the least square procedure.

There are also geometrical reason for avoiding the use of the linear least square scheme, as we shall see in the following brief study of the distortion properties under the linear least square scheme.

In the lateral motion case, the iso-distortion factor for the linear least square reconstruction scheme can be expressed as:

$$D = \frac{(\hat{U}, \hat{V}) \cdot (U + Z\beta_e, V - Z\alpha_e)}{(U + Z\beta_e)^2 + (V - Z\alpha_e)^2} \tag{19}$$

where we have made the same assumptions as in Section 2. Figure 11 (a) shows that the iso-distortion surfaces are also planes parallel to the image plane, identical to the case of epipolar reconstruction scheme. However, equation (19) shows that the distortion transformation does not belong to the class of relief transformation; thus there is no guarantee that the depth orders will be preserved.

In the forward motion case, the equation for the iso-distortion factor is:

$$D = \frac{(x, y) \cdot (x - \beta_e f Z, y + \alpha_e f Z)}{(x - \beta_e f Z)^2 + (y + \alpha_e f Z)^2}$$

Figures 11 (b) and (c) show the iso-distortion contours for the forward motion case. Different y values will yield different distributions of iso-distortion contours. These contours are more or less similar to the contours we obtained using the epipolar reconstruction scheme. For instance, the $D = 1$ contour still lies on the x-axis, and depth orders will in general not be preserved.

In conclusion, it is evident that different (n_x, n_y) would result in different distortion geometries. We are not trying to argue for any particular approach of recovering depth, but whichever scheme is adopted, it is the attendant distortion geometry that we are interested

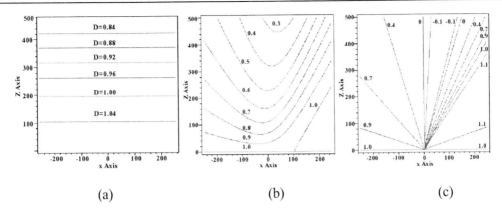

(a)   (b)   (c)

Figure 11: Families of iso-distortion contours using the linear least square scheme for depth reconstruction. (a) Lateral motion with $U = 0.81$, $\hat{U} = 1.0$, $V = 0.58$ and $\hat{V} = 0.4$; (b) forward motion case with $y = 0$; (c): forward motion case with $y = 100$. All the other parameters are identical: $\alpha_e = 0.001$, $\beta_e = 0.001$ and $f = 309.0$.

in. From this brief analysis, it seems that, geometrically, the "epipolar reconstruction approach" has certain favourable properties; it leads to the ordinal depth being preserved in the case of lateral motion, which is not so for the least square approach.

References

Adiv, G. 1989. Inherent ambiguities in recovering 3-D motion and structure from a noisy flow field. *IEEE Trans. PAMI* **11**: 477–489.

Bougnoux, S. 1998. From Projective to Euclidean Space under any practical situation, a criticism of self-calibration. In *Proc. Sixth Int. Conf. on Computer Vision* , pp. 790–796.

Chaumette, F., Boukir, S., Bouthemy, P. and Juvin, D. 1994. Optimal estimation of 3D structures using visual servoing. In *Proc. IEEE Conf. Computer Vision and Pattern Recognition* , pp. 347–354.

Cheong, L-F., Fermüller, C. and Aloimonos, Y. 1998. Effects of errors in the viewing geometry on shape estimation. *Computer Vision and Image Understanding* , 71(3):356–372.

Cheong, L-F. and Ng, K. 1999. Geometry of Distorted Visual Space and Cremona Transformation. *International Journal of Computer Vision* , 32(2):195–212.

Cheong, L-F. and Peh, C-H. 2004. Depth distortion under calibration uncertainty. *Computer Vision and Image Understanding* , 93(3): 221–244.

Collett, T. S. 1978. Peering - a locust behavior pattern for obtaining motion parallax information. *Journal of Experimental Biology* **76**, 237-241.

Coombs, D. and Roberts. K. 1993. Centering behaviour using peripheral vision. In *Proc. Conf. Computer Vision and Pattern Recognition*. pp 440–445.

Cornilleau-Pérès, V. and Droulez, J. 1989. Visual perception of surface curvature: psychophysics of curvature detection induced by motion parallax. *Perception and Psychophysics*, **46**: 351–364.

Daniilidis, K. and Spetsakis, M.E. 1995. Understanding Noise Sensitivity in Structure from Motion. In *Visual Navigation: From Biological Systems to Unmanned Ground Vehicles* , Y. Aloimonos (Ed.), Lawrence Erlbaum Assoc., Pub.

Darrell, T. and Pentland, A. 1991. Robust estimation of a multi-layered motion representation. In *Proc. IEEE Workshop on Visual Motion* , pp. 173–178.

Droulez, J. and Cornilleau-Pérès, V. 1990. Visual perception of surface curvature. The spin variation and its physiological implications. *Biological Cybernetics*, **62**: 211–224.

Dutta, R. and Snyder, M.A. 1990. Robustness of correspondence-based structure from motion. In *Proc. Int'l Conf. on Computer Vision* ,106–110.

Eriksson, E. S. 1980. Movement parallax and distance perception in the grasshopper *Phaulacridium vittatum* (Sjöstedt). *Journal of Experimental Biology* **86**, 337–341.

Fermüller, C. Passive navigation as a pattern recognition Problem. *Int'l Journal of Computer Vision*, **14**:147–158.

Fermüller, C., Cheong, L.F. and Aloimonos, Y. 1997. Visual Space Distortion. *Biological Cybernetics*, **77**:323–337.

J.M. Foley, 1980. "Binocular Distance Perception," *Psychological Review*, **87**(5), 411–434.

Grossmann, E. and Victor, J.S. 2000. Uncertainty analysis of 3D reconstruction from uncalibrated views. *Image Vision Computing*, **18**(9): 686–696.

Horn, B.K.P. 1987. Motion fields are hardly ever ambiguous. *Int'l Journal of Computer Vision*, **1**:259–274, 1987.

Kahl, F. 1995. Critical motions and ambiguous Euclidean reconstructions in autocalibration. In *Proc. Int. Conf. on Computer Vision*, pp. 469–475.

Koenderink, J.J. and van Doorn, A.J. 1995. Relief: Pictorial and Otherwise *Image and Vision Computing,* **13**(5):321–334.

Longuet-Higgins, H.C. 1981. A computer algorithm for reconstruction of a scene from two projections. *Nature* **293**, 133–135.

Lucas, B.D. 1984. Generalized Image Matching by the Method of Differences. PhD Dissertation, Department of Computer Science, Carnegie-Mellon University.

Negahdaripour, S. 1989. Critical surface pairs and triplets. *Int'l Journal of Computer Vision*, **3**:293–312.

Norman, J. F. and Lappin, J. S. 1992. The detection of surface curvatures defined by optical motion. *Perception and Psychophysics*, **51**:386-396.

K.N. Ogle, 1964. *Researches in Binocular Vision.* New York: Hafner.

Oliensis, J. 2000. A critique of structure-from-motion algorithms. *Computer Vision and Image Understanding*, **80**:172–214.

Oliensis, J. 2000. A New Structure From Motion Ambiguity. *IEEE Trans. PAMI*, **22**(7):685–700.

Poteser M. and Kral K. 1995. Visual distance discrimination in praying mantis larvae: An index of the use of motion parallax. *Journal of Experimental Biology* **198**: 2127–2137.

Santos-Victor, J., Sandini, G, Curotto, F. and Garibaldi. S. 1993. Divergent stereo for robot navigation: Learning from bees. *Proc. Conf. Computer Vision and Pattern Recognition*, New York, pp. 434–439.

Soatto, S. and Brockett, R. 1998. Optimal structure from motion: local ambiguities and global estimates. In *Proc. Conf. Computer Vision and Pattern Recognition*, pp. 282–288.

Sobel, E. C. 1990. The locust's use of motion parallax to measure distance. *Journal of Comparative Physiology* **167**, 579–588.

Srinivasan, M. V., Lehrer, M. and Horridge, G. A. 1990. Visual figure–ground discrimination in the honeybee: the role of motion parallax at boundaries. *Proc. R. Soc. Lond. B*, **238**, 331–350.

Sturm, P. 1997. Critical motion sequences for monocular self-calibration and uncalibrated Euclidean reconstruction. In *Proc. Conf. Computer Vision and Pattern Recognition*, pp. 1100–1105.

Sturm, P. 2002. Critical motion sequences for the self-calibration of cameras and stereo systems with variable focal length. *Image and Vision Computing*, **20**(5-6): 415–426.

Szeliski, R. and Kang, S.B. 1997. Shape ambiguities in structure from motion. *IEEE Transactions on Pattern Analysis and Machine Intelligence*, **19**(5): 506–512.

Thomas, J.I., Hanson, A. and Oliensis, J. 1993. Understanding noise: The critical role of motion error in scene reconstruction. In *Proc. DARPA Image Understanding Workshop*, pp. 691–695.

Todd, J.T. and Reichel, F.D. 1989. Ordinal structure in the visual perception and cognition of smoothly curved surfaces. *Psychological Review*, **96**(4):643–657.

Ullman, S. 1979. *The Interpretation of Visual Motion,* MIT Press, Cambridge and London.

Voss, R. and Zeil, J. 1998. Active vision in insects: An analysis of object-directed zig-zag flights in a ground-nesting wasp (*Odynerus spinipes*, Eumenidae). *Journal of Comparative Physiology A*, **182**: 377–387.

Wagner, H. 1989. Peering in barn owls. In *Neural mechanisms of behavior*, Erber, J., Menzel, R., Pflüger, H-J., Todt, D. (eds), Thieme, Stuttgart, pp. 238–239.

Weng, J., Huang, T.S and Ahuja, N. 1991. *Motion and Structure from Image Sequences,* Springer-Verlag.

Young, G.S. and Chellapa, R. 1992. Statistical analysis of inherent ambiguities in recovering 3-D motion from a noisy flow field. *IEEE Trans. PAMI* , **14**:995–1013.

In: Computer Vision and Robotics
Editor: John X. Liu, pp. 135-188

ISBN 1-59454-357-7
© 2006 Nova Science Publishers, Inc.

Chapter 5

MULTIPERSPECTIVE PANORAMIC DEPTH IMAGING

*Peter Peer*and Franc Solina†*
Computer Vision Laboratory, Faculty of Computer and Information Science,
University of Ljubljana, Tr"za"ska 25, 1001 Ljubljana, Slovenia

Abstract

In this chapter we present a stereo panoramic depth imaging system, which builds depth panoramas from multiperspective panoramas while using only one standard camera.

The basic system is mosaic-based, which means that we use a single standard rotating camera and assemble the captured images in a multiperspective panoramic image. Due to a setoff of the camera's optical center from the rotational center of the system, we are able to capture the motion parallax effect, which enables the stereo reconstruction.

The system has been comprehensively analysed. The analyses include the study of influence of different system parameters on the reconstruction accuracy, constraining the search space on the epipolar line, meaning of error in estimation of corresponding point, definition of the maximal reliable depth value, contribution of the vertical reconstruction and influence of using different cameras. They are substantiated with a number of experiments, including experiments addressing the baseline, the repeatability of results in different rooms, by using different cameras, influence of lens distortion presence on the reconstruction accuracy and evaluation of different models for estimation of system parameters. The analyses and the experiments revealed a number of interesting properties of the system.

According to the basic system accuracy we definitely can use the system for autonomous robot localization and navigation tasks.

Keywords: Computer vision, Stereo vision, Reconstruction, Depth image, Multiperspective panoramic image, Mosaicing, Motion parallax effect, Standard camera, Depth sensor

---

*E-mail address: peter.peer@fri.uni-lj.si
†E-mail address: franc.solina{@fri.uni-lj.si

1   Introduction

1.1   Motivation

A computer vision is a special kind of scientific challenge as we are all users of our own vision systems. Our vision is definitely a source of the major part of information we acquire and process each second. A stereo vision is perhaps even greater challenge, since our own vision system is a stereo one and it performs a complex task, which supplies us with 3D information on our surroundings in a very effective way.

Making machines see is a difficult problem. On one side we have psychological aspects of human visual perception, which try to explain how the visual information is processed in the human brain. On the other side we have technical solutions, which try to imitate human vision. Normally, it all starts with capturing digital images that store the basic information about the scene in a similar way that humans see. But this information represents only the beginning of a difficult process. By itself it does not reveal the information about the objects on the scene, their color, distances etc. to the machine. For humans, visual recognition is an easy task, but the human brain processing methods are still a mistery to us.

One part of the human visual perception is estimating the distances to the objects on the scene. This information is also needed by robots if we want them to be completely autonomous.

In this chapter we present a stereo panoramic depth imaging system.

Standard cameras have a limited field of view, which is usually smaller than the human field of view. Because of that people have always tried to generate images with a wider field of view, up to full 360 degree panorama [16].

One way to build panoramic images is by taking one column out of a captured image and mosaicing the columns. Such panoramic images are called multiperspective panoramic images. The crucial property of two or more multiperspective panoramic images is that they capture the information about the motion parallax effect, since the columns forming the panoramic images are captured from different perspectives.

Under the term stereo reconstruction we understand the generation of depth images from two or more captured images. A depth image is an image that stores distances to points on the scene. The stereo reconstruction procedure is based on relations between points and lines on the scene and images of the scene. If we want to get a linear solution of the reconstruction procedure then the images can interact with the procedure in pairs, triplets or quadruplets, and relations are named accordingly to the number of images as epipolar constraint, trifocal constraint or quadrifocal constraint [22]. We want the images to have the property that the same points and lines are visible in all images of the scene, which facilitate stereo reconstruction. This is the property of panoramic cameras and it presents our fundamental motivation. We do the stereo reconstruction from two symmetric multiperspective panoramic images.

In this chapter we address only the issue how to enlarge the horizontal field of view of images. The vertical field of view of panoramic images can be enlarged by using wide angle camera lenses [44], by using mirrors [25, 32] or by moving the camera also in the vertical direction and not only in the horizontal direction [16].

If we tried to build two panoramic images simultaneously by using two standard cameras which are mounted on two rotational robotic arms, we would have problems with non-static scenes. Clearly, one camera would capture the motion of the other camera. So we have decided to use one camera only. Accordingly, in this chapter we present a mosaic-based panoramic depth imaging system using only one standard camera and analyze its performance to see if it can be used for robot localization and navigation in a room.

1.2 Basics about the System

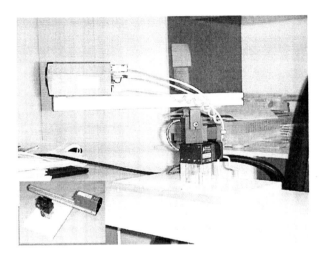

Figure 1: Hardware part of our system.

In Fig. 1 the hardware part of our system can be seen: a color camera is mounted on a rotational robotic arm so that the optical center of the camera is offset from the vertical axis of rotation. The camera is looking outward from the system's rotational center. Panoramic images are generated by repeatedly shifting the rotational arm by an angle which corresponds to a single pixel column of the captured image. By assembling the center columns of these images, we get a mosaic panoramic image. One of the drawbacks of mosaic-based panoramic imaging is that dynamic scenes are not well captured.

It can be shown that the epipolar geometry is very simple if we perform the reconstruction based on a symmetric pair of stereo panoramic images. We get a symmetric pair of stereo panoramic images when we take symmetric columns on the left and on the right hand side from the captured image center column. These columns are assembled in a mosaic stereo pair. The column from the left hand side of the captured image is mosaiced in the right eye panoramic image and the column from the right hand side of the captured image is mosaiced in the left eye panoramic image.

1.3   Structure of the Chapter

In the next section we compare different panoramic cameras with emphasis on mosaicing. In Sec. 3 we give an overview of related work and briefly present the contribution of our work towards the discussed subject. Sec. 4 describes the geometry of our system, Sec. 5 is devoted to the epipolar geometry and Sec. 6 describes the procedure of stereo reconstruction. The focus of this chapter is on the analysis of system capabilities, given in Sec. 7. In Sec. 8 we present experimental results. In the very end of this chapter we summarize the main conclusions.

2   Panoramic Cameras

Every panoramic camera belongs to one of three main groups of panoramic cameras: catadioptric cameras, dioptric cameras and cameras with moving parts. The basic property of a catadioptric camera is that it consists of a mirror (or mirrors [18]) and a camera. The camera captures the image which is reflected from the mirror. A dioptric camera is using a special type of lens, e.g. fish-eye lens, which increases the size of the camera's field of view. A panoramic image can also be generated by moving the camera along some path and mosaicing together the images captured in different locations on the path.

Type of panoramic camera	Number of images	Resolution of panoramic images	Real time	References
catadioptric camera	1	low	yes	[15, 18, 25, 28, 29, 33, 52]
dioptric camera	1	low	yes	[3, 7]
moving parts	a lot	high	no	[1, 8, 9, 10, 12, 13, 14, 16] [17, 19, 20, 21, 23, 25, 26] [27, 32, 33, 35, 36, 39, 43] [44]

Table 1: Comparison of different types of panoramic cameras with respect to the number of standard images needed to build a panoramic image, the resolution of panoramic images and the capability of building a panoramic image in real time.

The comparison of different types of panoramic cameras is shown in Tab. 1.

All types of panoramic cameras enable 3D reconstruction. The camera has a single viewpoint or a projection center if all light rays forming the image intersect in a single point. Cameras with this property are also called central cameras. Rays forming a non-central image do not pass through a single point, but rather intersect a line [10], a conic [25, 39, 40, 49], do not intersect at all [46] or are bound by other constraints suiting the practical or the theoretical demands [13, 17].

Mosaic-based procedures can be marked as non-central (we do not deal with a single center of projection), they do not execute in real time, but they give high resolution results. High resolution images enable effective depth reconstruction, since by increasing the resolution the number of possible depth estimates is also increasing. Thus mosaicing is not appropriate for capturing dynamic scenes and consequently not for reconstruction of dynamic scenes. The systems described in [1, 16] are exceptions because the light rays forming the mosaic panoramic image intersect in the rotational center of the system. These two systems are central systems. The system presented in [30, 41, 42] could also be treated as mosaic-based procedure, though its concept for generating panoramic depth images is very different from our concept.

Dioptric panoramic cameras with wide angle lenses can be marked as non-central [29], they build a panoramic image in real time and they give low resolution results. Cameras with wide angle lenses are appropriate for fast capturing of panoramic images and processing of captured images, e.g. for detection of obstacles or for localization of a mobile robot, but are less appropriate for reconstruction. Please note that we are talking about panoramic cameras here. Generally speaking, dioptric cameras can be central.

Only some of the catadioptric cameras have a single viewpoint. Cameras with a mirror (or mirrors) work in real time and they give low resolution results. Only two mirror shapes, namely hyperbolic and parabolic mirrors, can be used to construct a central catadioptric panoramic camera [29, 52]. Such panoramic cameras are appropriate for low resolution reconstruction of dynamic scenes and for motion estimation. It is also true that only for panoramic systems with hyperbolic and parabolic mirrors the epipolar geometry can be simply generalized [29, 52].

Since dioptric and catadioptric cameras give low resolution results, they are more appropriate for use with view-based systems [59] and less for use with reconstruction systems.

Of course, combinations of different cameras exist: e.g. a combination of the mosaicing camera and the catadioptric camera [25, 32] or a combination of the mosaicing camera and the wide angle camera [44]. Their main purpose is to enlarge the camera's vertical field of view.

3   Related Work

We can generate panoramic images either with the help of special panoramic cameras or with the help of a standard camera and with mosaicing standard images into panoramic images. If we want to generate mosaic 360 degree panoramic images, we have to move the camera on a closed path, which is in most cases a circle.

One of the best known commercial packages for creating mosaic panoramic images is QTVR (QuickTime Virtual Reality). It works on the principle of sewing together a number of standard images captured while rotating the camera [8]. Peleg et al. [27] introduced the method for creation of mosaiced panoramic images from standard images captured with a handheld video camera. A similar method was suggested by Szeliski and Shum [12], which also does not strictly constraint the camera path but assumes that a great motion parallax effect is not present. All methods mentioned so far are used only for visualization purposes since the authors did not try to reconstruct the scene.

The crossed-slits (X-slits) projection [53, 56, 61] uses a similar mosaicing technique with one important difference: the mosaiced strips are sampled from varying positions in the captured images. This makes the generation of virtual walkthroughs possible, i.e. we are again dealing with the visualization with the help of image-based rendering or new view synthesis.

Ishiguro et al. [1] suggested a method which enables scene reconstruction. They used a standard camera rotating on a circular path. The scene is reconstructed by means of mosaicing panoramic images together from the central column of the captured images and moving the system to another location where the task of mosaicing is repeated. The two created panoramic images are then used as the input to a stereo reconstruction procedure. The depth of an object was first estimated using projections in two images captured in different locations of the camera on the camera path. But since their primary goal was to create a global map of the room, they preferred to move the system attached to the robot about the room. Clearly, by moving the robot to another location and producing the second panoramic image of a stereo pair in this location rather than producing a stereo pair in a single location, they enlarged the disparity of the system. But this decision also has a few drawbacks: we cannot estimate the depth for all points on the scene, the time of capturing a stereo pair is longer and we have to search for the corresponding points on the sinusoidal epipolar curves. The depth was then estimated from two panoramic images taken at two different locations of the robot in the room.

Peleg and Ben-Ezra [19, 26] introduced a method for creation of stereo panoramic images without actually computing the 3D structure — the depth effect is created in the viewer's brain.

In [20], Shum and Szeliski described two methods used for creation of panoramic depth images, which use standard procedures for stereo reconstruction. Both methods are based on moving the camera on a circular path. Panoramic images are built by taking one column out of a captured image and mosaicing the columns. The authors call such panoramic images *multiperspective panoramic images*. The crucial property of two or more multiperspective panoramic images is that they capture the information about the motion parallax effect, since the columns forming the panoramic images are captured from different perspectives. The authors use such panoramic images as the input in a stereo reconstruction procedure. In [21], Shum et al. proposed a non-central camera called an omnivergent sensor in order to reconstruct scenes with minimal reconstruction error. This sensor is equivalent to the sensor presented in this chapter.

However, multiperspective panoramic images are not something new to the vision community [20]: they are a special case of *multiperspective panoramic images for cel animation* [13], a special case of *crossed-slits (X-slits) projection* [53, 56, 61], they are very similar to images generated by a procedure called *multiple-center-of-projection* [17], by the *manifold projection* procedure [27] and by the *circular projection* procedure [19, 26]. The principle of constructing multiperspective panoramic images is also very similar to the *linear pushbroom camera* principle for creating panoramic images [10].

The papers closest to our work [1, 20, 21] seem to lack two things: a comprehensive analysis of 1) the system's capabilities and 2) the corresponding points search using the epipolar constraint. Therefore, the focus of this chapter is on these two issues. While in [1] the authors searched for corresponding points by tracking the feature from the column building the first panorama to the column building the second panorama, the authors in [20] used an upgraded *plane sweep stereo* procedure. A key idea behind the approach in [21] is that it enables optimizing the input to traditional computer vision algorithms for searching the correspondences in order to produce superior results.

Further details about the related work are revealed in in the following sections, where we discuss specifics of our system.

4 System Geometry

Let us begin this section with description of how the stereo panoramic pair is generated. From the captured images on the camera's circular path we always take only two columns, which are equally distant from the middle column. We assume that the middle column that we are referring to in this chapter, is the middle column of the captured image, if not mentioned otherwise. The column on the right hand side of the captured image is then mosaiced in the left eye panoramic image and the column on the left hand side of the captured image is mosaiced in the right eye panoramic image. So, we are building each panoramic image from just a single pixel column of the captured image. Thus, we get a symmetric pair of stereo panoramic images, which yields a reconstruction with optimal characteristics (simple epipolar geometry and minimal reconstruction error) [21].

The geometry of our system for creating multiperspective panoramic images is shown in Fig. 2. The panoramic images are then used as the input to create panoramic depth images. Point C denotes the system's rotational center around which the camera is rotated. The offset of the camera's optical center from the rotational center C is denoted as r, describing the radius of the circular path of the camera. The camera is looking outward from the rotational center. The optical center of the camera is marked with O. The column of pixels that is sewn in the panoramic image contains the projection of point P on the scene. The distance from point P to point C is the depth l, while the distance from point P to point O is denoted by d. Further, θ is the angle between the line defined by points C and O and the line defined by points C and P. In the panoramic image the horizontal axis represents the path of the camera. The axis is spanned by μ and defined by point C, a starting point O_0, where we start capturing the panoramic image, and the current point O. φ denotes the angle between the line defined by point O and the middle column of pixels of the image captured by the physical camera looking outward from the rotational center (the latter column contains the projection of the point Q), and the line defined by point O and the column of pixels that will be mosaiced into the panoramic image (the latter column contains the projection of the point P). Angle φ can be thought of as a reduction of the camera's horizontal view angle α.

The geometry of capturing multiperspective panoramic images can be described with a pair of parameters (r, φ). By increasing (decreasing) each of them, we increase (decrease)

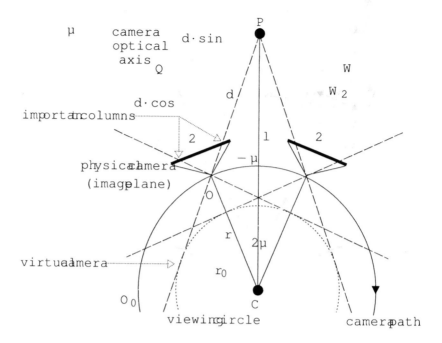

Figure 2: Geometry of our system for constructing multiperspective panoramic images. Note that a ground-plan is presented. The optical axis of the camera is kept horizontal.

the baseline ($2r_0$ [39], $r_0 = r \cdot \sin \varphi$ (Fig. 2)) of our stereo system.

Wei et al. [43] proposed an approach to solve the parameter (r, φ) determination problem for a symmetric stereo panoramic camera. The image acquisition parameters (r, φ) are calculated based on (subjectively) given parameters: the nearest and the furthest distances of the region of interest, the height of the region of interest and the width of the angular disparity interval. They conclude that neither the parameter r nor φ can satisfactorily match application requirements on their own and report that a general study of relations among parameters is in progress as they have discovered certain exceptions in experiments that require further researches.

The system in Fig. 2 is obviously a non-central since the light rays forming the panoramic image do not intersect in one point called the viewpoint, but instead are tangent ($\varphi \neq 0$) to a cylinder with radius r_0, called the viewing cylinder (Fig. 3). Thus, we are dealing with panoramic images formed by a projection from a number of viewpoints. This means that a captured point on the scene is seen in the panoramic image from one viewpoint only. This is why the panoramic images captured in this way are called multiperspective panoramic images.

For stereo reconstruction we need two images. If we look at only one circle on the viewing cylinder (Fig. 2) then we can conclude that our system is equivalent to a system with two cameras. In our case, two virtual cameras are rotating on a circular path, i.e. a

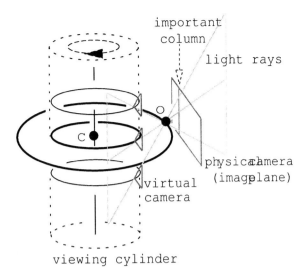

Figure 3: All the light rays forming the panoramic image are tangent to the viewing cylinder.

viewing circle, with radius r_0. The optical axis of a virtual camera is always tangent to the viewing circle. The panoramic image is generated from only one pixel from the middle column of each image captured by a virtual camera. This pixel is determined by the light ray which describes the projection of a scene point on the physical camera image plane. If we observe a point on the scene P, we see that both virtual cameras, which see this point, form a traditional stereo system of converging cameras.

Obviously, a symmetric pair of panoramic images used in the stereo reconstruction process could be captured also with a bunch of cameras rotating on a circular path with radius r_0, where the optical axis of each camera is tangent to the circular path (Fig. 3).

Two images differing in the angle of rotation of the physical camera setup (for example, two image planes marked in Fig. 2) are used to simulate a bunch of virtual cameras on the viewing cylinder. Each column of the panoramic image is obtained from a different position of the physical camera on a circular path. In Fig. 4 we present two symmetric pairs of panoramic images.

To automatically register captured images directly from the knowledge of the camera's viewing direction, the camera lens' horizontal view angle α and vertical view angle β are required. If we know this information, we can calculate the resolution of one angular degree, i.e. we can calculate how many columns and rows are within an angle of one degree. The horizontal view angle is especially important in our case, since we move the rotational arm only around it's vertical axis. To calculate these two parameters, we use an algorithm described in [16]. It is designed to work with cameras whose zoom settings and other internal camera parameters are unknown. The algorithm is based on the mechanical accuracy of the rotational arm. The basic step of our rotational arm corresponds to an angle of $0.051428\overline{5}°$. In general, this means that if we tried to turn the rotational arm for 360 degrees, we would

$$2\varphi = 29.9625°$$

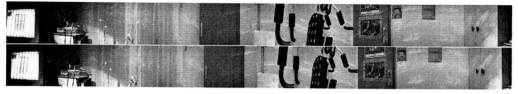

$$2\varphi = 3.6125°$$

Figure 4: Two symmetric pairs of panoramic images generated using different values of the angle φ. In Sec. 7.1 we explain where these values for the angle φ come from. Each symmetric pair of panoramic images comprises the motion parallax effect. This fact enables the stereo reconstruction.

perform 7000 steps. Unfortunately, the rotational arm that we use cannot turn for 360 degrees around it's vertical axis. The basic idea of the algorithm is to calculate the translation dx (in pixels) between two images, while the camera is rotated for a known angle $d\gamma$ in the horizontal direction. Since we know the exact angle by which we move the camera, we can calculate the horizontal view angle of the camera:

$$\alpha = \frac{W}{dx} \cdot d\gamma, \qquad (1)$$

where W is the width of the captured image in pixels.

The major drawback of this method is that it relies on the accuracy of the rotational arm. Because of that we rechecked the values of the view angles by calibrating the camera using a static camera and a checkboard pattern [11, 31, 54]. The input into the calibration procedure is a set of images with varying position of the pattern in each image. The results obtained were very similar, though the second method should be more reliable as it reveals more information about the camera model and also uses sub-pixel accuracy procedure. The latter calibration estimates the focal length, the principal point, the skew coefficient and distortions, to name just the most important parameters for us. It also reveals the errors of all estimated parameters. If we assume that the principal point is in the middle of the captured image, we can calculate the horizontal view angle of the camera from the estimated parameters:

$$\alpha = 2 \arctan \frac{W/2}{f}, \qquad (2)$$

where f is the estimated focal length.

Distortion parameters are also important, because we also investigate the influence of distortion on the system's results.

In any case, now that we know the value of α, we can calculate the resolution of one angular degree x_0:

$$x_0 = \frac{W}{\alpha}.$$

This equation enables us to calculate the width of the stripe W_s that will be mosaiced in the panoramic image when the rotational arm moves for an angle θ_0:

$$W_s = x_0 \cdot \theta_0. \tag{3}$$

From the above equation we can also calculate the angle of the rotational arm for which we have to move the rotational arm if the stripe is only one pixel column wide.

We used three different cameras in the experiments:

- a camera with the horizontal view angle $\alpha = 34°$ and the vertical view angle $\beta = 25°$,

- a camera with the horizontal view angle $\alpha = 39.72°$ and the vertical view angle $\beta = 30.54°$,

- a camera with the horizontal view angle $\alpha = 16.53°$ and the vertical view angle $\beta = 12.55°$.

In the process of the panoramic image construction we did not vary these two parameters. From here on, the first camera is used in the calculations and the experiments, if not stated differently.

5   Epipolar Geometry

Searching for the corresponding points in two images is a difficult problem. Generally speaking, the corresponding point can be anywhere in the second image. That is why we would like to constrain the search space as much as possible. Using the epipolar constraint we reduce the search space from 2D to 1D, i.e. to an epipolar line [4]. In Sec. 7.3 we prove that in our system we can effectively reduce the search space even on the epipolar line.

In this section we will only illustrate the procedure of the proof that the epipolar lines of the symmetric pair of panoramic images are image rows. This statement is true for our system geometry. For proof see [20, 23, 35, 51].

The proof in [23] is based on radius r_0 of the viewing cylinder (Figs. 2 and 3). We can express r_0 in the terms of known parameters r and φ as:

$$r_0 = r \cdot \sin\varphi.$$

We carry out the proof in three steps: *first*, we have to execute the projection equation for the line camera, *then* we have to write the projection equation for a multiperspective panoramic image and, in the *final* step, we prove the property of the epipolar lines for the case of a symmetric pair of panoramic images. In the first step, we are interested in how the point on the scene is projected to the camera's image plane [4], which is of dimension $n \times 1$ pixels in our case, since we are dealing with a line camera. In the second step, we have to write the relation between different notations of a point on the scene and the projection of this point on the panoramic image: notation of the scene point in Euclidean coordinates of the world coordinate system and in cylindric coordinates of the world coordinate system, notation of the projected point in angular coordinates of the (2D) panoramic image coordinate system and in pixel coordinates of the (2D) panoramic image coordinate system. When we know the relations between the above-mentioned coordinate systems, we can write the equation for projection of scene points on the cylindric image plane of the panorama. Based on the angular coordinates of the panoramic image coordinate system property, we can in the third step show that the epipolar lines of the symmetric pair of panoramic images are actually rows of panoramic images. The basic idea for the last step of the proof is as follows: If we are given an image point in one panoramic image, we can express the optical ray defined by a given point and the optical center of the camera in 3D world coordinate system. If we project this optical ray described in world coordinate system on the second panoramic image, we get an epipolar line corresponding to the given image point in the first panoramic image. After introducing proper relations valid for the symmetric case into the obtained equation, our hypothesis is confirmed.

The same result can be found in [20], where the authors proved the property of symmetric pair of panoramic images by directly investigating the presence of the vertical motion parallax effect in the panoramic images captured from the same rotational center. The generalization to the non-symmetric case for the camera looking inward and outward can be found in [51]. Even a more general case, in some respect, where the panoramic images can be captured from different rotational centers, is discussed in [35].

It was shown that the notion of the epipolar geometry, well known for both central perspective cameras [4, 22, 34] and central catadioptric cameras [28, 29, 52], can be generalized to some non-central cameras [37, 40, 46, 49]. The epipolar surfaces extend from planes to double-ruled quadrics: planes, rotational hyperboloids and hyperbolic paraboloids.

6   Stereo Reconstruction

Let us go back to Fig. 2. Using trigonometric relations evident from the sketch, we can write the equation for the depth estimation l of a point P on the scene. By the basic law of sines for triangles, we have:

$$\frac{r}{\sin(\varphi - \theta)} = \frac{d}{\sin\theta} = \frac{l}{\sin(180° - \varphi)}. \tag{4}$$

From this equation we can express the equation for depth estimation l as:

$$l = \frac{r \cdot \sin(180° - \varphi)}{\sin(\varphi - \theta)} = \frac{r \cdot \sin\varphi}{\sin(\varphi - \theta)}. \tag{5}$$

Eq. (5) implies that we can estimate depth l only if we know three parameters: r, φ and θ. r is given. Angle φ can be calculated on the basis of the camera's horizontal view angle α (Eq. (1)) as:

$$2\varphi = \frac{\alpha}{W} \cdot W_{2\varphi}, \tag{6}$$

where W is the width of the captured image in pixels and $W_{2\varphi}$ is the width of the captured image between columns forming the symmetric pair of panoramic images, given also in pixels. To calculate the angle θ, we have to find corresponding points on panoramic images. Our system works by moving the camera for the angle corresponding to one pixel column of the captured image. If we denote this angle by θ_0, we can express the angle θ as:

$$\theta = dx \cdot \frac{\theta_0}{2}, \tag{7}$$

where dx is the absolute value of difference between the corresponding points image coordinates on the horizontal axis x of the panoramic images.

Note that Eg. (5) does not contain the focal length f explicitly, but since the relationships between α and f on one side (Eq. (2)) and α and φ on the other side (Eq. (6)) exist, φ also depends upon f (the two models for estimating angle φ (Eqs. (6) and (8)) are discussed in Sec. 7.2):

$$\varphi = \arctan\frac{W_{2\varphi}/2}{f}. \tag{8}$$

Eq. (5) estimates the distance l to the perpendicular projection of the scene point P on the plane defined by the camera's circular (planar) path. The projection of the scene point P is marked with P' in Fig. 5. Since this estimation is an approximation of the real l, we have to improve the estimation by addressing the vertical reconstruction, i.e. by incorporating the vertical view angle β into Eq. (5).

Let us here adopt the following notation to introduce the influence of β on estimation of l: if a variable l or d depends on α only, we mark that as $l(\alpha)$ and $d(\alpha)$ (until now, these variables were marked simply l and d), but if a variable l or d depends on α and β, we mark that as $l(\alpha, \beta)$ and $d(\alpha, \beta)$. According to Fig. 5 the distance to the point P on the scene can be calculated as:

$$l(\alpha, \beta) = \sqrt{l(\alpha)^2 + Y^2} = \sqrt{l(\alpha)^2 + (l(\alpha) \cdot \tan\omega_2)^2}.$$

Because the value of ω_2 is unknown, we have to express it in terms of known parameters. We can do that, while Y can also be written as:

$$Y = d(\alpha) \cdot \tan\omega_1.$$

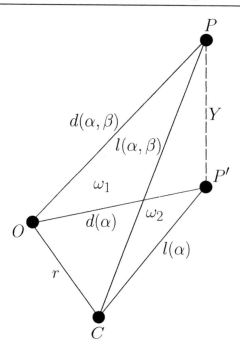

Figure 5: Important relations between system parameters for addressing the vertical reconstruction.

We can calculate ω_1 similarly as we calculated φ (Eqs. (6) and (8)):

$$2\omega_1 = \frac{\beta}{H} \cdot H_{2\omega_1} \quad \text{or} \quad \omega_1 = \arctan\frac{H_{2\omega_1}/2}{f},$$

where H is the height of the captured image in pixels and $H_{2\omega_1}$ is the height of the captured image between the image row that contains the projection of the scene point P and the symmetric row on the other side from the middle row, given also in pixels. And $d(\alpha)$ follows from Eq. (4):

$$d(\alpha) = \frac{l(\alpha) \cdot \sin\theta}{\sin\varphi}.$$

Now, we can write the equation for $l(\alpha, \beta)$ as:

$$l(\alpha, \beta) = \sqrt{l(\alpha)^2 + \left(\frac{l(\alpha) \cdot \sin\theta}{\sin\varphi} \cdot \tan\omega_1\right)^2}. \tag{9}$$

From now on, $l = l(\alpha)$ and when $l(\alpha, \beta)$ is used, this is explicitly stated.

The influence of addressing the vertical reconstruction on the reconstruction accuracy is discussed in Secs. 7.6 and 8.4.

7   Analysis of the System's Capabilities

7.1   Time Complexity of Panoramic Image Creation

The biggest disadvantage of our system is that it cannot produce panoramic images in real time since we create them stepwise by rotating the camera for a very small angle. Because of mechanical vibrations of the system, we also have to ensure to capture an image when the system is completely still. The time that the system needs to create a panoramic image is much too long to allow it work in real time.

In a single circle around the system's vertical axis our system constructs 11 panoramic images: 5 symmetric pairs and a panoramic image from the middle columns of the captured images. It captures and saves 1501 images with resolution of 160×120 pixels, where radius is $r = 30$ cm and the shift angle is $\theta_0 = 0.205714°$. We have choosen the resolution of 160×120 pixels because it represents a good compromise between overall time complexity of the system and its accuracy, as it is shown in the following sections. We cannot capture $360/\theta_0$ images because of the limitation of the rotational arm. Namely, the rotational arm cannot turn for 360 degrees around its vertical axis.

The middle column of the captured image was in our case the 80th column. The distances between the columns building up symmetric pairs of panoramic images were 141, 125, 89, 53 and 17 columns. These numbers include two columns building up each pair. In consequence the values of the angle 2φ (Eq. (6)) are $29.9625°$ (141 columns), $26.5625°$ (125 columns), $18.9125°$ (89 columns), $11.2625°$ (53 columns) and $3.6125°$ (17 columns), respectively. (Here we used the camera with the horizontal view angle $\alpha = 34°$.)

The acquisition process takes little over 15 minutes on a 350 MHz Intel PII PC. The steps of the acquisition process are as follows:

1. Move the rotational arm to its initial position.

2. Capture and save the image.

3. Contribute image parts to the panoramic images.

4. Move the arm to the new position.

5. Check in the loop if the arm is already in the new position. The communication between the program and the arm is written in the file for debugging purposes. After the program exits the loop, it waits for 300 ms in order to stabilize the arm in the new position.

6. Repeat steps 2 to 5 until the last image is captured.

7. When the last image has been captured, contribute image parts to the panoramic images and save them.

We could achieve faster execution since our code is not optimized. For example, we did not optimize the waiting time (300 ms) after the arm is in the new position. No computations are done in parallel.

7.2   Influence of Parameters r, φ and θ_0 on the Reconstruction Accuracy

In order to estimate the depth as precisely as possible, the parameters involved in the calculation also have to be estimated precisely. In this section we reveal the methods used for estimation of parameters r, φ and θ_0.

θ_0 denotes the angle corresponding to one pixel column of the captured image, for which we rotate the camera. It can be calculated from Eq. (3):

$$\theta_0 = \frac{\alpha}{W}. \tag{10}$$

For $\alpha = 34°$ and W=160 pixels, we get $\theta_0 = 0.2125°$. On the other hand, we know that the accuracy of our rotational arm is $\varepsilon = 0.0514285^{\overline{}°}$, so the best possible approximate value is $\theta_0 = 0.205714°$. Since each column in the panoramic image in reality describes the latter angle θ_0, we always use in calculations $\theta_0 = n \cdot \varepsilon$, $n \in I\!N$, which is closest to the result obtained from Eq. (10). The experiment in Sec. 8.5 confirms that this decision is correct. To discriminate the two values between each other, let us mark them as $\theta_0(\alpha)$ (Eq. (10)) and $\theta_0(\varepsilon)$ (the estimation based on the accuracy of our rotational arm). We use them from now on, but where only θ_0 is given, then $\theta_0 = \theta_0(\varepsilon)$.

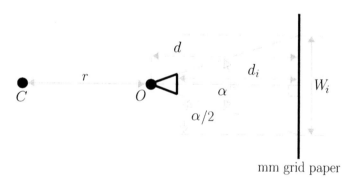

Figure 6: The relation between the parameters, which are important for determining the radius r.

r represents the distance between the rotational center of the system and the optical center of the camera. Since the exact position of the optical center is normally not known (not given by the manufacturer), we have to estimate its position. Optical firms with their special equipment would do the best job, but since this has not been an option for us, we have used a simple method, which has been proved quite useful (Fig. 6): First the camera horizontal view angle α has been estimated. Then we have captured a few images of the mm grid paper from known distances d_i from one point on the camera to the paper. The optical axis has been assumed to be perpendicular to the paper surface. From each image we have read the width W_i of it in mm and used all now known values (α, d_i and W_i) to estimate the distance d from the paper to the optical center by manually drawing a geometrically precise relation between the parameters. More distances d_i have been used to check the consistency of all estimates. At the end the position of the optical center has been calculated as an average over all estimated values. Because we know the distances d_i and d, we also know the position of the optical center with respect to the point on the camera from which we have measured the distances d_i. Finally, we can measure the distance r. Nevertheless, this is a rough estimation of the optical center position, but it can be optimized as shown in the experiment in Sec. 8.9.

φ determines the column of each captured image, which is mosaiced into the panoramic image. The two models for estimating angle φ (Eqs. (6) and (8)) differ from one another: the first one is linear, while the second one is not. But since we use cameras with the maximal horizontal view angle $\alpha = 39.72°$, the biggest possible difference between the models is only $0.3137°$ (at the point, where ratio $W_{2\varphi}/W = 91/160$). In the experiments we use such values of φ that the difference is very small, i.e. the biggest difference is lower than $0.1°$. The experiment in Sec. 8.6 shows that we obtain slightly better results with the linear model for a given (estimated) set of parameters. This is why the linear model was used in all other experiments.

We discuss the angle θ_0 and the radius r in relation with the one-pixel error in estimation of the angle φ in the end of Sec. 7.4.

7.3 Constraining the Search Space on the Epipolar Line

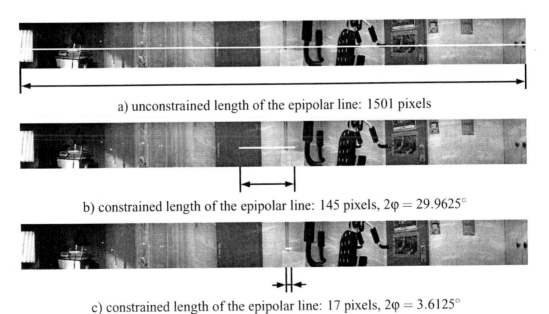

a) unconstrained length of the epipolar line: 1501 pixels

b) constrained length of the epipolar line: 145 pixels, $2\varphi = 29.9625°$

c) constrained length of the epipolar line: 17 pixels, $2\varphi = 3.6125°$

Figure 7: We can effectively constrain the search space on the epipolar line.

Knowing that the width of the panoramic image is much bigger than the width of the captured image, we would have to search for a corresponding point along a very long epipolar line (Fig. 7a). Therefore we would like to constraint the search space on the epipolar line as much as possible. This means that the stereo reconstruction procedure executes faster. A side effect is also an increased confidence in the estimated depth.

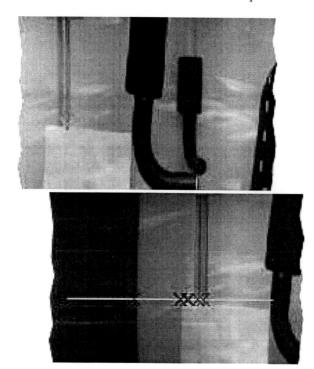

Figure 8: Constraining the search space on the epipolar line in case of $2\varphi = 29.9625°$. In the left eye panorama (top image) we have denoted the point for which we are searching the corresponding point with a green cross. In the right eye panorama (bottom image) we have used green color to mark the part of the epipolar line on which the corresponding point must lie. The best corresponding point is marked with a red cross. With blue crosses we have marked a number of points which presented temporary best corresponding point before we actually found the point with the maximal correlation.

From Eq. (5) we can derive two conclusions, which nicely constraint the search space:

1. Theoretically, the minimal possible estimation of depth is $l_{min} = r$. This is true for $\theta = 0°$. However, this is impossible in practice since the same point on the scene cannot be seen in the column that will be mosaiced in the panorama for the left eye and at the same time in the column that will be mosaiced in the panorama for the right eye. If we observe the horizontal axis of the panoramic image with respects to the direction of the rotation, we can see that every point on the scene that is shown on both panoramic images (Fig. 4) is first imaged in the panorama for the left eye and then in the panorama for the right eye. Therefore, we have to wait until the point imaged in the column building up the left eye panorama moves in time to the column building up the right eye panorama. If θ_0 presents the angle by which the camera is shifted, then $2\theta_{min} = \theta_0$. In consequence, we have to make at least one basic shift of

the camera to enable a scene point projected in a right column of the captured image forming the left eye panorama to be seen in the left column of the captured image forming the right eye panorama.

Based on this fact, we can search for the corresponding point in the right eye panorama starting from the horizontal image coordinate $x + \frac{2\theta_{\min}}{\theta_0} = x + 1$ forward, where x is the horizontal image coordinate of the point in the left eye panorama for which we are searching the corresponding point. Thus, we get the value $+1$ since the shift for the angle θ_0 describes the shift of the camera for a single column of the captured image.

In our system, the minimal possible depth estimation $l_{\min}$ depends on the value of the angle φ:

$$l_{\min}(2\varphi = 29.9625°) \quad = \quad 302 \text{ mm}$$
$$\ldots$$
$$l_{\min}(2\varphi = 3.6125°) \quad = \quad 318 \text{ mm}.$$

2. Theoretically, the estimation of depth is not constrained upwards, but from Eq. (5) it is evident that the denominator must be non-zero. Practically, this means that for the maximal possible depth estimation $l_{\max}$ the difference $\varphi - \theta_{\max}$ must be equal to the value in the interval $(0, \frac{\theta_0}{2})$. We can write this fact as: $\theta_{\max} = n \cdot \frac{\theta_0}{2}$, where $n = \varphi \text{ div } \frac{\theta_0}{2}$ and $\varphi \text{ mod } \frac{\theta_0}{2} \neq 0$.

If we write the constraint for the last point, which can be a corresponding point on the epipolar line, in analogy with the case of determining the starting point that can be a corresponding point on the epipolar line, we have to search for the corresponding point in the right eye panorama to including the horizontal image coordinate $x + \frac{2\theta_{\max}}{\theta_0} = x + n$. Here x is the horizontal image coordinate of the point on the left eye panorama for which we are searching the corresponding point.

Equivalently, like in case of the minimal possible depth estimation $l_{\min}$, the maximal possible depth estimation $l_{\max}$ also depends upon the value of the angle φ:

$$l_{\max}(2\varphi = 29.9625°) \quad = \quad 54687 \text{ mm}$$
$$\ldots$$
$$l_{\max}(2\varphi = 3.6125°) \quad = \quad 86686 \text{ mm}.$$

In the following sections we show that we cannot trust the depth estimates near the last point of the epipolar line search space, but we have proven that we can effectively constrain the search space.

To illustrate the use of specified constraints on real data, let us present the following example which describes the working process of our system: while the width of the panorama is 1501 pixels, when searching for a corresponding point, we have to check only $\varphi \text{ div } \frac{\theta_0}{2} = 145$ pixels in case of $2\varphi = 29.9625°$ (Figs. 7b and 8) and only 17 in case of $2\varphi = 3.6125°$ (Fig. 7c).

From the last paragraph we could conclude that the stereo reconstruction procedure is much faster for a smaller angle φ. However, in the next section we show that a smaller angle φ, unfortunately, has also a negative property.

7.4   Meaning of the One-pixel Error in Estimation of the Angle θ

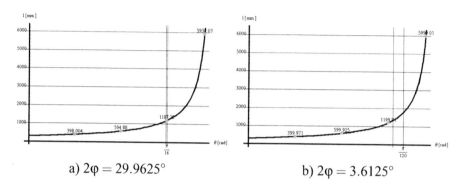

a) $2\varphi = 29.9625°$ b) $2\varphi = 3.6125°$

Figure 9: The dependence of depth l on angle θ (Eq. (5), $r = 30$ cm and two different values of φ are used). To visualize the one-pixel error in estimation of the angle θ, we have marked the interval of width $\frac{\theta_0}{2} = 0.102857°$ between the vertical lines near the third point.

Let us first define what we mean under the term one-pixel error. As the images are discrete, we would like to know what is the value of the error in the depth estimation if we miss the right corresponding point for only one pixel. And we would like to have this information for various values of the angle φ.

Before we illustrate the meaning of the one-pixel error in estimation of the angle θ, let us take a look at the graphs in Fig. 9. The graphs show the dependence of the depth function l on the angle θ when two different values of the angle φ are used. It is evident that the depth function l rises slower in case of a bigger angle φ. This property decreases the error in the depth estimation l when a bigger angle φ is used and this decrease in the error becomes even more evident if we know that the horizontal axis is discrete and the intervals on the axis are $\frac{\theta_0}{2}$ degrees wide (see Fig. 9). If we compare the width of the interval in both graphs with respect to the width of the interval that θ is defined in (θ ∈ [0, φ]), we can see that the interval with the width of $\frac{\theta_0}{2}$ degrees is much smaller when a bigger angle φ is used. This subsequently means that the one-pixel error in estimation of the angle θ is much smaller when a bigger angle φ is used, since a shift for the angle θ_0 describes the shift of the camera for a single column of pixels.

Because of a discrete horizontal axis θ (Fig. 9), with intervals $\frac{\theta_0}{2}$ degrees wide (in our case $\theta_0 = 0.205714°$), the number of possible depth estimates is proportional to the angle φ: we can calculate φ div $\frac{\theta_0}{2} = 145$ different depth values (Eq. (5)) if we use the angle $2\varphi = 29.9625°$ (Fig. 10a) and only 17 different depth values if we use the angle $2\varphi = 3.6125°$ (Fig. 10b). This is the disadvantage of small angles φ (see the experiment in Sec. 8.1).

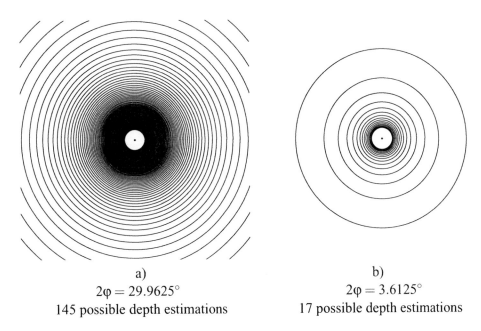

a)
$2\varphi = 29.9625°$
145 possible depth estimations

b)
$2\varphi = 3.6125°$
17 possible depth estimations

Figure 10: The number of possible depth estimates is proportional to the angle φ. Each circle denotes a possible depth estimation value.

	$\theta - \frac{\theta_0}{2}$	θ	$\theta + \frac{\theta_0}{2}$
l [mm]	394.5	398	401.5
Δl [mm]	3.5		
(error)		3.5	

a) $\theta = \theta_1 = \frac{\varphi}{4}, 2\varphi = 29.9625°$

	$\theta - \frac{\theta_0}{2}$	θ	$\theta + \frac{\theta_0}{2}$
l [mm]	372.5	400	431.8
Δl [mm]	27.5		
(error)		31.8	

b) $\theta = \theta_1 = \frac{\varphi}{4}, 2\varphi = 3.6125°$

	$\theta - \frac{\theta_0}{2}$	θ	$\theta + \frac{\theta_0}{2}$
l [mm]	2252.9	2373.2	2507
Δl [mm]	120.3		
(error)		133.8	

c) $\theta = \theta_2 = \frac{7\varphi}{8}, 2\varphi = 29.9625°$

	$\theta - \frac{\theta_0}{2}$	θ	$\theta + \frac{\theta_0}{2}$
l [mm]	1663	2399.6	4307.4
Δl [mm]	736.6		
(error)		1907.8	

d) $\theta = \theta_2 = \frac{7\varphi}{8}, 2\varphi = 3.6125°$

Table 2: The one-pixel error Δl in estimation of the angle θ, where $r = 30$ cm (Eq. (5)).

Let us illustrate the meaning of the one-pixel error in estimation of the angle θ: We would like to know what is the error of the angle θ at the beginning of the interval over which θ is defined ($\theta \in [0, \varphi]$) and what is the error of the angle θ near the end of this interval?

For this purpose we choose angles $\theta_1 = \frac{\varphi}{4}$ and $\theta_2 = \frac{7\varphi}{8}$. We are also interested in the nature of the error for different values of the angle φ. In this example we use our already standard values for the angle φ: $2\varphi = 29.9625°$ and $2\varphi = 3.6125°$. The results in Tab. 2 give the values of the one-pixel error in estimation of the angle θ for different values of parameters θ and φ.

From the results in Tab. 2 we can conclude that the error is much bigger in case of a smaller angle φ than in case of a bigger angle φ. The second conclusion is that the value of the error increases as the value of the angle θ gets closer to the value of the angle φ. This is true regardless of the value of the angle φ. This two conclusions are also evident from Fig. 10: possible depth estimations lie on concentric circles centered in the center of the system, with the distance between circles increasing the further away they lie from the center (see also the experiment in Sec. 8.3). The figure nicely illustrates the fact that in case of a small angle φ, we can estimate only a few different depths and the fact that the one-pixel error in estimation of the angle θ increases if we move away from the center of the system.

We would like to get reliable depth estimates, but at the same time we would like the reconstruction procedure to execute fast. Here, we are faced with two contradicting requirements, since we have to make a compromise between the accuracy of the system and the speed of the reconstruction procedure. Namely, if we wanted to achieve the maximal possible accuracy, then we would use the maximal possible angle φ. But this means that we would have to conduct a search for the corresponding points on a larger segment of the epipolar line. Consequently, the speed of the reconstruction process would be lower. We would come to the same conclusion if we wanted to achieve a higher speed of the reconstruction procedure, since the speed of the reconstruction process is inversely proportional to its accuracy.

By varying the parameters θ_0 and r we change the size of the error:

- By increasing the resolution of captured images, we decrease the angle θ_0 (Eq. (10)) and subsequently decrease the rotational angle of the camera between two successively captured images forming the stereo panoramic images. By nearly the same factor that we increase (decrease) the resolution of captured images, we decrease (increase) the value of the error Δl, while the reconstruction process takes more (less) time by nearly the same factor. By decreasing (increasing) the value θ_0 we are able to calculate more (less) depth values and consequently, we achieve bigger (lower) accuracy. Another way to influence the parameter θ_0 is to vary the horizontal view angle α. This influence is presented separately in Sec. 7.7.

- By the same factor that we increase (decrease) the radius r, we increase (decrease) the (biggest possible and sensible) depth estimation l and the size of error Δl. Obviously, if the camera optical center is at the same distance from one really close object for different r, we achieve bigger accuracy by using smaller r. The behavior of Δl_{min} given in the next section nicely illustrates this fact. If we vary the parameter r, the process of reconstruction is not any faster or slower. In practice, a bigger r means that we can reconstruct bigger scenes (rooms). The geometry of our system is adequate of reconstructing (smaller) rooms and is not really suitable for reconstruction of an outdoor scene. This is due to the inherent property of the system: we do not trust in the estimated depth l of far-away objects on the scene if the size of the error Δl is too big. If we vary the parameter r, the number of possible depth estimates naturally stays the same.

7.5   Definition of the Maximal Reliable Depth Value

In Sec. 7.3 we have defined the minimal possible depth estimation l_{min} and the maximal possible depth estimation l_{max}, but we have not said anything about the meaning of the one-pixel error in estimation of the angle θ for these two estimated depths. Let us examine the size of the error Δl for these two estimated depths. We calculate Δl_{min} as the absolute value of difference between the depth l_{min} and the depth l for which the angle θ is bigger than the angle θ_{min} by the angle $\frac{\theta_0}{2}$:

$$\Delta l_{min} = |l_{min}(\theta_{min}) - l(\theta_{min} + \frac{\theta_0}{2})| = |l_{min}(\frac{\theta_0}{2}) - l(\theta_0)|.$$

Similarly, we calculate the error Δl_{max} as the absolute value of difference between the depth l_{max} and the depth l for which the angle θ is smaller than the angle θ_{max} by the angle $\frac{\theta_0}{2}$:

$$\Delta l_{max} = |l_{max}(\theta_{max}) - l(\theta_{max} - \frac{\theta_0}{2})| = |l_{max}(n\frac{\theta_0}{2}) - l((n-1)\frac{\theta_0}{2})|,$$

where the variable n denotes a positive number in equation: $n = \varphi \text{ div } \frac{\theta_0}{2}$.

	$2\varphi = 29.9625°$	$2\varphi = 3.6125°$
Δl_{min}	2 mm	19 mm
Δl_{max}	30172 mm	81587 mm

Table 3: The one-pixel error Δl in estimation of the angle θ for the minimal possible depth estimation l_{min} and the maximal possible depth estimation l_{max} with respect to the angle φ and the radius $r=30$ cm.

In Tab. 3 we have gathered the error sizes for different values of the angle φ. The results confirm statements in Sec. 7.4. We can add one additional conclusion: The value of error Δl_{max} is unacceptably high and this is true regardless of the value of the angle φ. This is why we have to sensibly decrease the maximal possible depth estimation l_{max}. In practice, this leads us to defining the upper boundary of the allowed error size (Δl) for a single pixel in the estimation of the angle θ. Using it, we subsequently define the maximal reliable depth value (see the example in the next section).

7.6   Contribution of the Vertical Reconstruction

Addressing the vertical reconstruction is essential for getting as accurate results as possible. In this section we investigate how big is the difference between the depths estimated without (Eq. (5)) and with (Eq. (9)) addressing the vertical reconstruction.

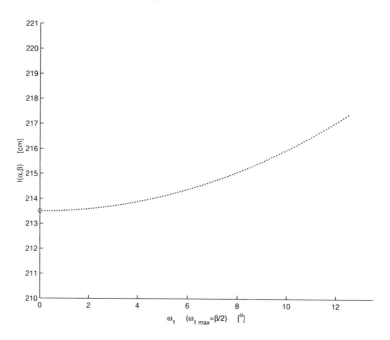

Figure 11: The contribution of the vertical reconstruction is small for the camera with the horizontal view angle $\alpha = 34°$ and the vertical view angle $\beta = 25°$ (Eq. (9)). The diamond marks the depth l_{max} estimated without addressing the vertical reconstruction (Eq. (5)). For detailed description see Sec. 7.6.

Let us first define the maximal reliable depth value l_{max} as suggested in the previous section for the camera with the horizontal view angle $\alpha = 34°$ and the vertical view angle $\beta = 25°$. If we do not allow the error size Δl to be more than 10 cm for $r = 30$ cm, $2\varphi = 29.9625°$ and $\theta_0 = 0.205714°$, then, consequently, $l_{max} = 213.5$ cm. By introducing the influence of the vertical view angle β into Eq. (9):

$$\omega_{1\ max} = \frac{\beta}{2},$$

we get $l_{max}(\alpha, \beta) = 217.4$ cm (Fig. 11). This means that the contribution of the vertical reconstruction is small ($l_{max}(\alpha, \beta) - l_{max} = 3.9$ cm, which is 1.8% of $l_{max}(\alpha, \beta)$), but as expected it has a positive influence on the overall results as shown in the experiment in

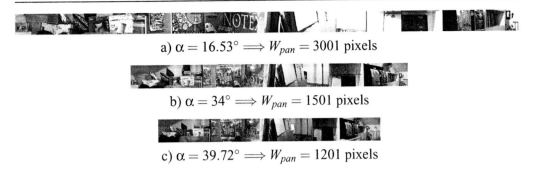

a) $\alpha = 16.53° \Longrightarrow W_{pan} = 3001$ pixels

b) $\alpha = 34° \Longrightarrow W_{pan} = 1501$ pixels

c) $\alpha = 39.72° \Longrightarrow W_{pan} = 1201$ pixels

Figure 12: Different cameras characterized by the horizontal view angle α give panoramic images with different horizontal resolution W_{pan}.

Sec. 8.4. By increasing (decreasing) the angle β (using different cameras) we also increase (decrease) the contribution of the vertical reconstruction.

7.7 Influence of Using Different Cameras

Each camera can be characterized by its horizontal view angle α. According to Eq. (10), the angle θ_0 gets bigger with bigger α, having in mind that the width W (the resolution) of the captured images stays the same. This means that we have to capture less images with the camera characterized by bigger α in order to generate the panoramic image of the same scene. Let us illustrate this fact by presenting generated panoramic images of the same scene, where we varied α (we used different cameras). In Fig. 12 we can see that for the cameras mentioned in the end of Sec. 4, we get panoramic images of different horizontal resolution, while the vertical resolution is equal for all cameras:

a) the camera with the horizontal view angle $\alpha = 16.53°$ gives a panoramic image with the width $W_{pan} = 3001$ pixels,

b) the camera with the horizontal view angle $\alpha = 34°$ gives a panoramic image with the width $W_{pan} = 1501$ pixels and

c) the camera with the horizontal view angle $\alpha = 39.72°$ gives a panoramic image with the width $W_{pan} = 1201$ pixels.

So, by enlarging the camera horizontal field of view the width of the panorama gets lower, while the height of the panorama stays the same. But the same height captures more scene, since also the camera vertical field of view is enlarged.

This means that we generate a panoramic image faster if we use a camera with a wider view angle. But the drawback here is that the horizontal angular resolution (the number of possible depth estimates per one degree) gets lower. As already (implicitly) mentioned in the end of Sec. 7.4, by varying the resolution of the captured images we vary the horizontal and the vertical resolution of the generated panoramic image at fixed α. But now we can

		θ	$\theta + \frac{\theta_0}{2}$
l [mm]	$2\varphi = 29.9625°$ ($\alpha = 34°$)	398	401.5
	$2\varphi = 38.47875°$ ($\alpha = 39.72°$)	396.7	400.2
Δl [mm] (error)	$2\varphi = 29.9625°$ ($\alpha = 34°$)	3.5	
	$2\varphi = 38.47875°$ ($\alpha = 39.72°$)	3.5	

a) $\theta = \theta_1 = \frac{\varphi}{4}$

		θ	$\theta + \frac{\theta_0}{2}$
l [mm]	$2\varphi = 29.9625°$ ($\alpha = 34°$)	2373.2	2507
	$2\varphi = 38.47875°$ ($\alpha = 39.72°$)	2355.8	2488.8
Δl [mm] (error)	$2\varphi = 29.9625°$ ($\alpha = 34°$)	133.8	
	$2\varphi = 38.47875°$ ($\alpha = 39.72°$)	133	

b) $\theta = \theta_2 = \frac{7\varphi}{8}$

Table 4: The one-pixel error Δl in estimation of the angle θ for different cameras (α) at the similar number of the possible depth estimates following from φ, where $r = 30$ cm (Eq. (5)). For $2\varphi = 29.9625°$ the results are the same as in Tab. 2. Consequently, the same θ values are used in this table.

add one more conclusion, namely, if we vary α then we vary only the horizontal resolution of the generated panoramic image at fixed resolution of the captured images.

For each camera (and not only for the cameras that we use) the maximal number of possible depth estimates depends on the horizontal resolution W of captured images. From $\varphi_{max} = \alpha/2$, θ_0 and the equation for determining the number of possible depth estimates φ_{max} div $\frac{\theta_0}{2}$, we get very similar results for different cameras (α) at fixed W. (All the results are equal to W if $\theta_0(\varepsilon) = \theta_0(\alpha)$ (see the discussion on estimation of the angle θ_0 ($\theta_0(\alpha)$, $\theta_0(\varepsilon)$) in Sec. 7.2).) This means that the comparison of results gained using different cameras should not be done at the similar φ, but rather at the similar number of the possible depth estimates. This fact is used in the experiment in Sec. 8.8.

The size of the one-pixel error Δl in estimation of the angle θ (Sec. 7.4), for φ's defined in described way, is also similar. This is evident from Tab. 4.

8 Experimental Results

In the experiments the following cameras were used:

- *camera #1* with parameters:
  - $\alpha = 34°$
  - $\beta = 25°$
  - $r = 30$ cm
  - $2\varphi = 29.9625°$
  - $\theta_0 = 0.205714°$

- *camera #2* with parameters:
  - $\alpha = 39.72°$
  - $\beta = 30.54°$
  - $r = 31$ cm
  - $2\varphi = 38.47875°$
  - $\theta_0 = 0.257143°$

- *camera #3* with parameters:
  - $\alpha = 16.53°$
  - $\beta = 12.55°$
  - $r = 35.6$ cm
  - $2\varphi = 15.3935625°$
  - $\theta_0 = 0.102857°$.

All the panoramic images are generated from images with resolution of 160×120 pixels.

Correspondences for each feature point on the scene used in the evaluation have been determined with a *normalized correlation* procedure [4] and rechecked manually for consistency. If the difference between the manually and the automatically determined correspondence has been more than one pixel, such feature has not been used in the evaluation, otherwise we believe in the automatically obtained result rather than in the manually obtained result, because the latter is a subjective result, while the other is an objective result. Fact is that it is hard to manually determine the corresponding point due to the discrete nature of images. Nevertheless, in more than 75% the two results have been the same.

We use the normalized correlation procedure to search for corresponding points because it is one of the most commonly used technique employed for that purpose. On the other hand, correlation-based stereo algorithms are the only ones that can produce sufficiently dense depth images with an algorithmic structure which lends itself nicely to fast implementations because of the simplicity of the underlying computation [5]. Various improvements to real time correlation-based stereo vision are discussed in [5, 45]. To improve the results we could also employ multiple-baseline approach [6, 36]. It has been shown that by using multiple-baseline stereo, match ambiguities can be reduced and the reconstruction precision can be improved as well. Other interesting methods than just those based on correlation are described in [2]. A nice survey about a taxonomy and evaluation of dense two-frame stereo correspondence algorithms is given in [48]. In [55], the authors review recent advances in computational stereo, focusing primarily on correspondence methods, methods for occlusion and real time implementations.

The normalized correlation procedure uses the principle of similarity of scene parts within two scene images. The basic idea of the procedure is to find the part of the scene in the second image which is most similar to a given part of the scene in the first image. The procedure uses a window, within which the similarity is measured with help of the correlation technique.

We use this procedure also when we generate depth images. Additionally, to increase the confidence in the estimated depth, we employ a procedure called *back-correlation* [4]. The main idea of this procedure is to first find a point $\mathbf{m}_2$ in the second image which corresponds to a point $\mathbf{m}_1$ given in the first image. Then we have to find the point corresponding to the point $\mathbf{m}_2$ in the first image. Let us denote this corresponding point by $\mathbf{m}'_1$. If the point $\mathbf{m}_1$ is equal to the point $\mathbf{m}'_1$ then we keep the estimated depth value. Otherwise, we do not keep the estimated depth value. This means that the point $\mathbf{m}_1$, for which the back-correlation was not successful, has no depth estimation associated with it in the depth image. Using the back-correlation procedure we also solve the problem of occlusions. On the other hand, the normalized correlation score can also be used for estimating the confidence in the estimated depth.

All results were generated by using a correlation window of size $2n+1 \times 2n+1$, $n=4$, if not mentioned otherwise. We searched for corresponding points only in the panoramic image row determined by the epipolar geometry.

The primary evaluation of the system is based on mentioned feature points on the scene. The quantitative measure, which gives the average error of the estimated depth (l (Eq. (5)) or $l(\alpha, \beta)$ (Eq. (9))) in comparison to the actual distance (d) over n scene points, is calculated as:

$$AVG_\% = \frac{\sum_{i=1}^{n} |l_i - d_i|/d_i}{n} \cdot 100\%.$$

The second measure, which is in the results written right beside the first one, is the standard deviation following from:

$$SD_\% = \sqrt{\frac{\sum_{i=1}^{n} \left(\frac{|l_i - d_i|}{d_i} \cdot 100\% - AVG_\%\right)^2}{n-1}},$$

which reveals how tightly all the various estimated depths are clustered around the average error in the set of data.

On the other hand, the evaluation is also given qualitatively, i.e. visually, where this is needed.

Note that all the presented results are rounded upon their calculations and not before.

Every time we refer to the features on the scene in tables or figures, the appropriate features are also marked for better orientation in the panoramic image given at the bottom of tables and figures.

In the first three experiments (Secs. 8.1, 8.2 and 8.3) we use l (Eq. (5)) rather than $l(\alpha, \beta)$ (Eq. (9)), so that afterwards we are able to demonstrate the influence of the vertical reconstruction on the reconstruction accuracy.

8.1   Influence of Different φ Values on the Reconstruction Accuracy — The Quantitative Evaluation

Experiment background: See the discussion on the number of possible depth estimates with respect to the angle φ in Sec. 7.4. The results were obtained with *camera #1*.

Results: The comparison of results using $2\varphi = 3.6125°$ and $2\varphi = 29.9625°$ (see Sec. 7.1 about how these values were obtained) is presented in Tab. 5.

feature	d [cm]	$2\varphi = 3.6125°$		$2\varphi = 29.9625°$	
		l [cm]	$l - d$ [cm (% of d)]	l [cm]	$l - d$ [cm (% of d)]
1	111.5	89.4	-22.1 (-19.8%)	109	-2.5 (-2.3%)
2	95.5	76.7	-18.8 (-19.6%)	89.3	-6.2 (-6.5%)
3	64	53.8	-10.2 (-15.9%)	59.6	-4.4 (-6.9%)
4	83.5	76.7	-6.8 (-8.1%)	78.3	-5.2 (-6.2%)
5	92	89.4	-2.6 (-2.8%)	89.3	-2.7 (-2.9%)
6	86.5	76.7	-9.8 (-11.3%)	82.7	-3.8 (-4.4%)
7	153	133.4	-19.6 (-12.8%)	159.8	6.8 (4.5%)
8	130.5	133.4	2.9 (2.2%)	135.5	5 (3.8%)
9	88	76.7	-11.3(-12.8%)	87.6	-0.4 (-0.5%)
10	92	89.4	-2.6 (-2.8%)	89.3	-2.7 (-2.9%)
11	234.5	176.9	-57.6 (-24.6%)	213.5	-21 (-8.9%)
12	198	176.9	-21.1 (-10.7%)	179.1	-18.9 (-9.5%)
13	177	176.9	-0.1 (-0.1%)	186.7	9.7 (5.5%)
		AVG%=11% ± 7.7%		AVG%=5% ± 2.6%	

Table 5: The comparison of results for two different values of φ.

Conclusion: As expected, the results with $2\varphi = 29.9625°$ are much better, since this angle ensures many more possible depth estimates.

8.2 Time Analysis of the Stereo Reconstruction Process

Experiment background: Searching for the corresponding point presents the most expensive part of the stereo reconstruction process. In this section we present some time results, given in hours, minutes and seconds, though the ratios between these results are more important, since the measured times depend on the code itself (optimized or unoptimized, sequential or parallel processing), the stereo-matching algorithm, the speed of the processor, the number of processors etc. As already mentioned, our code is not optimized, no processing is done in parallel, we use normalized correlation algorithm and all the calculations are done on a 350 MHz Intel PII PC (in C++ programming language). For better illustration we have run the reconstruction process over the whole generated pair of stereo panoramic images.

On one side, we have constructed dense panoramic images, which means that we have tried to find the corresponding point in the right eye panorama for every point in the left eye panorama.

On the other side, the sparse depth images have been created by searching only for the correspondences of feature points in input panoramic images. The feature points used have been vertical edges on the scene, derived by filtering the panoramic images with the Sobel filter for searching the vertical edges [1, 4]. The time needed for locating the features on the scene reconstructed in the sparse depth image is included in the presented times. But the time needed for acquisition of panoramic images is not included in the reconstruction time.

Some of the generated depth images are presented in the next section.

The results were obtained with *camera #1*.

Results: The comparison of results using $2\varphi = 3.6125°$, $2\varphi = 29.9625°$ and the back-correlation algorithm ($BC = true$ or $false$), while building dense and sparse depth images, is given in Tab. 6.

Conclusion: As expected, the time needed for the reconstruction with the back-correlation search is approximately twice the time needed for the reconstruction without it, while the back-correspondence search algorithm has the same complexity as the correspondence search algorithm (because the basic algorithm is the same in both cases, just the role of the stereo images are swapped). And if we use the smaller angle φ, the reconstruction times are up to approximately eight times smaller from presented ones. This is due to the fact that in case of smaller angle φ we have to check only 17 pixels on the epipolar line, while in case of bigger angle φ we have to check 145 pixels on the epipolar line. The ratio between these two numbers is approximately equal to the speed-up factor.

As mentioned, all results have been generated by using a correlation window of size $2n + 1 \times 2n + 1$, $n=4$. For comparison, if $n=3$ then the time needed to create the dense panoramic depth image, while $2\varphi = 29.9625°$ and $BC = true$, is 4 hours, 20 minutes and 55 seconds. The ratio between the window areas is again approximately equal to the speed-up

	sparse depth image reconstruction time [min./sec.]	dense depth image reconstruction time [hours/min./sec.]
$2\varphi = 29.9625°$ $BC = true$	1/10	6/42/20
$2\varphi = 29.9625°$ $BC = false$	0/38	3/21/56
$2\varphi = 3.6125°$ $BC = true$	0/33	0/52/56
$2\varphi = 3.6125°$ $BC = false$	0/21	0/29/6

Table 6: The comparison of the stereo reconstruction times.

factor. On the other hand, if we run the same process on the faster computer (PC Intel PIV/2.0 GHz), the time needed to gain the same result is 1 hour, 1 minute and 29 seconds. The speed-up factor could again be attributed to the ratio between the processor frequencies. Nevertheless, the newer processor is approximately 4 times faster, which means that after optimizing the code, introducing Intel's MMX SIMD (Single Instruction Multiple Data) instruction set [15, 55] etc., we would gain the sparse panoramic depth image for $2\varphi = 29.9625°$ and $BC = true$ in real time. At this point, real time to us means one stereo reconstruction per second. For autonomous robot navigation the sparse depth image based on vertical edges already contains important information about the environment.

Further stereo reconstruction process speed-up could be achieved by processing 8-bit grayscale images with lower resolution, by doing the reconstruction of only part of the scene in which we are interested, using the property of successive pixels in the panoramic images to constrain the search space on the epipolar line even more, using different stereo-matching algorithm etc. But the most efficient way to ensure the real time reconstruction (at video rate) is to employ cluster of computers, doing real parallel processing [57]. Until very recently, all truly real time implementations made use of special purpose hardware, like digital signal processors (DSP) or field programmable gate arrays (FPGA) [5, 55].

On the other hand, real time correlation based stereo algorithms are discussed in [5, 45, 60]. In the latter, i.e. [60], the real time dense reconstruction is performed on symmetric multiperspective panoramic images with resolution of 1324×120 pixels. The reconstruction is done in 0.34 seconds on a 1.7 GHz PC.

According to Sec. 7.7, the speed-up could also be achieved if we use a camera with a wider field of view, since this means that the width of the generated panoramic images is lower. Consequently, the speed of the reconstruction process is higher. If we generate the sparse depth image then the speed-up is not that noticeable, since the number of pixels presenting edges is more or less the same. But in case of dense depth image the speed-up factor can be substantial: The basic speed-up factor is given by the ratio between the widths of panoramic images.

Real time, on the other hand, is a wide term, as it has different meanings in relation with different applications and consequently, in our case, with demanded reconstruction accuracy.

Let us at the end of this section also touch the storage requirements. Our panoramic images are each of approximately 0.5 MB in size (bmp format), while in [43] the size of each panoramic image is approximately 3400 MB (format is not specified). Their images are really of hyper-resolution (19478×5184 pixels), but acquisition requirements (time, storage, processing, cost) are obviously of great pretension.

8.3 Influence of Different φ Values on the Reconstruction Accuracy — The Qualitative Evaluation

Experiment background: We have used a simple stereo-matching algorithm based on the correlation technique. In spite of that, we are interested in how good the obtained results, i.e. depth images, are visually. Since it is hard to evaluate the quality of generated depth images, we present four reconstructions of the room from generated depth images. In this way, we are able to evaluate the quality of generated depth images and consequently the quality of the system. The plan of the room that we have reconstructed is given in Fig. 13. In the sketch we have marked the features on the scene that help us evaluate the quality of generated depth images. The result of the (3D) reconstruction process is a ground-plan of the scene. The goal of the experiment is to see how well the reconstruction fits the real room. The results were obtained with *camera #1*.

Results: Fig. 14 shows some results of our system. In case denoted with b), we have constructed the dense panoramic image. Black color marks the points on the scene with no depth estimation associated. Otherwise, the nearer the point on the scene is to the rotational center of the system, the lighter the point appears in the depth image.

In case denoted with d), we have used the information about the confidence in the estimated depth (case c), which we get from the normalized correlation estimations. In this way, we have eliminated from the dense depth image all depth estimates which do not have a high enough associated confidence estimation. The lighter the point appears in case c), the more we trust in the estimation of the normalized correlation for this point. In case marked with e), we have created a sparse depth image by searching only for the correspondences of feature points (vertical edges) in input panoramic images.

The following properties are common to the (3D) reconstructions in Figs. 15, 16, 17 and 18:

- Big dots denote the actual positions of features on the scene (measured by hand).

- A big dot near the center of the reconstruction shows the position of the center of our system.

- Small black dots represent reconstructed points on the scene.

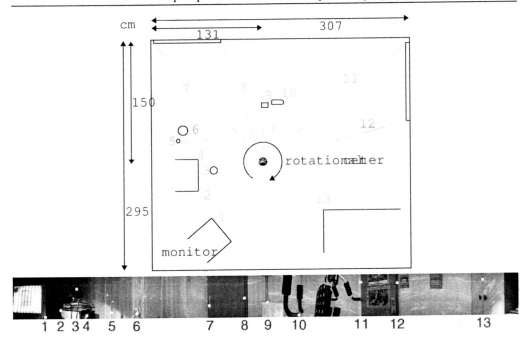

Figure 13: The top picture contains the plan of the reconstructed room. In the bottom picture we have marked the features on the scene that help us evaluate the quality of generated depth images.

- Lines between black dots denote links between two successively reconstructed points.

The result of the reconstruction process based on the 68th row of the dense depth image is given in Fig. 15 for the angle $2\varphi = 29.9625°$ and in Fig. 16 for the angle $2\varphi = 3.6125°$. We have used back-correlation and weighting. In Figs. 15 and 16 black dots are reconstructed on the basis of the estimated depth values, which are stored in the same row of the depth image. The features on the scene marked with big dots are not necessarily visible in the same row.

We have built sparse depth images by first detecting vertical edges in panoramic images. We have made an assumption that points on vertical edges have the same depth which is approximately true in the examples shown here. The results of the reconstruction shown in Figs. 17 and 18 are based on information within the entire sparse depth image: first, we calculate the average depth within each column of the depth image and then we show this average depth value in the ground-plan of the scene. In Figs. 17 and 18 the results have been derived from the sparse depth image gained by using back-correlation. The result in Fig. 17 is given for the angle $2\varphi = 29.9625°$ and the result in Fig. 18 is given for the angle $2\varphi = 3.6125°$. We imposed one additional constraint on the reconstruction process: each column in the depth image must contain at least four points with associated depth estimates or the average depth is not shown in the ground-plan of the scene.

Conclusion: Although the correlation technique has been used the presented results are

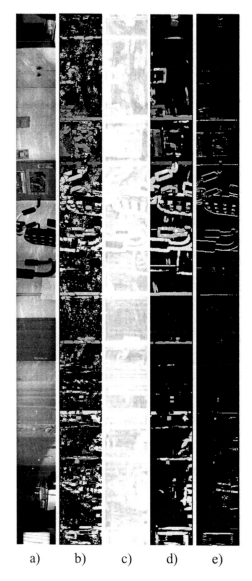

a)   b)   c)   d)   e)

Figure 14: Some stereo reconstruction results when creating the depth image for the left eye at the angle $2\varphi = 29.9625°$: a) the left eye panorama, b) a dense depth image / using back-correlation / reconstruction time: 6 hours, 42 min., 20 sec., c) confidence in the estimated depth, d) the dense depth image after weighting / without back-correlation / reconstruction time: 3 hours, 21 min., 56 sec., e) a sparse depth image / without back-correlation / reconstruction time: 38 seconds. The number of pixels for which we searched for the correspondences in case of b) was 147840 ($\times 2$ due to employed back-correlation) and only 4744 in case of e). In case of b) we calculated 21436800 ($\times 2$) correlation scores and in case of e) only 67110 scores.

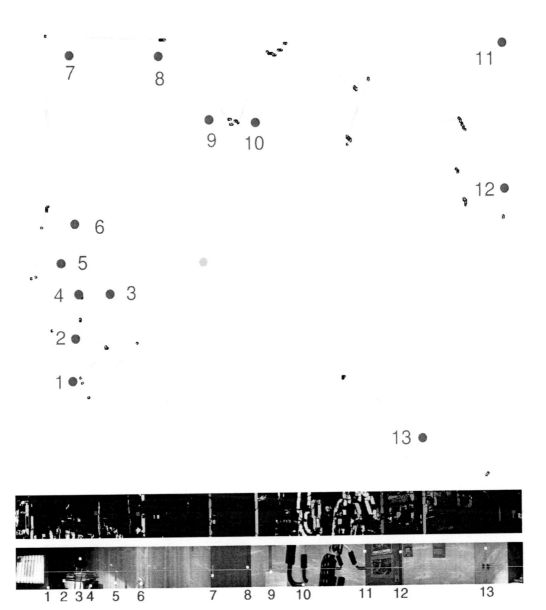

Figure 15: The top picture is a ground-plan showing the result of the reconstruction process based on the 68th row of the dense depth image. We have used back-correlation and weighting for the angle $2\varphi = 29.9625°$. The corresponding depth image is shown in the middle picture. For better orientation, the reconstructed row and the features on the scene for which we have measured the actual depth by hand are shown in the bottom picture. The features on the scene marked with big dots and associated numbers are not necessarily visible in this row.

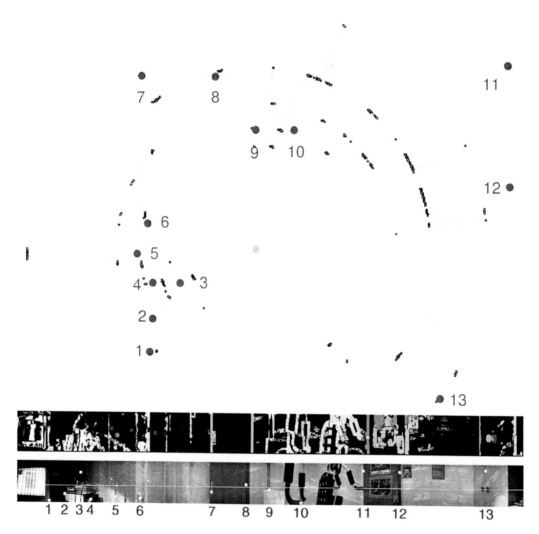

Figure 16: The top picture is a ground-plan showing the result of the reconstruction process based on the 68th row of the dense depth image. We have used back-correlation and weighting for the angle $2\varphi = 3.6125°$. The corresponding depth image is shown in the middle picture. For better orientation, the reconstructed row and the features on the scene for which we have measured the actual depth by hand are shown in the bottom picture. The features on the scene marked with big dots and associated numbers are not necessarily visible in this row.

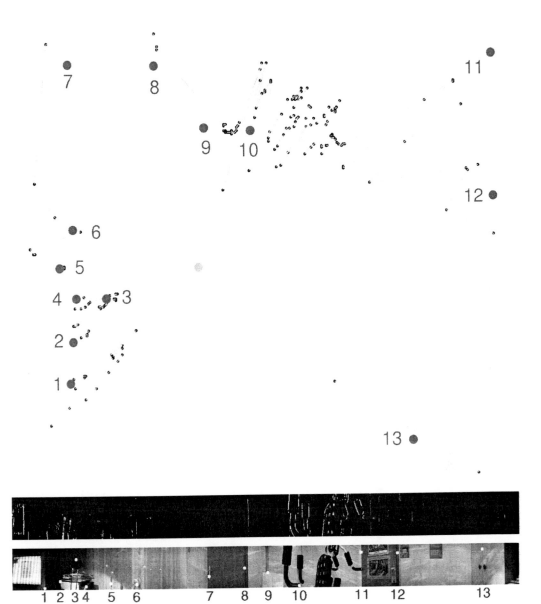

Figure 17: The top picture is a ground-plan showing the result of the reconstruction process based on the average depth within each column of the sparse depth image. We have used back-correlation for the angle $2\varphi = 29.9625°$. The corresponding sparse depth image is shown in the middle picture.

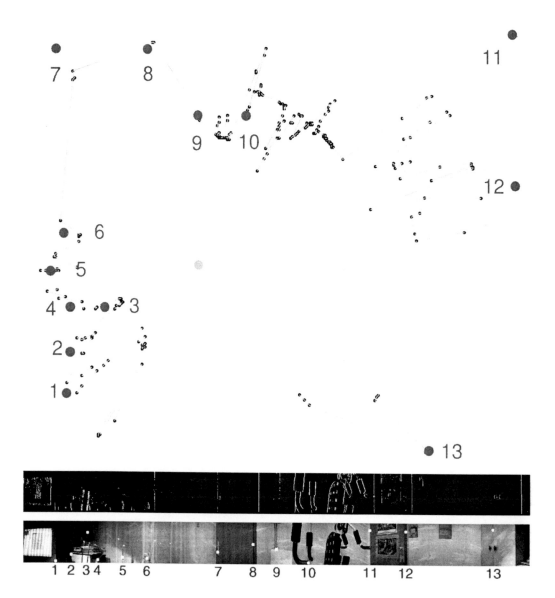

Figure 18: The top picture is a ground-plan showing the result of the reconstruction process based on the average depth within each column of the sparse depth image. We have used back-correlation for the angle $2\varphi = 3.6125°$. The corresponding sparse depth image is shown in the middle picture.

good: We can see that the reconstructions correspond to the outline of the room and that the reconstructions support well the statements made throughout Sec. 7 — and this was exactly the point of this experiment.

In Fig. 16 we can observe two properties of the system (Sec. 7.4): the reconstructed points are on concentric circles centered in the center of the system and the distance between the circles increases the further away they lie from the center. The figure nicely illustrates the fact that in case of a small angle φ, we can estimate only a few different depths and the fact that the one-pixel error in estimation of the angle θ increases as we move away from the center of the system.

As expected, the correlation technique had performed badly on uniform parts of the scene (e.g. walls), while the edges on the scene are well exposed and assessed in the depth images.

8.4 Influence of Addressing the Vertical Reconstruction

Experiment background: In Sec. 7.6 we have showed that the contribution of the vertical reconstruction is small. Here, we want to prove its positive influence on the overall accuracy of the system. The results were obtained with *camera #1*.

Results: The comparison of depth estimates l (Eq. (5)) and $l(\alpha, \beta)$ (Eq. (9)) is presented in Tab. 7.

feature	d [cm]	l [cm]	$l - d$ [cm (% of d)]	$l(\alpha, \beta)$ [cm]	$l(\alpha, \beta) - d$ [cm (% of d)]
1	111.5	109	-2.5 (-2.3%)	110.0	-1.5 (-1.3%)
2	95.5	89.3	-6.2 (-6.5%)	89.6	-5.9 (-6.1%)
3	64	59.6	-4.4 (-6.9%)	59.6	-4.4 (-6.8%)
4	83.5	78.3	-5.2 (-6.2%)	78.8	-4.7 (-5.6%)
5	92	89.3	-2.7 (-2.9%)	89.8	-2.2 (-2.4%)
6	86.5	82.7	-3.8 (-4.4%)	83.1	-3.4 (-3.9%)
7	153	159.8	6.8 (4.5%)	160.2	7.2 (4.7%)
8	130.5	135.5	5 (3.8%)	135.5	5.0 (3.8%)
9	88	87.6	-0.4 (-0.5%)	87.6	-0.4 (-0.4%)
10	92	89.3	-2.7 (-2.9%)	90.1	-1.9 (-2.1%)
11	234.5	213.5	-21 (-8.9%)	215.0	-19.5 (-8.3%)
12	198	179.1	-18.9 (-9.5%)	180.4	-17.6 (-8.9%)
13	177	186.7	9.7 (5.5%)	188.5	11.5 (6.5%)
			AVG$_\%$=5% $\pm$ 2.6%		AVG$_\%$=4.7% $\pm$ 2.7%

Table 7: The comparison of results without and with addressing the vertical reconstruction.

Conclusion: As expected, addressing the vertical reconstruction brings better results. This was observed also in other cases: reconstructions of different rooms using different cameras.

8.5 Influence of Different θ_0 Values on the Reconstruction Accuracy

Experiment background: See the discussion on estimation of the angle θ_0 ($\theta_0(\alpha)$, $\theta_0(\varepsilon)$) in Sec. 7.2. The results were obtained with *camera #1*.

Results: The comparison of results using $\theta_0(\alpha) = 0.2125°$ (Eq. (10)) and $\theta_0(\varepsilon) = 0.205714°$ (the estimation based on the accuracy of our rotational arm) is presented in Tab. 8.

feature	d [cm]	$\theta_0(\varepsilon) = 0.205714°$		$\theta_0(\alpha) = 0.2125°$	
		$l(\alpha,\beta)$ [cm]	$l(\alpha,\beta) - d$ [cm (% of d)]	$l(\alpha,\beta)$ [cm]	$l(\alpha,\beta) - d$ [cm (% of d)]
1	111.5	110.0	-1.5 (-1.3%)	132.2	20.7 (18.6%)
2	95.5	89.6	-5.9 (-6.1%)	102.5	7.0 (7.3%)
3	64.0	59.6	-4.4 (-6.8%)	63.6	-0.4 (-0.6%)
4	83.5	78.8	-4.7 (-5.6%)	87.9	4.4 (5.2%)
5	92.0	89.8	-2.2 (-2.4%)	102.7	10.7 (11.6%)
6	86.5	83.1	-3.4 (-3.9%)	93.6	7.1 (8.2%)
7	153.0	160.2	7.2 (4.7%)	220.7	67.7 (44.2%)
8	130.5	135.5	5.0 (3.8%)	174.3	43.8 (33.6%)
9	88.0	87.6	-0.4 (-0.4%)	99.7	11.7 (13.3%)
10	92.0	90.1	-1.9 (-2.1%)	103.1	11.1 (12.1%)
11	234.5	215.0	-19.5 (-8.3%)	351.3	116.8 (49.8%)
12	198.0	180.4	-17.6 (-8.9%)	263.4	65.4 (33.0%)
13	177.0	188.5	11.5 (6.5%)	281.8	104.8 (59.2%)
		AVG%=4.7% ± 2.7%		AVG%=22.8% ± 19%	

Table 8: The comparison of results for two different values of θ_0.

Conclusion: As expected, the results with $\theta_0(\varepsilon) = 0.205714°$ are much better, since it presents the angle for which the robotic arm is rotated in reality. The fact that even a small deviation from real $\theta_0(\varepsilon)$ brings much worse results is also obvious from these results.

8.6 Linear Versus Non-linear Model for Estimation of Angle φ

Experiment background: See the discussion on estimation of the angle φ in Sec. 7.2. The results were obtained with *camera #1*.

Results: The comparison of results using $2\varphi = 29.9625°$ (Eq. (6)) and $2\varphi = 30.15774565°$ (Eq. (8)) is presented in Tab. 9.

feature	d [cm]	$2\varphi = 29.9625°$		$2\varphi = 30.15774565°$	
		$l(\alpha,\beta)$ [cm]	$l(\alpha,\beta) - d$ [cm (% of d)]	$l(\alpha,\beta)$ [cm]	$l(\alpha,\beta) - d$ [cm (% of d)]
1	111.5	110.0	-1.5 (-1.3%)	108.1	-3.4 (-3.0%)
2	95.5	89.6	-5.9 (-6.1%)	88.5	-7.0 (-7.4%)
3	64.0	59.6	-4.4 (-6.8%)	59.2	-4.8 (-7.4%)
4	83.5	78.8	-4.7 (-5.6%)	78.0	-5.5 (-6.6%)
5	92.0	89.8	-2.2 (-2.4%)	88.6	-3.4 (-3.7%)
6	86.5	83.1	-3.4 (-3.9%)	82.2	-4.3 (-5.0%)
7	153.0	160.2	7.2 (4.7%)	155.7	2.7 (1.8%)
8	130.5	135.5	5.0 (3.8%)	132.4	1.9 (1.5%)
9	88.0	87.6	-0.4 (-0.4%)	86.5	-1.5 (-1.7%)
10	92.0	90.1	-1.9 (-2.1%)	88.9	-3.1 (-3.4%)
11	234.5	215.0	-19.5 (-8.3%)	206.7	-27.8 (-11.9%)
12	198.0	180.4	-17.6 (-8.9%)	174.6	-23.4 (-11.8%)
13	177.0	188.5	11.5 (6.5%)	180.4	182.1 (3.4%)
		AVG$_\%$=4.7% $\pm$ 2.7%		AVG$_\%$=5.3% $\pm$ 3.5%	

1 2 3 4   5   6   7   8 9 10   11   12   13

Table 9: The comparison of results for two different values of φ: the first one is gained from the linear and the second one from the non-linear model for estimation of angle φ.

Conclusion: The results are not much different, though the results obtained with the linear model are better. Similar results were obtained with *camera #2* in a different room (using again $l(\alpha,\beta)$; see the experiment in Sec. 8.8): For $2\varphi = 38.47875°$ (the linear model) the results were AVG$_\%$=2.7% $\pm$ 2.3%, while for $2\varphi = 38.57170666°$ (the non-linear model) the results were AVG$_\%$=3.1% $\pm$ 2.6%. Based on these results we can conclude that the linear model is better, at least for a given (estimated) set of parameters.

8.7 Repeatability of Results — Different Room

Experiment background: We want to see if we can achieve similar results as in Sec. 8.4, using the same camera (*camera #1*) in a different room?
Results: The results obtained in the different room are presented in Tab. 10.
Conclusion: The overall results are very similar. We can conclude that we can achieve similar accuracy in different rooms. This is also evident from the next experiment, where

feature	d [cm]	$l(\alpha,\beta)$ [cm]	$l(\alpha,\beta)-d$ [cm (% of d)]
1	63.2	61.5	-1.7 (-2.7%)
2	51.5	50.8	-0.7 (-1.3%)
3	141.0	147.3	6.3 (4.5%)
4	142.0	158.0	16.0 (11.3%)
5	216.0	220.4	4.4 (2.0%)
6	180.0	182.7	2.7 (1.5%)
7	212.0	248.8	36.8 (17.4%)
8	49.0	45.4	-3.6 (-7.4%)
9	49.0	45.4	-3.6 (-7.4%)
10	97.0	95.1	-1.9 (-2.0%)
11	129.5	142.2	12.7 (9.8%)
12	134.0	136.6	2.6 (1.9%)
13	119.0	118.4	-0.6 (-0.5%)
14	156.0	162.5	6.5 (4.2%)
15	91.0	91.2	0.2 (0.2%)
16	97.7	99.3	1.6 (1.6%)
17	111.0	109.3	-1.7 (-1.6%)
18	171.5	175.7	4.2 (2.4%)
19	171.5	182.9	11.4 (6.7%)

$$AVG_\% = 4.5\% \pm 4.5\%$$

1 2   3   4   5   6 7   8   9 10   11 12/13 14 15   16   17/18/19

Table 10: The results obtained in the different room, but with the same camera as in Sec. 8.4.

we have reconstructed the third room with three different cameras. One of them is again *camera #1*. Small differences in results are expected, since each room has its own shape, i.e. the depth distribution around the center of the system is different. And we know how this influences the accuracy, while we are limited with the number of possible depth estimates, which are approximations of the real distances (Sec. 7.4).

8.8 Repeatability of Results — Different Cameras

Experiment background: We want to see if we can achieve similar results as in Secs. 8.4 and 8.7, using different cameras in a different, third room? As mentioned in Sec. 7.7, the comparison of results gained using different cameras should not be done at the similar φ, but rather at the similar number of possible depth estimates. This fact is used in this experiment.

Results: The comparison of results for three different cameras is given in Tab. 11. Note that for features marked 3, 5, 6, 7, 15, 19, 20 and 21 the real distance d in case of *camera #3* is different from the presented one. The reason for this lies in the vertical view angle of the camera β, which is smaller in comparison to other two cameras. This means that some marked feature points are not seen in the panoramic images generated with *camera #3*, so we have chosen a nearby features with similar distances (see the panoramic image in Tab. 12). By all means, in the calculations we have used the correct distances.

feature	d [cm]	camera #1		camera #2		camera #3	
		$l(\alpha,\beta)$ [cm]	$l(\alpha,\beta)-d$ [cm (% of d)]	$l(\alpha,\beta)$ [cm]	$l(\alpha,\beta)-d$ [cm (% of d)]	$l(\alpha,\beta)$ [cm]	$l(\alpha,\beta)-d$ [cm (% of d)]
1	165.0	162.3	-2.7 (-1.6%)	159.1	-5.9 (-3.6%)	149.0	-16.0 (-9.7%)
2	119.0	118.0	-1.0 (-0.9%)	118.0	-1.0 (-0.8%)	114.1	-4.9 (-4.2%)
3	128.0	133.7	5.7 (4.4%)	130.1	2.1 (1.7%)	119.2	-6.3 (-5.0%)
4	126.5	125.6	-0.9 (-0.7%)	118.3	-8.2 (-6.5%)	114.0	-12.5 (-9.9%)
5	143.0	146.7	3.7 (2.6%)	141.4	-1.6 (-1.1%)	127.8	-13.2 (-9.3%)
6	143.0	151.9	8.9 (6.2%)	141.5	-1.5 (-1.1%)	130.9	-10.6 (-7.5%)
7	142.5	152.7	10.2 (7.2%)	145.0	2.5 (1.7%)	130.9	-11.1 (-7.8%)
8	136.5	141.0	4.5 (3.3%)	135.6	-0.9 (-0.7%)	131.0	-5.5 (-4.0%)
9	104.5	106.8	2.3 (2.2%)	104.6	0.1 (0.1%)	99.1	-5.4 (-5.2%)
10	81.7	79.6	-2.1 (-2.5%)	79.4	-2.3 (-2.8%)	78.6	-3.1 (-3.7%)
11	84.5	80.6	-3.9 (-4.6%)	80.6	-3.9 (-4.6%)	82.3	-2.2 (-2.6%)
12	83.5	82.7	-0.8 (-0.9%)	83.8	0.3 (0.4%)	83.6	0.1 (0.1%)
13	97.0	94.9	-2.1 (-2.2%)	95.7	-1.3 (-1.3%)	93.8	-3.2 (-3.3%)
14	110.0	114.9	4.9 (4.5%)	109.5	-0.5 (-0.5%)	104.9	-5.1 (-4.6%)
15	180.0	191.1	11.1 (6.2%)	165.8	-14.2 (-7.9%)	158.1	-12.9 (-7.5%)
16	124.5	129.9	5.4 (4.3%)	125.2	0.7 (0.6%)	119.2	-5.3 (-4.2%)
17	132.5	132.4	-0.1 (-0.1%)	127.9	-4.6 (-3.5%)	121.8	-10.7 (-8.0%)
18	134.5	136.6	2.1 (1.5%)	131.6	-2.9 (-2.2%)	124.7	-9.8 (-7.3%)
19	113.0	109.4	-3.6 (-3.2%)	107.8	-5.2 (-4.6%)	101.1	-6.9 (-6.4%)
20	125.0	121.6	-3.4 (-2.8%)	118.7	-6.3 (-5.0%)	111.6	-7.4 (-6.3%)
21	130.0	128.8	-1.2 (-1.0%)	121.8	-8.2 (-6.3%)	116.6	-8.4 (-6.8%)
		AVG%=3% $\pm$ 2%		AVG%=2.7% $\pm$ 2.3%		AVG%=5.9% $\pm$ 2.5%	

Table 11: The results obtained in the third room with three different cameras.

Conclusion: The results show that similar overall accuracy can be achieved if we use different cameras. The reason for somewhat worse results in case of *camera #3* could be attributed to the systematic error presence in the estimation of parameter r, as investigated in the next experiment.

8.9   Possibility of Systematic Error Presence in the Estimation of r

Experiment background: In Sec. 7.2 we have described how the estimation of parameter r is performed. Obviously, it is harder to estimate the location of the optical center in this way, if the view angle is smaller. The problem is even bigger if the camera cannot focus well on the near objects. Let us say that this estimation process is a good starting point for the estimation of system accuracy. We can optimize the estimation of r by minimizing AVG%: Simply, by letting r go through an interval of possible values around the estimated value, we can calculate AVG% for each value of r and, in the end, assign to r the value which minimizes AVG%. The results were obtained with *camera #3*.

Results: Tab. 12 compares the accuracy results before ($r = 356$ mm) and after ($r = 376$ mm) the optimization of parameter r.

Conclusion: We see that the results obtained after the optimization are much better. That r has been underestimated is also obvious from the results of the difference $l(\alpha, \beta) - d$ before the optimization, while they are bigger than normal and they are all, except one, negative.

The same optimization process could of course be used with all other cameras.

The remaining error in accuracy could be attributed to:

- the fact that we are limited with the number of possible depth estimates, which are approximations of the real distances (Sec. 7.4),

- the error in estimations of other parameters (e.g. α),

- the error due to the lens distortion presence (this matter is addressed in Sec. 8.10),

- the human factor (e.g. the distances to the features on the scene are measured manually) and/or

- the possible errors in robotic arm movement.

8.10   Influence of Lens Distortion Presence on the Reconstruction Accuracy

Experiment background: Lens distortion is a well known property of camera lens, which causes images to be spherised at their center. This basically means that the pixels that should be on the image edge are actually moved more towards the center of the image. How much they are moved towards the center depends on the camera field of view. Bigger is the field of view, bigger is the error due to the lens distortion. For *camera #3* the maximal error due to the lens distortion is small, only 0.8 pixel in 160×120 pixel images. On the other hand, for *camera #2* the maximal error due to the lens distortion is already 5 pixels in 160×120 pixel images. Fig. 19 nicely illustrates this fact. (The error in 640×480 pixel images is 4 times bigger.)

Since the best results obtained with *camera #3* are already very good (Sec. 8.9) and the size of distortion here is very small, we use the camera with the widest view angle, i.e. *camera #2*, in this experiment. Fig. 20 shows the camera model gained after the calibration process over a set of 640×480 pixel images [54]. We have used this model to undistort the captured images before they were merged into the panoramic images.

As we mentioned, the pixels that should be on the image edge are actually moved more towards the center of the image, which means that after the distortion is corrected the camera field of view gets smaller (Fig. 19). The new vertical field of view α_{new} was estimated using a simple observation about the part of the scene (number of pixels n) that disappeared from the image due to the distortion correction (similar to Eq. (6)):

$$\alpha_{new} = \frac{W - n}{W} \cdot \alpha; \text{ for } n = 9 \Rightarrow \alpha_{new} = 37.48575°.$$

Table 12: The comparison of results before and after the optimization of parameter r.

feature	d [cm]	before optimization $l(\alpha,\beta)$ [cm]	before optimization $l(\alpha,\beta)-d$ [cm (% of d)]	after optimization $l(\alpha,\beta)$ [cm]	after optimization $l(\alpha,\beta)-d$ [cm (% of d)]
1	165.0	149.0	-16.0 (-9.7%)	157.4	-7.6 (-4.6%)
2	119.0	114.1	-4.9 (-4.2%)	120.5	1.5 (1.2%)
3	125.5	119.2	-6.3 (-5.0%)	125.9	0.4 (0.3%)
4	126.5	114.0	-12.5 (-9.9%)	120.4	-6.1 (-4.8%)
5	141.0	127.8	-13.2 (-9.3%)	135.0	-6.0 (-4.2%)
6	141.5	130.9	-10.6 (-7.5%)	138.3	-3.2 (-2.3%)
7	142.0	130.9	-11.1 (-7.8%)	138.2	-3.8 (-2.7%)
8	136.5	131.0	-5.5 (-4.0%)	138.4	1.9 (1.4%)
9	104.5	99.1	-5.4 (-5.2%)	104.6	0.1 (0.1%)
10	81.7	78.6	-3.1 (-3.7%)	83.1	1.4 (1.7%)
11	84.5	82.3	-2.2 (-2.6%)	86.9	2.4 (2.9%)
12	83.5	83.6	0.1 (0.1%)	88.3	4.8 (5.8%)
13	97.0	93.8	-3.2 (-3.3%)	99.1	2.1 (2.1%)
14	110.0	104.9	-5.1 (-4.6%)	110.8	0.8 (0.7%)
15	171.0	158.1	-12.9 (-7.5%)	167.0	-4.0 (-2.4%)
16	124.5	119.2	-5.3 (-4.2%)	125.9	1.4 (1.2%)
17	132.5	121.8	-10.7 (-8.0%)	128.7	-3.8 (-2.9%)
18	134.5	124.7	-9.8 (-7.3%)	131.7	-2.8 (-2.1%)
19	108.0	101.1	-6.9 (-6.4%)	106.8	-1.2 (-1.2%)
20	119.0	111.6	-7.4 (-6.3%)	117.8	-1.2 (-1.0%)
21	125.0	116.6	-8.4 (-6.8%)	123.1	-1.9 (-1.5%)
		AVG%=5.9% $\pm$ 2.5%		AVG%=2.2% $\pm$ 1.5%	

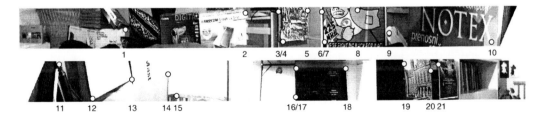

Figure 19: Each row of images shows distorted image (left) and undistorted image (right), after the distortion has been suppressed. The images in the top row have been taken with *camera #2*, while the images in the bottom row have been taken with *camera #3*. The resolution of all presented images is 640×480 pixels.

W is again the width of the captured image.

We also know that different α brings different r (Sec. 7.2), so we have to correct the size of parameter r as well:

$$r_{new} = \frac{\alpha_{new}}{\alpha} \cdot r = 293 \text{ mm}.$$

Similarly, all other parameters could be estimated if they are needed, e.g. the focal length f could be estimated from Eq. (2). But θ_0 stays the same as it still (even after the distortion correction) represents the angle for which the rotational arm has been moved between each two successively captured images.

Results: By using undistorted images to generate panoramic images and the new values of parameters, we obtain results presented on the right side in Tab. 13. For comparison, on the left side the results using distorted sequence are presented. In case of undistorted sequence, we have used $2\varphi = 37.04933695°$ as it ensures a similar number of possible depth estimates as the basic settings of *camera #2* (Sec. 7.7). Note that for the feature marked 11 the real

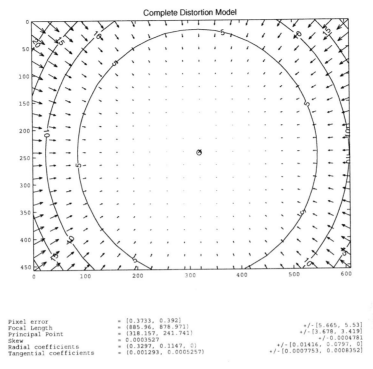

Figure 20: The camera model gained after the calibration process over a set of 640×480 pixel images [54]. Note that the values of estimated parameters (that are given in pixels) are $4\times$ smaller for 160×120 pixel images, which we use for generating panoramic images, and that the errors given in the right bottom side of the figure are approximately three times the standard deviations (for reference). Having that in mind, we see that the principal point in 160×120 pixel images is right in the middle of the images. The values on the curves in the figure present the errors in pixels due to the lens distortion.

distance d in case of undistorted sequence is different from the presented one. The reason for this lies in the fact that the normalized correlation procedure has been unable to find the appropriate corresponding point, so we have chosen a nearby feature with a similar distance. By all means, in the calculations we have used the correct distance.

Conclusion: We can conclude that processing undistorted images brings better results, though quite comparable. Having in mind that undistorting the sequence means that more processing time is needed (for instance, in Matlab (running on a 2.0 GHz Intel PIV PC) it takes a few hours to process 1501 images of size 160×120 pixels), perhaps we should be satisfied with the results gained using the distorted sequence. Another drawback of undistorted images is that they are more blurred in comparison to distorted originals (Fig. 19). Nevertheless, by using cameras with even wider field of view the distortion gets more obvious, and consequently we cannot always neglect its presence.

feature	d [cm]	distorted sequence		undistorted sequence	
		$l(\alpha,\beta)$ [cm]	$l(\alpha,\beta)-d$ [cm (% of d)]	$l(\alpha,\beta)$ [cm]	$l(\alpha,\beta)-d$ [cm (% of d)]
1	165.0	159.1	-5.9 (-3.6%)	165.6	0.6 (0.4%)
2	119.0	118.0	-1.0 (-0.8%)	118.4	-0.6 (-0.5%)
3	128.0	130.1	2.1 (1.7%)	123.7	-4.3 (-3.4%)
4	126.5	118.3	-8.2 (-6.5%)	122.3	-4.2 (-3.3%)
5	143.0	141.4	-1.6 (-1.1%)	139.5	-3.5 (-2.5%)
6	143.0	141.5	-1.5 (-1.1%)	139.5	-3.5 (-2.4%)
7	142.5	145.0	2.5 (1.7%)	143.8	1.3 (0.9%)
8	136.5	135.6	-0.9 (-0.7%)	133.7	-2.8 (-2.0%)
9	104.5	104.6	0.1 (0.1%)	103.8	-0.7 (-0.7%)
10	81.7	79.4	-2.3 (-2.8%)	77.1	-4.6 (-5.6%)
11	84.5	80.6	-3.9 (-4.6%)	78.4	-5.1 (-6.1%)
12	83.5	83.8	0.3 (0.4%)	81.7	-1.8 (-2.1%)
13	97.0	95.7	-1.3 (-1.3%)	96.5	-0.5 (-0.6%)
14	110.0	109.5	-0.5 (-0.5%)	106.3	-3.7 (-3.4%)
15	180.0	165.8	-14.2 (-7.9%)	173.0	-7.0 (-3.9%)
16	124.5	125.2	0.7 (0.6%)	122.7	-1.8 (-1.4%)
17	132.5	127.9	-4.6 (-3.5%)	129.4	-3.1 (-2.3%)
18	134.5	131.6	-2.9 (-2.2%)	133.6	-0.9 (-0.7%)
19	113.0	107.8	-5.2 (-4.6%)	109.8	-3.2 (-2.8%)
20	125.0	118.7	-6.3 (-5.0%)	122.4	-2.6 (-2.0%)
21	130.0	121.8	-8.2 (-6.3%)	126.1	-3.9 (-3.0%)
		$\text{AVG}_\% = 2.7\% \pm 2.3\%$		$\text{AVG}_\% = 2.4\% \pm 1.6\%$	

Table 13: The comparison of results obtained without and with the lens distortion correction.

9   Summary

We have presented a comprehensive analysis of our mosaic-based system for construction of depth panoramic images using only one standard camera.

The conclusions about the system effectiveness and its accuracy are well exposed in Secs. 7 and 8. Nevertheless, let us summarize the main conclusions made throughout the chapter and indicate in which section of the chapter each conclusion has been made:

- The geometry of capturing multiperspective panoramic images can be described with a pair of parameters (r, φ) (Sec. 4). By increasing (decreasing) each of them, we increase (decrease) the baseline $(2r_0)$ of our stereo system.

- The stereo pair acquisition procedure with only one standard camera cannot be executed in real time (Sec. 7.1).

- The epipolar geometry in case of symmetric stereo pair of panoramic images, which we use in the reconstruction process, is very simple: epipolar lines are image rows (Sec. 5).

- The parameters of the system should be estimated as precisely as possible, since already a small difference can cause a big difference in the reconstruction accuracy of the system (Secs. 7.2, 8.5, 8.6 and 8.9).

- We can effectively constrain the search space on the epipolar line (Sec. 7.3). This follows directly from the interpretation of the equation for depth estimation l (Eq. (5)), while other rules for constraining the search space, known from traditional stereo vision systems, can also be applied in addition to the basic constraint. An example of such rule is to seek for the neighboring pair of corresponding points only from the previously found correspondence on.

- The confidence in the estimated depth is variable: 1) the bigger the slope of the function l, the smaller the confidence in the estimated depth (one-pixel error Δl gets bigger) and 2) the bigger the value φ for each camera (α), the bigger the number of possible depth estimates and consequently the bigger the confidence (Secs. 7.4, 8.1 and 8.3).

- We can influence the parameter θ_0 by varying the resolution of captured images or by varying the horizontal view angle α (Secs. 7.4 and 7.7).

- By varying the radius r, we vary the biggest possible and sensible depth estimation l and the size of the one-pixel error Δl (Sec. 7.4).

- The bigger the value α, the smaller the horizontal resolution of panoramic images at fixed resolution of captured images (Sec. 7.7). Consequently, the number of possible depth estimates per one degree gets lower.

- In practice, from the autonomous robot localization and navigation system point of view, we should define the upper boundary of the allowed one-pixel error size Δl (Sec. 7.5).

- The contribution of the vertical reconstruction is small in general, but has a positive influence on the overall results (Secs. 7.6 and 8.4).

- The numbers of possible depth estimates are very similar for different cameras (α) at fixed resolution of the captured images (Sec. 7.7).

- The size of the one-pixel error Δl is also similar at similar number of possible depth estimates for different cameras (Sec. 7.7).

- The reconstruction process can execute in real time (Sec. 8.2).

- The reconstructed points lie on concentric circles centered in the center of rotation and the distance between circles (the one-pixel error Δl) increases the further away they lie from the center (Secs. 7.4 and 8.3).

- The linear model for estimation of angle φ have been proved better for a given set of parameters in comparison to the non-linear model (Sec. 8.6).

- We can achieve similar reconstruction accuracy with panoramas build from only one-pixel column ($W_s = 1$) of the captured images in different rooms, even with different cameras (Secs. 8.7 and 8.8).

- The remaining error in accuracy could be attributed to a number of possible reasons (Sec. 8.9).

- Processing undistorted images in general brings better though comparable results, but undistorting the sequence can be time expensive task and we are forced to re-estimate some parameters of the system after the distortion is corrected (Sec. 8.10).

All this is true for the cameras used in the dissertation, while for really wide angle cameras some conclusions perhaps demand further investigation in direction presented by the conclusion.

We should also expose the fact that we have developed few other simple procedures along the way, which have been proved useful in various aspects. Like the method for estimating the position of the optical center (Sec. 7.2) or the method for defining the maximal reliable depth value (Sec. 7.5).

In the end, let us write only the conclusion of all conclusions, which answers to the question written at the very beginning of this chapter (Sec. 1.1): Can the system be used for robot localization and navigation in a room? According to the accuracy achieved the answer is: Yes!

Our future work is directed primarily in the development of an application for the real time autonomous localization and navigation of a mobile robot in a room.

Acknowledgements

A preliminary and much shorter version of this chapter was published in Kluwer's International Journal of Computer Vision [47].

This work was supported by the Ministry of education, science and sport of Republic of Slovenia (project Computer vision P2-0214).

References

1992

[1] Ishiguro H., Yamamoto M., Tsuji S.: Omni-directional stereo. *IEEE Trans. PAMI*, **14**(2), 257–262.

[2] Paar G., Pölzleitner W.: Robust disparity estimation in terrain modeling for spacecraft navigation. Proc. *IEEE ICPR*, The Hague, The Netherlands, August 30 – September 3, I:738–741.

[3] Weng J., Cohen P., Herniou M.: Camera calibration with distortion models and accuracy evaluation. *IEEE Trans. PAMI*, **14**(10), 965–980.

1993

[4] Faugeras O.: *Three-Dimensional Computer Vision: A Geometric Viewpoint*. MIT Press, Cambridge, Massachusetts, London, England.

[5] Faugeras O., Hotz B., Mathieu H., Viéville T., Zhang Z., Fua P.,Théron E., Moll L., Berry G., Vuillemin J., Bertin P., Proy C.: Real time correlation based stereo: algorithm implementations and applications. *Technical Report 2013*, INRIA Sophia-Antipolis, France.Available at:ftp://ftp-sop.inria.fr/pub/rapports/RR-2013.ps.gz   .

[6] Okutomi M., Kanade T.: A Multiple-Baseline Stereo. *IEEE Trans. PAMI*,**15**(4), 353–363.

1995

[7] Basu A., Licardie S.: Alternative models for fish-eye lenses.*Pattern Recognition Letters*,**16**(4), 433–441.

[8] Chen S.: Quicktime VR — an image-based approach to virtual environment navigation. Proc. *ACM SIGGRAPH*, Los Angeles, USA, August 6–11, 29–38.

1996

[9] Szeliski R.: Video Mosaics for Virtual Environments, *IEEE Computer Graphics and Applications*, **16**(2), 22–30.

1997

[10] Gupta R., Hartley R. I.: Linear pushbroom cameras.*IEEE Trans. PAMI*, **19**(9), 963–975.

[11] Heikkilä J., Silvén O.: A Four-Step Camera Calibration Procedure with Implicit Image Correction. Proc. *IEEE CVPR*, San Juan, Puerto Rico, June 17–19, 1106–1112.

[12] Szeliski R., Shum H. Y.: Creating full view panoramic image mosaics and texture-mapped models.*Computer Graphics (ACM SIGGRAPH)*, Los Angeles, USA, August 3–8, 251–258.

[13] Wood D., Finkelstein A., Hughes J., Thayer C., Salesin D.:Multiperspective panoramas for cel animation.*Computer Graphics (ACM SIGGRAPH)*, Los Angeles, USA, August 3–8, 243–250.

1998

[14] Benosman R., Maniere T., Devars J.: Panoramic stereovision sensor. Proc. *IEEE ICPR*, Brisbane, Australia, August 16–20, I:767–769.

[15] Gluckman J., Nayar S. K., Thorek K. J.: Real-time omnidirectional and panoramic stereo. Proc. *DARPA Image Understanding Workshop*, Monterey,USA, November.

[16] Prihavec B., Solina F.: User interface for video observation over the internet. *Journal of Network and Computer Applications*, **21**, 219–237.

[17] Rademacher P., Bishop G.: Multiple-center-of-projection images. *Computer Graphics (ACM SIGGRAPH)*, Orlando, USA, July 19–24, 199–206.

1999

[18] Nayar S. K., Peri V.: Folded Catadioptric Camera. Proc. *IEEE CVPR*, Fort Collins, USA, June 23–25, II:217–223.

[19] Peleg S., Ben-Ezra M.: Stereo panorama with a single camera. Proc. *IEEE CVPR*, Fort Collins, USA, June 23–25, I:395–401.

[20] Shum H. Y., Szeliski R.: Stereo Reconstruction from Multiperspective Panoramas. Proc. *IEEE ICCV*, Kerkyra, Greece, September 20–25, I:14–21.

[21] Shum H. Y., Kalai A., Seitz S. M.: Omnivergent Stereo. Proc. *IEEE ICCV*, Kerkyra, Greece, September 20–25, I:22–29.

2000

[22] Hartley R., Zisserman A.: *Multiple View Geometry in Computer Vision.* Cambridge University Press, Cambridge, UK.

[23] Huang F., Pajdla T.: Epipolar geometry in concentric panoramas. *Technical Report CTU-CMP-2000-07*, Center for Machine Perception, Czech Technical University, Prague, Czech Republic. Available at: ftp://cmp.felk.cvut.cz/pub/cmp/articles/pajdla/Huang-TR-2000-07.ps.gz .

[24] Jogan M., Leonardis A.: Robust localization using the eigenspace of spinning-images. Proc. *IEEE Workshop on Omnidirectional Vision*, Hilton Head Island, USA, June 12, 37–44.

[25] Nayar S. K., Karmarkar A.: 360×360 Mosaics. Proc. *IEEE CVPR*, Hilton Head Island, USA, June 13–15, II:388–395.

[26] Peleg S., Pritch Y., Ben-Ezra M.: Cameras for stereo panoramic imaging. Proc. *IEEE CVPR*, Hilton Head Island, USA, June 13–15, I:208–214.

[27] Peleg S., Rousso B., Rav-Acha A., Zomet A.: Mosaicing on adaptive manifolds. *IEEE Trans. PAMI*, **22**(10), 1144–1154.

[28] Svoboda T.: Central Panoramic Camera Design, Geometry, Egomotion. *Ph.D. Thesis*, Center for Machine Perception, Czech Technical University, Prague,Czech Republic. Available at:ftp://cmp.felk.cvut.cz/pub/cmp/articles/svoboda/phdthesis.ps.gz .

[29] Svoboda T., Pajdla T.: Panoramic cameras for 3D computation. Proc. *Czech Pattern Recognition Workshop*, Prague, Czech Republic, 63–70.

[30] Tanahashi H., Yamamoto K., Wang C., Niwa Y.: Development of a Stereo Omnidirectional Imaging System (SOS). Proc. *IEEE International Conference on Industrial Electronics, Control and Instrumentation*, Nagoya, Japan, October 22–28, 289–294.

[31] Zhang Z.: A flexible new technique for camera calibration. *IEEE Trans. PAMI*, **22**(11), 1330-1334.

2001

[32] Bakstein H., Pajdla T.: 3D Reconstruction from 360×360 Mosaics. Proc. *IEEE CVPR*, Kauai, Hawaii, USA, December 8–14, I:72–77.

[33] Benosman R., Kang S. B. (Eds.): *Panoramic Vision: Sensors, Theory and Applications*. Springer-Verlag, New York, USA.

[34] Faugeras O., Luong Q.-T.: *The Geometry of Multiple Images.* MIT Press, Cambridge, Massachusetts, London, England.

[35] Huang F., Wei S. K., Klette R.: Geometrical Fundamentals of Polycentric Panoramas. Proc. *IEEE ICCV*, Vancouver, Canada, July 9–12, I:560–565.

[36] Li Y., Tang C. K., Shum H. Y.: Efficient Dense Depth Estimation from Dense Multiperspective Panoramas. Proc. *IEEE ICCV*, Vancouver, Canada, July 9–12, I:119–126.

[37] Pajdla T.: Epipolar geometry of some non-classical cameras. Proc. *Computer Vision Winter Workshop (CVWW)*, Bled, Slovenia, February 7–9, 223–233.

[38] Peer P., Solina F.: Capturing mosaic-based panoramic depth images with a single standard camera. *International Journal of Machine Graphics and Vision,***10**(3), 369–397.

[39] Peleg S., Ben-Ezra M., Pritch Y.: Omnistereo: Panoramic Stereo Imaging. *IEEE Trans. PAMI,* **23**(3), 279–290.

[40] Seitz S. M.: The Space of All Stereo Images. Proc. *IEEE ICCV*, Vancouver, Canada, July 9–12, I:26–33.

[41] Shimada D., Tanahashi H., Kato K., Yamamoto K.: Extract and Display Moving Object in All Direction by Using Stereo Omnidirectional System (SOS). Proc. *IEEE International Conference on 3-D Digital Imaging and Modeling,*Quebec City, Canada, May 28 – June 1, 42–47.

[42] Tanahashi H., Shimada D., Yamamoto K., Niwa Y.: Acquisition of Three-Dimensional Information in Real Environment By Using Stereo Omni-directional System (SOS). Proc. *IEEE International Conference on 3-D Digital Imaging and Modeling,*Quebec City, Canada, May 28 – June 1, 365–371.

[43] Wei S. K., Huang F., Klette R.: Determination of geometric parameters for stereoscopic panorama cameras. *International Journal of Machine Graphics and Vision,***10**(3), 399–427.

2002

[44] Bakstein H., Pajdla T.: Panoramic Mosaicing with a 180° Field of View Lens. Proc. *IEEE Workshop on Omnidirectional Vision*, Copenhagen, Denmark, June, 60–67.

[45] Hirschmüller H., Innocent P. R., Garibaldi J.: Real-Time Correlation-Based Stereo Vision with Reduced Border Errors. *International Journal of Computer Vision,***47**(1/2/3), 229–246.

[46] Pajdla T.: Stereo with Oblique Cameras. *International Journal of Computer Vision,* **47**(1/2/3), 161–170.

[47] Peer P., Solina F.: Panoramic Depth Imaging: Single Standard Camera Approach. *International Journal of Computer Vision,* **47**(1/2/3), 149–160.

[48] Scharstein D., Szeliski R.: A Taxonomy and Evaluation of Dense Two-Frame Stereo Correspondence Algorithms. *International Journal of Computer Vision,***47**(1/2/3), 7–42.

[49] Seitz S. M., Kim J.: The Space of All Stereo Images. *International Journal of Computer Vision,***48**(1), 21–38.

[50] Shah M.: Guest Introduction: The Changing Shape of Computer Vision in the Twenty-First Century. *International Journal of Computer Vision,* **50**(2), 103–110.

[51] Sivic J.: Geometry of Concentric Multiperspective Panoramas. *M.Sc. Thesis,* Center for Machine Perception, Czech Technical University, Prague, Czech Republic.

[52] Svoboda T., Pajdla T.: Epipolar Geometry for Central Catadioptric Cameras. *International Journal of Computer Vision,* **49**(1), 23–37.

2003

[53] Bakstein H., Pajdla T.: Ray space volume of omnidirectional 180° × 360° images. Proc. *Computer Vision Winter Workshop (CVWW)*, Valtice, Czech Republic, February 3–6, 39–44.

[54] Bouguet J.-Y.: Camera Calibration Toolbox for Matlab. California Institute of Technology. Available at:http://www.vision.caltech.edu/bouguetj/calib_doc/index.html   .

[55] Brown M. Z., Burschka D., Hager G. D.: Advances in Computational Stereo. *IEEE Trans. PAMI*, **25**(8), 993–1008.

[56] Feldman D., Zomet A., Weinshall D., Peleg S.: New view synthesis with non-stationary mosaicing. Proc. *Computer Vision / Computer Graphics Collaboration forModel-based Imaging, Rendering, image Analysis and Graphical special Effects (MIRAGE)*, INRIA Rocquencourt, France, March 10–11, 48–56.

[57] Matsuyama T., Wu X., Takai T., Nobuhara S.: Real-Time Generation and High Fidelity Visualization of 3D Video. Proc. *Computer Vision / Computer Graphics Collaboration forModel-based Imaging, Rendering, image Analysis and Graphical special Effects (MIRAGE)*, INRIA Rocquencourt, France, March 10–11, 1–10.

[58] Peer P., Solina F.: Towards a Real Time Panoramic Depth Sensor. Proc. *International Conference on Computer Analysis of Images and Patterns (CAIP)*, Groningen, The Netherlands, August 25–27, 107–115.

[59] Skočaj D.: Robust subspace approaches to visual learning and recognition. *Ph.D. Thesis*, University of Ljubljana, Faculty of Computer and Information Science. Available at: http://eprints.fri.uni-lj.si   .

[60] Sun C., Peleg S.: Fast Panoramic Stereo Matching Using Cylindrical Maximum Surfaces. *IEEE Trans. on Systems, Man and Cybernetics – Part B*, Accepted for publication.

[61] Zomet A., Feldman D., Peleg S., Weinshall D.: Mosaicing New Views: The Crossed-Slits Projection. *IEEE Trans. PAMI*, **25**(6), 741–754.

In: Computer Vision and Robotics
Editor: John X. Liu, pp. 189-243

ISBN 1-59454-357-7
© 2006 Nova Science Publishers, Inc.

Chapter 6

A GEOMETRIC ERROR ANALYSIS OF 3-D RECONSTRUCTION ALGORITHMS

Tao Xiang [1]*and Loong-Fah Cheong* [2] †
[1]Department of Computer Science
Queen Mary, University of London, London E1 4NS, UK
[2]Department of Electrical and Computer Engineering
National University of Singapore, Singapore 119260

Abstract

The estimation of 3-D motion and structure is one of the most important function-alities of an intelligent vision system. In spite of the best efforts of a generation of computer vision researchers, we still do not have a practical and robust system for accurately estimating motion and structure from a sequence of moving imagery under all motion-scene configurations. We put forth in this study a geometrically motivated 3-D motion and structure error analysis which is capable of shedding light on global effect such as inherent ambiguities. This is in contrast with the usual statistical kinds of error analyses which can only deal with local effect such as noise perturbations, and in which much of the results regarding global ambiguities are empirical in nature. The error expression that we derive allows us to predict the exact conditions likely to cause ambiguities and how these ambiguities vary with motion types such as lateral or forward motion. Such an investigation may alert us to the occurrence of ambiguities under different conditions and be more careful in picking the solution. Our formu-lation, though geometrically motivated, was also put to use in modeling the effect of noise and in revealing the strong influence of feature distribution. Given the erroneous 3-D motion estimates caused by the inherent ambiguities, it is also important to un-derstand the impact such motion errors have on the structure reconstruction. In this

---
*E-mail address: txiang@dcs.qmul.ac.uk; Tel: (+44)-(0)20-7882-5230; Fax: (+44)-(0)20-8980-6533
†E-mail address: eleclf@nus.edu.sg

study, various robustness issues related to the different types of second order shape recovered from motion cue are addressed. Experiments on both synthetic and real image sequences were conducted to verify the various theoretical predictions.

This study would be most beneficial for an intelligent vision system that needs to have an estimate of the robustness of the 3-D motion and structure information recovered from the world. Such information would allow the system to carry out its tasks more effectively and to seek more information if necessary.

Keywords: Structure from motion, error analysis, epipolar constraint, inherent ambiguity, depth distortion, shape recovery, iso-distortion framework.

1   Introduction

In spite of the best efforts of a generation of computer vision researchers, we still do not have a practical and robust system for accurately estimating 3-D motion and structure from a sequence of moving imagery under all motion-scene configuration. The estimation of 3-D motion and structure is notorious for its noise sensitivity; a small amount of error in the image measurements can lead to very different solutions. Structure from motion (SFM) algorithms proposed in the past two decades faced this problem to varying extent which has led to many error analyses (Adiv, 1989; Daniilidis and Spetsakis, 1997; Weng et al., 1991; Young, 1992). To date, however, few of them have ever attempted to give a topological characterization of the residuals associated with different optimization criteria which would make explicit the configuration of the error surface, especially the distribution of the local minima of the cost functions. Ideally, such a characterization should consider the ambiguities under a full range of motion-scene configurations. The rationale for such a comprehensive description of the ambiguities is that since most SFM algorithms perform well only in restricted domains, it was important to evaluate the limits of applicability of these algorithms. That is, each algorithm should be evaluated specifically against likely problem conditions. If such understanding could be achieved, it then becomes possible to fuse the results of several SFM algorithms or to fuse the visual motion cues with other cues such as vestibular signals. This viewpoint has been expressed by (Oliensis, 2000a).

In this study, we propose an approach that lends itself towards understanding the full behavior of SFM algorithms. Instead of dealing with specific algorithms each using different optimization techniques, we study one class of algorithms based on a weighted differential epipolar constraint. This class includes most of the existing differential SFM algorithms using optical flow as input. What permits an unifying view of these different algorithms is a new optimization criterion to be presented in this study. It is based on the difference between the original optical flow and the reprojected flow obtained via a backprojection of the reconstructed depth, analogous to the distance between the observation and the reprojection of the reconstructed point in the discrete case (Zhang, 1998). We showed that different weighted differential epipolar constraints used in the literature correspond to the different ways of reconstructing depth using the optimization criterion presented in this study. Thus this criterion also lends a geometric interpretation to the various weights used. More importantly, it allows us to develop a simple and explicit expression for the residual errors of the optimization functions in terms of the errors in the 3-D motion estimates and enables us to predict the exact conditions likely to cause ambiguities. The result is that inherent ambiguities in both translation and rotation estimates are identified; how the likelihood of these ambiguities varies with the scene and the motion types such as lateral or forward motion is also made apparent. To round off the investigation of motion ambiguity, we extend our analysis to include the effect of noise in the image measurements, using both the isotropic and the anisotropic noise models. Our investigation unravels the impact a realistic anisotropic noise distribution can have on the topology of the cost functions.

An error analysis of SFM algorithms would not be complete without saying something about how the depths would be recovered given such motion ambiguities. As a consequence of the motion ambiguities, the estimated 3-D motion parameters contain errors; thus the reconstructed depth would be a distorted version of the physical depth. The need to characterize such depth distortion arising from errors in the motion estimates prompted the work of Cheong et al. (Cheong et al., 1998), which gave an account of the systematic nature of the errors in the depth estimates via the so-called iso-distortion framework. It showed that the most general description of such a transformation from the physical to the perceived space is very complicated, belonging to the family of Cremona transformations. The work of Cheong and Xiang (Cheong and Xiang, 2001) built upon that framework and considered the depth distortion under two generic types of motion, namely, lateral and forward motion, with a view to obtain robust recovery of depth information. In this study, the iso-distortion framework is employed to analyze the errors in second order shape recovery given an erroneous 3-D motion estimation. We elucidate the impact of errors in 3-D motion estimates on second order shape descriptors such as normal curvatures and shape index (Koenderink and Van Doorn, 1992). In particular, our findings show that the second order shape is recovered with varying degrees of uncertainty depending on the types of 3-D motion executed. Furthermore, we make clear that different shapes exhibit different sensitivities in their recovery to errors in 3-D motion estimates. Shape reconstruction typically occurs in a context where different cues such as shading, contour, range data and motion are available (Cutting and Vishton, 1995). It has been stated that different depth cues are suited for recovering different kinds of depth representation. If the confidence level of the shape recovered from motion can be ascertained in details against various motion-scene configurations, it then becomes possible to adopt a proper fusion mechanism to combine the depths recovered from different cues to produce better results. The work presented here has taken a step towards this direction.

The understanding of the distortion in 3-D motion and structure estimation enables us to explain some well-known human perceptual illusions such as the "rotating cylinder illusion". Correlatively, if human suffers from such distortion and yet can perform many tasks efficiently, it is hoped that a deeper understanding of the distortion would help an intelligent vision system emulate human in these performances. The understanding of such distortion may also be useful for other purposes such as alerting one to the occurrence of ambiguities, thereby allowing us to pick up the true solution.

Relation to Previous Work

The SFM problem is usually treated as two subproblems, namely, the measurement of 2-D image displacement (correspondences) or velocity (optical flow), and the extraction of 3-D relative motion and structure information using as input the 2-D image measurements. Due to the ill-conditioned nature of the first subproblem, the input to the 3-D motion estimation algorithms inevitably contains errors. In view of such errors, most of the previous error analysis on 3-D motion estimation (Adiv, 1989; Daniilidis and Spetsakis, 1997; Weng et al., 1991; Young, 1992; Heeger and Jepson, 1992; Maybank, 1993) related the errors of the estimated 3-D motion parameters to the measurement errors in the first subproblem. The errors are typically expressed as a high variance or a bias in the motion parameters through some statistical analysis (Adiv, 1989; Daniilidis and Spetsakis, 1997; Weng et al., 1991; Young, 1992; Heeger and Jepson, 1992; Maybank, 1993), or given as empirical figures (Dutta and Snyder, 1990) through some simulations. A comprehensive survey of such analysis was given by Daniilidis and Spetsakis (Daniilidis and Spetsakis, 1997). Several results have been established by such analysis:

- Maybank (Maybank, 1993), Jepson and Heeger (Heeger and Jepson, 1992) established the result that the plane defined by the true translation and the optical axis can be determined by most SFM algorithms reliably. They obtained this result based on strict assumptions such as infinitesimal field of view (FOV). The finding is closely related to the bas-relief ambiguity obtained in this study, although we do not need the field of view assumption.

- If the field of view is small or depth variation is insufficient, rotation about an axis parallel to the image plane can easily be confounded with lateral translations. This has been demonstrated through both theoretical work and experimental study (Daniilidis and Spetsakis, 1997).

- The estimated translation is biased towards the viewing direction if the error metric is not appropriately normalized.

Little work has been contributed to a systematical characterization of the topology of the cost functions. Recently, however, several studies have emerged in this direction. Soatto et al. and Chiuso et al. (Soatto and Brockett, 1998; Chiuso et al., 2000) attempted to achieve optimal SFM (in differential approaches) by understanding the error surface configuration of the cost functions. They noted the existence of a minimum at the opposite end of the bas-relief valley (termed as rubbery ambiguity in their studys), and attributed it to the presence of noise, although the simulation results showed that the minimum persisted with noiseless input. Ma et al. (Ma et al., 2001), adopting the discrete approach, unified different optimization criteria under an "optimal triangulation" procedure and analyzed the impact of noise on the ambiguities for different optimization criteria. They characterized the behavior of the critical points under noise by making use of the properties of the so-called "essential manifold".

The major difference between our work and the preceding work lies in the fact that we highlight the importance of the inherent ambiguities of the SFM problem itself, without considering the effect of noise initially. Indeed, all the major ambiguities identified in the literature can be accounted for by such noiseless consideration. We argue that while dealing with the statistical adequacy of the various criteria is important, it is equally important to understand the detailed nature of the inherent ambiguities which is caused by the geometry of the problem itself and thus cannot be removed by any statistical schemes (relieved, yes). In this respect, the work of Fermüller and Aloimonos (Fermüller and Aloimonos, 2000) and Oliensis (Oliensis, 2000b, 2004) are the closest in spirit to our work. Not surprisingly, there are many common findings, though there are some aspects that are different too.

The work of Fermüller and Aloimonos (Fermüller and Aloimonos, 2000) presented a geometrical-statistical investigation of the observability of 3-D motion. They studied the conditions on the errors in the motion estimates for the local minima on error surface to arise. The cost functions are expressed in terms of the true motion parameters and the errors in the estimated motion parameters. Our work adopted similar notations but used very different method of analysis. Various assumptions were required in their work such as random distribution of feature points over the image plane and random depths over the 3-D space. They also assumed small FOV and neglected all the second order flow terms caused by the rotational parameters. The epipolar constraint considered in their work was unweighted, which, together with those assumptions led to some results that were different from ours. In particular, it was shown that when all the motion parameters are estimated simultaneously, the solution for the focus of expansion (FOE) can have a local minima at the image center, which is obviously due to the unweighted epipolar constraint. Our result shows that if the epipolar constraint is properly weighted, this minima should not occur, unless some specific motion-scene configuration such as forward motion arises, in which case the true minimum will also be at the image center. Indeed, our study considers a variety of motion-scene configurations which are not studied in (Fermüller and Aloimonos, 2000). Finally, the kind of noise considered in (Fermüller and Aloimonos, 2000) was found to have no influence on the overall structure of the cost functions. Our theoretical analysis and experimental results show that a more realistic noise model often has significant impact on the cost function behavior.

The objective of Oliensis' work (Oliensis, 2000b, 2004) is very similar to ours, that is, to characterize the overall error surface via an explicit analytical model. However, the means through which we achieve the end are quite different. Our approach is much simpler, allowing us to achieve an intuitive grasp of the geometric nature of the ambiguities. For instance, we explain the formation of the local minimum at the opposite end of the bas-relief valley (termed as flipped minimum in Oliensis' work) through the coupling of the rotational and the translational motions. Our more intuitive formulation renders it more suitable for analysis under a wider range of motion-scene configurations; specifically, we focus on different types of translational motions, ranging from purely forward motion to purely lateral motion. Other factors that influence the error surface, such as the distribution of feature points and the scene structure, are also studied in a more systematic and detailed manner. Oliensis' work concentrated only on the error surface for translation estimation; the corresponding ambiguities in the rotation estimates are not made explicit. Our work fully characterizes the error surfaces for both the rotational and the translational parameters.

Finally, in contrast to our work, all the preceding works focus on motion estimation and devote much less attention on the closely related problem of depth estimation. While some of the works (Weng et al., 1991; Szeliski and Kang, 1997; Grossmann and Victor, 2000) predicted the sensitivity of the depth estimates to small amounts of image noise, the situation where the errors in the depth estimates arise from the erroneous 3-D motion parameters has not been dealt with, except in the case of critical surface pairs (Horn, 1987; Negahdaripour, 1989).

Organization

The organization of this chapter is as follows. First, we briefly review in section 2 the optimization criteria in both the discrete and the differential cases. We then introduce the notions of the iso-distortion framework and discuss how it can be used to address the reliability of depth recovery. In section 3, the differential reprojection criterion is proposed to unify the various criteria based on differential epipolar constraints. We then seek to characterize the various inherent ambiguities in 3-D motion estimation and the corresponding depth distortion properties under these ambiguity configurations. We employ a cost function visualization method to visualize the topology of the cost functions, so as to both verify the various theoretical predictions and to reveal further properties of the cost functions. Based on such understanding, we are able to explain some well-known human visual illusions. We characterize the role of the measurement noise on the behavior of SFM algorithms in section 4. In particular, the global effects of isotropic and anisotropic noise distribution are studied. These are followed by experiments on real images to verify the various predictions made and to study the feasibility of a more robust algorithm based on the topology of the cost functions. In Section 5, various robustness issues surrounding different types of second order shape estimates recovered from motion cue are addressed. The chapter ends with the conclusions of the work.

2   Background and Prerequisite

2.1   Model and Notations

In this study, we denote the estimated parameters with the hat symbol ($\hat{}$) and errors in the estimated parameters with the subscript e (where error of any estimate s is defined as $s_e = s - \hat{s}$). We use bold lower-case character to denote vector and bold upper-case character to denote matrix. Unless otherwise stated, vectors are column vectors. Given a n-vector $\mathbf{s}$, $[\mathbf{s}]_m$ is defined as the m-vector which consist of the first m ($m < n$) components of $\mathbf{s}$, $\underline{\mathbf{s}}$

is defined as the (n+1)-vector with 0 added as the last component, and $\bar{\mathbf{s}}$ is the associated skew-symmetric matrix of $\mathbf{s}$. The symbol $(\cdot)$ represents the dot product of vectors. For any vector $\mathbf{s} = (s_1, s_2)^T$, $\mathbf{s}^\perp$ represents the vector $(s_2, -s_1)^T$ which is perpendicular to $\mathbf{s}$ with the same magnitude. The symbol $(\|\,\|)$ represents the Euclidean norm of a vector and the symbol $(|\,|)$ the absolute value of a variable.

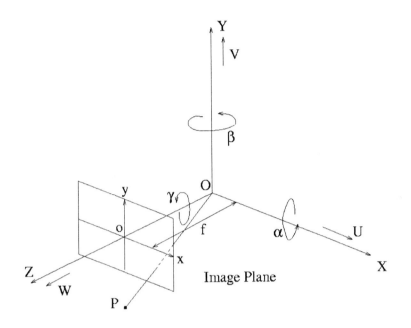

Figure 1: The image formation model.

A pinhole camera model with perspective projection is assumed as shown in Figure 1; it is moving with a translational velocity $\mathbf{v} = (U, V, W)^T$ and a rotational velocity $\mathbf{w} = (\alpha, \beta, \gamma)^T$. A point P in the world produces an image point p in the image plane which is f pixels away from the optical center; if $\mathbf{P} = (X, Y, Z)^T$ and $\mathbf{p} = (x, y, f)^T$ are the co-ordinates corresponding to P and p respectively, we have: $\mathbf{p} = f\frac{\mathbf{P}}{Z}$. The focal length f is assumed to be known since we are dealing with calibrated motion in this study.

The image velocity due to camera motion is given by the following familiar equation (Longuet-Higgins, 1981):

$$\dot{\mathbf{p}} = -\mathbf{Q_p}\left(\frac{\mathbf{v}}{Z} + \bar{\mathbf{w}}\mathbf{p}\right) \tag{1}$$

where $\dot{\mathbf{p}} = (u, v, 0)^T$, $\mathbf{Q_p} = \begin{bmatrix} f & 0 & -x \\ 0 & f & -y \\ 0 & 0 & 0 \end{bmatrix}$.

Equation (1) can alternatively be written in terms of its components as:

$$u = \frac{u_{tr}}{Z} + u_{rot}$$

$$= (x - x_0)\frac{W}{Z} + \frac{\alpha xy}{f} - \beta\left(\frac{x^2}{f} + f\right) + \gamma y$$

$$v = \frac{v_{tr}}{Z} + v_{rot}$$

$$= (y - y_0)\frac{W}{Z} + \alpha\left(\frac{y^2}{f} + f\right) - \frac{\beta xy}{f} - \gamma x$$

$$\tag{2}$$

where $(x_0, y_0) = (f\frac{U}{W}, f\frac{V}{W})$ is the focus of expansion (FOE). We define $\dot{\mathbf{p}}_{tr} = (u_{tr}, v_{tr})^T$ and $\dot{\mathbf{p}}_{rot} = (u_{rot}, v_{rot})^T$, where $\frac{\dot{\mathbf{p}}_{tr}}{Z}$ and $\dot{\mathbf{p}}_{rot}$ are the components of the flow due to translation and rotation respectively. Since only the direction of the translation can be recovered from the flow field, we can set $W = 1$ for the case of general motion; the case of pure lateral motion ($W = 0$) will be discussed separately where required.

2.2   3-D Motion Estimation

2.2.1   Discrete Case

The SFM problem in the discrete case amounts to the estimation of the fundamental matrix $\mathbf{F}$ (or the Essential matrix $\mathbf{E}$ if the camera is calibrated) based on a sufficiently large set of point correspondences. The geometry of the discrete two-image motion analysis has been well studied and is succinctly captured by the epipolar equation:

$$\mathbf{p}_1\mathbf{F}\mathbf{p}_2 = 0 \tag{3}$$

where $\mathbf{p}_1$ and $\mathbf{p}_2$ are the corresponding points on the two images and $\mathbf{F}$ is the fundamental matrix. The geometric meaning of the epipolar equation is that $\mathbf{p}_1$ must lie on the epipolar line of $\mathbf{p}_2$ given by $\mathbf{F}\mathbf{p}_2$. Directly minimizing equation (3) leads to a closed-form solution whose results are sensitive to noise. It was also argued (Zhang, 1998) that this optimization criterion does not totally reflect the epipolar geometry and thus is not physically meaningful. A couple of non-linear optimization criteria were thus proposed. Three criteria are derived based on: distance between points and epipolar lines, gradient-weighted epipolar errors and distances between points and their reprojections. The corresponding cost function are denoted as J_{D1}, J_{D2} and J_{D3} respectively. The geometric meanings of J_{D1} and J_{D3} are obvious from their names, whereas J_{D2} is obtained based on the following statistical consideration: When independent and identically distributed Gaussian noise is assumed, minimizing the Mahalanobis distance between points and epipolar lines gives rise to J_{D2}. While J_{D3} has obvious geometric meaning, it also possesses a statistical interpretation: It corresponds to the case of optimizing based on the maximum *a posterior* (MAP) principle under the same noise model. Zhang (Zhang, 1998) studied the relationship between these three criteria under different motion configurations. J_{D2} was recommended since it is equivalent to J_{D3} under most configurations and yet is computationally more efficient. Ma et al. (Ma et al., 2001) investigated the behavior of different criteria with and without noise.

Similar to Zhang's results, these criteria were shown to be intimately related and were unified under a new "optimal triangulation" procedure (our proposed unifying scheme for the differential case is similar in idea).

2.2.2 Differential Case

A motion estimation algorithm based on the differential epipolar constraint can be developed analogous to the discrete case, from which the following cost function can be obtained (Brooks et al., 1997):

$$J_{E1} = \sum_{i=1}^{n} \left(\mathbf{p}_i^T \bar{\tilde{\mathbf{v}}} \dot{\mathbf{p}}_i + \mathbf{p}_i^T \bar{\tilde{\mathbf{v}}} \bar{\tilde{\mathbf{w}}} \mathbf{p}_i \right)^2 \tag{4}$$

where n is the number of image velocity measurement. As discussed before, we first focus on the case where the optical flow input was noise-free; thus we have used the term $\dot{\mathbf{p}}_i$ in (4). The actual case of the optical flow containing noise component would be addressed in later sections.

The geometric meaning of J_{E1} is that in the image plane the vector $[\dot{\mathbf{p}}_i]_2 - \hat{\dot{\mathbf{p}}}_{rot_i}$ (the de-rotated flow) should be parallel to the vector $\hat{\dot{\mathbf{p}}}_{tr_i}$, or equivalently, perpendicular to $\hat{\dot{\mathbf{p}}}_{tr_i}^\perp$. Thus the cost function J_{E1} can also be expressed as:

$$J_{E1} = \sum_{i=1}^{n} \left(([\dot{\mathbf{p}}_i]_2 - \hat{\dot{\mathbf{p}}}_{rot_i}) \cdot \hat{\dot{\mathbf{p}}}_{tr_i}^\perp \right)^2 \tag{5}$$

Minimizing the preceding amounts to a linear optimization problem which can be solved by a linear least square method. However, it faces the same problem as its discrete counterpart. A well-studied bias of the linear algorithms is that the estimated translation will be biased towards the image center. In view of this bias, a statistically more adequate implementation of the differential epipolar constraint should be:

$$
\begin{aligned}
J_{E2} &= \sum_{i=1}^{n} \left(\frac{([\dot{\mathbf{p}}_i]_2 - \hat{\dot{\mathbf{p}}}_{rot_i}) \cdot \hat{\dot{\mathbf{p}}}_{tr_i}^\perp}{\| [\dot{\mathbf{p}}_i]_2 - \hat{\dot{\mathbf{p}}}_{rot_i} \| \| \hat{\dot{\mathbf{p}}}_{tr_i}^\perp \|} \right)^2 \\
&= \sum_{i=1}^{n} (\sin \theta_i)^2
\end{aligned}
\tag{6}
$$

where θ_i is the angle between $\hat{\dot{\mathbf{p}}}_{tr_i}$ and $[\dot{\mathbf{p}}_i]_2 - \hat{\dot{\mathbf{p}}}_{rot_i}$.

The counterpart of J_{E2} in the discrete case would be J_{D1}. Both of them are non-linear and involve heavy computation to obtain their solutions. Like the discrete case, there are a variety of other non-linear methods which are basically different weighted versions of J_{E1} each driven by slightly different considerations. Some of these methods derived the weight based on a statistical analysis of noise (Kanatani, 1993; Ma et al., 2000). For instance, Ma et al. (Ma et al., 2000) presented a normalized epipolar constraint which would yield MAP estimates given an independent and identically distributed Gaussian noise:

$$J_{E3} = \sum_{i=1}^{n} \left(\frac{\mathbf{p}_i^T \tilde{\mathbf{v}} \dot{\mathbf{p}}_i + \mathbf{p}_i^T \tilde{\mathbf{v}} \tilde{\mathbf{w}} \mathbf{p}_i}{\|\hat{\dot{\mathbf{p}}}_{tr_i}\|} \right)^2 \tag{7}$$

Others like Brooks et al. (Brooks et al., 1998) presented an algebraic-geometric point of view: the hyperpoint $[\mathbf{p}_i, \dot{\mathbf{p}}_i]^T$ should lie on the hypersurface $\mathbf{p}_i^T \tilde{\mathbf{v}} \dot{\mathbf{p}}_i + \mathbf{p}_i^T \tilde{\mathbf{v}} \tilde{\mathbf{w}} \mathbf{p}_i = 0$. Like the gradient-weighted epipolar error in the discrete case, they proposed a cost function which minimizes the first order approximation of the Euclidean distance between the hyperpoint and the hypersurface:

$$J_{E4} = \sum_{i=1}^{n} \frac{\left(\mathbf{p}_i^T \tilde{\mathbf{v}} \dot{\mathbf{p}}_i + \mathbf{p}_i^T \tilde{\mathbf{v}} \tilde{\mathbf{w}} \mathbf{p}_i \right)^2}{\|2 \tilde{\mathbf{v}} \tilde{\mathbf{w}} \mathbf{p}_i + \tilde{\mathbf{v}} \dot{\mathbf{p}}_i\|^2 + \|\tilde{\mathbf{v}} \mathbf{p}_i\|^2} \tag{8}$$

This cost function is very similar to Kanatani's renormalization criterion (Kanatani, 1993) which is based on statistical consideration.

As in the discrete case, one can ask what is the geometric meaning of these various criteria, beside their statistical interpretation? The differential reprojection criterion to be developed later allows us to answer this question.

2.3   Depth Estimation

3-D motion estimation is regarded as the first step towards the full recovery of 3-D shape information from 2-D measurements. Therefore any error in the 3-D motion estimates will systematically affect the perceived space. However, the reliability of the depth estimates could have quite different behavior from that of 3-D motion estimates. That is, motion-scene configuration that allows robust motion recovery may yield less than desirable depth estimates, and vice versa. Another substantive question is of course, whether there is any interaction between the errors in the motion estimates and the corresponding distortion in the recovered depth. That is, would the distortion in the perceived space in turn affect motion estimation? Partially to address these questions, the iso-distortion framework was introduced in (Cheong et al., 1998). Let us first revisit some notations that would be useful for this study.

2.3.1   Iso-distortion Framework

The iso-distortion framework seeks to understand the geometric laws under which the recovered scene is distorted due to some errors in the estimated motion parameters. The distortion in the perceived space is visualized by looking at the locus of constant distortion, known as the iso-distortion surfaces.

From equation (1), the reconstructed depth can be expressed using the estimated motion parameters:

$$\hat{Z} = -\frac{\hat{\mathbf{v}}^T \mathbf{Q_p}^T \underline{\mathbf{n}}}{\left(\dot{\mathbf{p}}^T - \mathbf{p}^T \tilde{\mathbf{w}} \mathbf{Q_p}^T \right) \underline{\mathbf{n}}} \tag{9}$$

where **n** is a unit vector in the image plane representing a direction. In general, when the estimated motion parameters contain errors, different choices of **n** will give rise to different reconstructions. One possibility is to recover depth by setting **n** to be along the estimated epipolar direction, which is the direction pointing from the image feature point to the estimated FOE; this scheme is heretoforth named as the "epipolar reconstruction" scheme. It is based on the intuition that the epipolar direction contains the strongest translational flow and hence represents the best direction for depth recovery. Another possibility is to let **n** be the image intensity gradient direction, based on the intuition that the normal flow can be recovered reliably. Various other possibilities exist, each can be given different geometric or statistical interpretation.

Substituting the expression for the true flow $\dot{\mathbf{p}}^T$ (using equation (1) again) into equation (9), we have:

$$\hat{Z} = Z \left(\frac{-\hat{\mathbf{v}}^T \mathbf{Q_p}^T \mathbf{n}}{-\mathbf{v}^T \mathbf{Q_p}^T \mathbf{n} + Z \left(\mathbf{p}^T \bar{\mathbf{w}}_e \mathbf{Q_p}^T \mathbf{n} \right)} \right) \tag{10}$$

From the above equation we can see that $\hat{Z}$ is related to Z through a multiplicative factor given by the terms inside the bracket, which we denote by D and term as the distortion factor:

$$D = \frac{-\hat{\mathbf{v}}^T \mathbf{Q_p}^T \mathbf{n}}{-\mathbf{v}^T \mathbf{Q_p}^T \mathbf{n} + Z \left(\mathbf{p}^T \bar{\mathbf{w}}_e \mathbf{Q_p}^T \mathbf{n} \right)} \tag{11}$$

For specific values of $\mathbf{v}$, $\hat{\mathbf{v}}$ and $\mathbf{w}_e$ and for any fixed distortion factor D, equation (11) describes a surface $g(x, y, Z) = 0$ in the xyZ-space, which we call an iso-distortion surface. This iso-distortion surface has the obvious property that points lying on it are distorted in depth by the same multiplicative factor D. The systematic nature of the distortion can then be made clear by looking at the organization of these iso-distortion surfaces.

The geometric laws for distortion can also be characterized algebraically as a distortion transformation from the physical space to the perceived space. In general, the resulting distortion transformation is a Cremona transformation (Cheong and Ng, 1999) whose properties are quite complex. However, under special cases, as are some of the ambiguity cases to be discussed later, the transformation reduces to that of the projective transformation with some nice depth properties.

2.3.2 Depth Error Sensitivity Under Different Motion Configurations

The iso-distortion framework has been used to seek some generic motion types that rendered depth recovery more robust and reliable (Cheong and Xiang, 2001). Lateral and forward motions were compared both under calibrated and uncalibrated scenarios. The fundamental conclusions are that under lateral movement (possibly coupled with rotation) and certain conditions, while it might be very difficult to resolve the ambiguity between translation and rotation, ordinal depth can be recovered with robustness, whereas for forward motion, the depth recovery is too sensitive to errors to admit meaningful scene reconstruction.

The preceding conclusions were established without imposing any constraints on the motion error. However, it is evident that the values of these motion errors are not arbitrary. Rather, ambiguities inherent in the SFM algorithms are likely to impose further constraints on these motion errors. Given these errors, what can be said about the distortion in depth given any types of translational motion? This will be addressed in the next section.

3 Differential Reprojection Criterion and its Error Surface

3.1 Differential Reprojection Criterion

From the geometric standpoint, the differential epipolar constraint in equation (4) is a "weak" constraint in the sense that it can be satisfied by any two vectors ($[\dot{\mathbf{p}}_i]_2 - \dot{\mathbf{p}}_{rot_i}$) and $\dot{\mathbf{p}}_{tr_i}$ that are parallel to each other. There is no requirement on the magnitudes of the two vectors although the true estimate should satisfy: $[\dot{\mathbf{p}}_i]_2 - \frac{\dot{\mathbf{p}}_{tr_i}}{Z} - \dot{\mathbf{p}}_{rot_i} = 0$. Thus a "stronger" and more adequate criterion based on the idea of reprojection is proposed; it is based on the difference between the original optical flow and the reprojected flow obtained via a backprojection of the reconstructed depth. It is thus analogous to J_{D3} in the discrete case. Furthermore, similar to Ma et al.'s "optimal triangulation" in the discrete case (Ma et al., 2001), we will see in this section that this criterion unifies the various weighted versions of the differential epipolar constraint and also lends a geometric interpretation to the weights used. More importantly it allows us to develop a geometric treatment of the motion ambiguity conditions, as we shall see in section 3.2.

Substituting the recovered depth in equation (9) into equation (1), we obtain the reprojected (estimated) flow field, denoted by $\hat{\dot{\mathbf{p}}}$, as follows:

$$\hat{\dot{\mathbf{p}}} = \frac{\mathbf{Q_p}\hat{\mathbf{v}}\left(\dot{\mathbf{p}}^T - \mathbf{p}^T\bar{\hat{\mathbf{w}}}\mathbf{Q_p}^T\right)\mathbf{n}}{\hat{\mathbf{v}}^T\mathbf{Q_p}^T\mathbf{n}} - \mathbf{Q_p}\bar{\hat{\mathbf{w}}}\mathbf{p} \tag{12}$$

The difference between the original optical flow and the reprojected flow can thus be expressed as:

$$\begin{aligned}
\dot{\mathbf{p}}_e &= \dot{\mathbf{p}} - \hat{\dot{\mathbf{p}}} \\
&= \frac{\left(\mathbf{C_1} - \mathbf{C_1}^T + \mathbf{C_2} - \mathbf{C_2}^T\right)\mathbf{n}}{\hat{\mathbf{v}}^T\mathbf{Q_p}^T\mathbf{n}} \\
&= \frac{\mathbf{C}\mathbf{n}}{\hat{\mathbf{v}}^T\mathbf{Q_p}^T\mathbf{n}} \tag{13}
\end{aligned}$$

where $\mathbf{C_1} = \dot{\mathbf{p}}\hat{\mathbf{v}}^T\mathbf{Q_p}$, $\mathbf{C_2} = \mathbf{Q_p}\bar{\hat{\mathbf{w}}}\mathbf{p}\hat{\mathbf{v}}^T\mathbf{Q_p}^T$, $\mathbf{C} = \left(\mathbf{C_1} - \mathbf{C_1}^T + \mathbf{C_2} - \mathbf{C_2}^T\right)$ and $\mathbf{C_1}, \mathbf{C_2}, \mathbf{C}$ are all 3×3 matrices. It can be easily shown that $\mathbf{C}$ is a skew-symmetrical matrix. Thus we have:

$$\mathbf{n}^T\mathbf{C}\mathbf{n} = 0$$

The above equation implies that along $\mathbf{n}$, the direction of depth recovery, the reprojected flow $\hat{\dot{\mathbf{p}}}$ has exactly the same component as the original optical flow $\dot{\mathbf{p}}$. As a consequence:

$$\|\dot{\mathbf{p}}_e\| = |\dot{\mathbf{p}}_e^T \underline{\mathbf{n}}^\perp| \tag{14}$$

If we define a cost function J_R based on the reprojected flow difference, it can be written as:

$$J_R = \sum_{i=1}^{n} \left(\frac{\|\mathbf{C}_i \underline{\mathbf{n}}_i\|}{\hat{\mathbf{v}}^T \mathbf{Q}_{\mathbf{p}_i}{}^T \underline{\mathbf{n}}_i} \right)^2$$

Using equation (14), J_R can be written as:

$$
\begin{aligned}
J_R &= \sum_{i=1}^{n} \left(\frac{\mathbf{n}_i{}^T \mathbf{C}_i \underline{\mathbf{n}}_i{}^\perp}{\hat{\mathbf{v}}^T \mathbf{Q}_{\mathbf{p}_i}{}^T \underline{\mathbf{n}}_i} \right)^2 \\
&= \sum_{i=1}^{n} \left(\frac{\mathbf{p}_i^T \tilde{\mathbf{v}} \dot{\mathbf{p}}_i + \mathbf{p}_i^T \tilde{\mathbf{v}} \tilde{\mathbf{w}} \mathbf{p}_i}{\hat{\mathbf{v}}^T \mathbf{Q}_{\mathbf{p}_i}{}^T \underline{\mathbf{n}}_i} \right)^2
\end{aligned} \tag{15}
$$

A comparison of equation (4) with equation (15) reveals the relationship between J_R and J_{E1}: J_R is a weighted version of J_{E1} with the weight given by the projection of $\hat{\dot{\mathbf{p}}}_{tr_i}$ on the direction $\mathbf{n}_i$. It follows that J_R can also be written as:

$$J_R = \sum_{i=1}^{n} \left(\frac{\hat{\dot{\mathbf{p}}}_{tr_i} \cdot ([\dot{\mathbf{p}}_i]_2 - \hat{\dot{\mathbf{p}}}_{rot_i})^\perp}{\hat{\dot{\mathbf{p}}}_{tr_i} \cdot \mathbf{n}_i} \right)^2 \tag{16}$$

Various weighted differential epipolar constraints differ mainly in the choice of $\mathbf{n}$. Possible choices of $\mathbf{n}$ include the "epipolar reconstruction" direction ($\mathbf{n} = \frac{\hat{\dot{\mathbf{p}}}_{tr}}{\|\hat{\dot{\mathbf{p}}}_{tr}\|}$), which results in J_{E3}, gradient direction (the normal flow approach), or the Linear Least Square Reconstruction (LLSR) direction[1]. Other more simplistic choices include constant direction, random direction, etc.

It follows that while the formulation of the differential reprojection criterion J_R is motivated by the need to have a stronger geometric constraint, it often has statistical meaning too. Furthermore J_R can be seen as a scheme unifying the various weighted epipolar constraints. It follows that to understand the behavior of these SFM algorithms based on weighing the epipolar constraint, one can focus on studying the differential reprojection criterion. All these algorithms inherit properties from the differential reprojection criterion; in particular, much of the ambiguity conditions of these algorithms are common and can be studied by looking at the numerator of J_R.

---

[1] LLSR direction refers to $\mathbf{n} = \frac{[\dot{\mathbf{p}}]_2 - \hat{\dot{\mathbf{p}}}_{rot}}{\|[\dot{\mathbf{p}}]_2 - \hat{\dot{\mathbf{p}}}_{rot}\|}$. Depth recovered along this direction is the standard linear least square estimate of depth from equation (2), which minimizes the "estimated measurement error" $\|\hat{\dot{\mathbf{p}}}_{tr} - \hat{Z}([\dot{\mathbf{p}}]_2 - \dot{\mathbf{p}}_{rot})\|$. Details of the properties of this depth reconstruction scheme can be found in (Cheong and Xiang, 2001).

3.2   Analyzing from a Geometric Point of View

To analyze how various factors, such as motion types, field of view, feature and depth distribution, govern the formation of motion ambiguities (or equivalently, the local minima in the error surface described by J_R), we need to express J_R in terms of the various component errors in the 3-D motion estimates. This allows us to obtain a more obliging form for analyzing in more specific details the ambiguity behavior over a wide range of conditions. Substituting $[\dot{\mathbf{p}}_i]_2 = (u_i, v_i)^T = (\frac{x_i - x_0}{Z_i} + u_{rot_i}, \frac{y_i - y_0}{Z_i} + v_{rot_i})^T$, $\dot{\hat{\mathbf{p}}}_{tr_i} = (x_i - \hat{x}_0, y_i - \hat{y}_0)^T$ and $\dot{\hat{\mathbf{p}}}_{rot_i} = (\hat{u_{rot_i}}, \hat{v_{rot_i}})^T$ into (16), we have:

$$J_R = \sum \left(\frac{(x - \hat{x}_0, y - \hat{y}_0) \cdot (v_{rot_e} - \frac{y_{0e}}{Z}, \frac{x_{0e}}{Z} - u_{rot_e})}{(x - \hat{x}_0, y - \hat{y}_0) \cdot \mathbf{n}} \right)^2 \tag{17}$$

where

$$
\begin{aligned}
(x_{0e}, y_{0e}) &= (x_0 - \hat{x}_0, y_0 - \hat{y}_0) \\
(u_{rot_e}, v_{rot_e}) &= (\frac{\alpha_e xy}{f} - \beta_e \left(\frac{x^2}{f} + f \right) + \gamma_e y, \\
& \quad \alpha_e \left(\frac{y^2}{f} + f \right) - \frac{\beta_e xy}{f} - \gamma_e x)
\end{aligned}
$$

For notational convenience, we omit the subscript i in the expression of J_R; it is understood that the summation runs over all feature points. To facilitate discussion, we also introduce the following notations. We denote the expression contained in the outer bracket of (17) as $\dot{\mathbf{p}}_e(\hat{\mathbf{v}}, \mathbf{v}, \mathbf{w}_e)$ (where the dependence of $\dot{\mathbf{p}}_e$ on the motion errors has been made explicit), and the vectors $(x - \hat{x}_0, y - \hat{y}_0)^T$ and $(v_{rot_e} - \frac{y_{0e}}{Z}, \frac{x_{0e}}{Z} - u_{rot_e})^T$ as $\mathbf{t}_1$ and $\mathbf{t}_2$ respectively (it is indeed the interaction between $\mathbf{t}_1$ and $\mathbf{t}_2$ that accounts for much of the inherent motion ambiguities). We also adopt the terminology that for the vectors $\mathbf{t}_1$ and $\mathbf{t}_2$, $\mathbf{t}_{1,n}$ and $\mathbf{t}_{2,n}$ denote the nth order component with respect to x and y; thus we have:

$$
\begin{cases}
\mathbf{t}_1 &= \mathbf{t}_{1,0} + \mathbf{t}_{1,1} \\
\mathbf{t}_2 &= \mathbf{t}_{2,0} + \mathbf{t}_{2,1} + \mathbf{t}_{2,2} + \mathbf{t}_{2,Z}
\end{cases}
\tag{18}
$$

where $\mathbf{t}_{1,0} = (-\hat{x}_0, -\hat{y}_0)^T$, $\mathbf{t}_{1,1} = (x, y)^T$, $\mathbf{t}_{2,0} = (\alpha_e f, \beta_e f)^T$, $\mathbf{t}_{2,1} = (-\gamma_e x, -\gamma_e y)^T$ and $\mathbf{t}_{2,2} = (\alpha_e \frac{y^2}{f} - \frac{\beta_e xy}{f}, -\frac{\alpha_e xy}{f} + \beta_e \frac{x^2}{f})^T$. The last item $\mathbf{t}_{2,Z}$ in the above equation denotes the depth dependent term $(-\frac{y_{0e}}{Z}, \frac{x_{0e}}{Z})^T$. The depth Z may be dependent on x and y in a complex manner; thus we use the notation $\mathbf{t}_{2,Z}$ and leave the order unspecified.

To visualize the residual error surface, it is easier to deal with a 3-dimensional surface. We use for this purpose the translation error surface, which is described parametrically with two free variables, the estimated FOE $(\hat{x}_0, \hat{y}_0)$. We know that given this hypothesized FOE, the rotation variables can be solved in terms of the estimated FOE so as to minimize J_R. The residual error J_R can thus be obtained for each FOE candidate, describing the entire residual surface completely. Unless otherwise stated, the error surface in this study refers to this type of error surface.

We first make some assumptions on the distribution of feature points and depth. We assume that the feature points are evenly distributed in the image plane, as is the distribution of the "depth-scaled feature points" $(\frac{x}{Z}, \frac{y}{Z})$. The latter assumption generally requires that the distribution of depths are independent of the corresponding image co-ordinates x and y. Later we will see how the error surface will be affected when these assumptions do not hold.

3.2.1   Several General Observations

Equation (17) shows that for any given data set (x, y, Z), the residual error is a function of the true FOE (x_0, y_0), the estimated FOE $(\hat{x}_0, \hat{y}_0)$ and the error in the rotation estimates $(\alpha_e, \beta_e, \gamma_e)$. Evidently, ambiguities would arise when the errors in the motion estimates satisfy the following conditions to make the numerator of $\dot{\mathbf{p}}_e(\hat{\mathbf{v}}, \mathbf{v}, \mathbf{w}_e)$ vanish: 1) making $\|\mathbf{t}_2\|$ small and 2) making $\mathbf{t}_1$ and $\mathbf{t}_2$ perpendicular to each other. The second condition is generally not satisfiable at all points of the image; thus making $\|\mathbf{t}_2\|$ small (condition one) contributes towards ambiguity. Making $\|\mathbf{t}_1\|$ small does not contribute towards ambiguity if we have suitably normalized J_R with the term in the denominator. We thus can make the following observations:

1. When the estimated FOE moves towards infinity, the direction of $\mathbf{t}_1$ approaches that of $\mathbf{t}_{1,0}$, which is constant. Pointing towards a constant direction represents a necessary condition for $\mathbf{t}_1$ and $\mathbf{t}_2$ to be perpendicular to each other.

2. From the expression of $\mathbf{t}_2$, we can see that $\mathbf{t}_{2,0}$ and $\mathbf{t}_{2,Z}$ are pointing towards constant directions for all the feature points. Intuitively, $\mathbf{t}_2$ will be more perpendicular to $\mathbf{t}_1$ when both $\mathbf{t}_{2,0}$ and $\mathbf{t}_{2,Z}$ are perpendicular to $\mathbf{t}_{1,0}$. This relationship can be illustrated with the diagram shown in Figure 2. The vector $\mathbf{t}_{1,1}$ can be regarded as a perturbation to the vector $\mathbf{t}_{1,0}$, and similarly, $\mathbf{t}_{2,1}$ and $\mathbf{t}_{2,2}$ can be regarded as perturbations to $\mathbf{t}_{2,0}$ and $\mathbf{t}_{2,Z}$. However, if the feature points are sufficiently evenly distributed (such that the vectors $\mathbf{t}_{1,1}$ are evenly spread on either side of $\mathbf{t}_{1,0}$ and the sum of vectors $\mathbf{t}_{2,1}$ and $\mathbf{t}_{2,2}$ are evenly spread on either side of $\mathbf{t}_{2,0}$ and $\mathbf{t}_{2,Z}$), and the distribution of depth Z is symmetrical with respect to the $\mathbf{t}_{1,0}$ direction, then making $\mathbf{t}_{2,0}$ and $\mathbf{t}_{2,Z}$ perpendicular to $\mathbf{t}_{1,0}$ is a reasonable choice for the minimization of J_R.

Thus we have

$$\frac{x_0}{y_0} = \frac{\hat{x}_0}{\hat{y}_0} \tag{19}$$

and

$$\frac{\alpha_e}{\beta_e} = -\frac{\hat{y}_0}{\hat{x}_0} \tag{20}$$

Equation (19) imposes a constraint on the direction of the estimated translation, namely, the three points $(\hat{x}_0, \hat{y}_0)$, (x_0, y_0) and $(0,0)$ should lie on a straight line. We henceforth refer to this constraint as the Translation Direction (TDir) constraint.

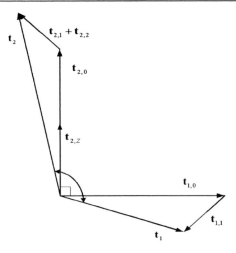

Figure 2: Geometry of $\mathbf{t_1}$ and $\mathbf{t_2}$.

Equation (20) imposes a constraint on the direction of $\mathbf{w}_e$, which shall be henceforth referred to as the Rotation Error Direction (RDir) constraint.

3. Since $\mathbf{t_1}$ cannot be made exactly perpendicular to $\mathbf{t_2}$, small $\|\mathbf{t_2}\|$ will help to reduce the numerator term of J_R. Obviously, small $\|\mathbf{t_2}\|$ can be achieved by having small errors in the motion estimates. Alternatively, since $\mathbf{t_{2,0}}$ and $\mathbf{t_{2,Z}}$ are pointing towards constant direction, they can be made to approximately cancel off each other by an appropriate choice in the errors in the motion estimates, which is:

$$\begin{cases} \alpha_e & = & \frac{y_{0_e}}{Z_{avg}f} \\ \beta_e & = & -\frac{x_{0_e}}{Z_{avg}f} \end{cases} \tag{21}$$

where Z_{avg} is the average scaled depth of the scene in view. In view of its constraint on the magnitude of $\mathbf{w}_e$, we refer to this constraint as the Rotation Magnitude (RMag) constraint, although it also implies directional constraint given by $\frac{\alpha_e}{\beta_e} = -\frac{y_{0_e}}{x_{0_e}}$. Comparing with (20), it is evident that RDir and RMag can hold simultaneously only if TDir also holds. Note that an exact cancelation of $\mathbf{t_{2,0}}$ and $\mathbf{t_{2,Z}}$ is impossible unless the scene can be modeled as a frontal-parallel plane.

4. $\mathbf{t_{2,1}}$ and $\mathbf{t_{2,2}}$ are determined by γ_e and α_e, β_e respectively. Under general scene, there is no way for $\mathbf{t_{2,1}}$ and $\mathbf{t_{2,2}}$ to be canceled off with other terms; $\|\mathbf{w}_e\|$ has to be small for $\mathbf{t_{2,1}}$ and $\mathbf{t_{2,2}}$ to be small. In this sense, $\mathbf{t_{2,1}}$ and $\mathbf{t_{2,2}}$ contribute to accurate estimation of rotation; unfortunately, their effects are weak unless the field of view is large. In view of the subsidiary role of the constraint it exerts on the magnitude of the rotation errors, we term it as RMag2 constraint. Another important fact about $\mathbf{t_{2,2}}$ is that it will be exactly perpendicular to $\mathbf{t_1}$ (independent of the feature points co-ordinates)

when $t_{1,0} = (0,0)$. Therefore, RMag2 constraint will be ineffective on the magnitude of α_e and β_e when the estimated FOE coincides with the origin.

From the preceding observations, we can establish the following conclusions:

1. **Translation estimates.** One of the well known phenomenon in motion perception is the bas-relief ambiguity. Basically it amounts to a valley on the translation error surface, along a straight line that is defined by the true FOE and the image center. We term this straight line the bas-relief line and this valley the bas-relief valley. TDir is the direct reason for the formation of such a valley.

2. **Rotation estimates.** Of the three rotational estimates, any error in $\hat{\gamma}$ would have purely deleterious effect on the minimization of $\|\dot{\mathbf{p}}_e(\hat{\mathbf{v}}, \mathbf{v}, \mathbf{w}_e)\|$. Thus, in the case of noiseless flow field, we expect accurate estimation of γ (those experienced in the art of the SFM algorithms will know that this is often not the case in numerical practice). The effects of α_e and β_e on the residual error are more complex. On the one hand, given a FOE error, α_e and β_e that satisfy the RMag constraint can make $\|\mathbf{t_2}\|$ small, thus leading to small residual error. On the other hand, the RMag2 constraint would prefer α_e and β_e to be small. Furthermore, a FOE estimate that is close to the origin weakens the RMag2 constraint on α_e and β_e. These effects will determine the values of the rotation estimates that minimize J_R, and the rotation estimates will in turn influence the shape of the bas-relief valley.

3.2.2   Error Profile along the Bas-relief Valley

We have established in the preceding section the existence of the bas-relief valley; the variation of the error along the bas-relief valley itself is the subject of this section. First and foremost, the location of the true FOE has a critical influence on the shape of the bas-relief valley and this in turn has implication for any motion algorithm trying to deal with a wide range of translational motion. We will use an example to elucidate the influence of the true FOE location. Figure 3(a) illustrates the case where the true FOE (x_0, y_0) lies somewhere in between the image center and infinity. The dotted line in the figure corresponds to the bas-relief valley. As the estimated FOE leaves the true FOE and moves along the bas-relief valley, we can identify several factors that influence the outcome of equation (17) for J_R.

1. **Translational error.** Consider first the effects of the translational terms by setting the rotational errors to zero. As we vary the estimated FOE along the bas-relief valley, we study the $\|\dot{\mathbf{p}}_e(\hat{\mathbf{v}}, \mathbf{v}, \mathbf{w}_e)\|^2$ curve for a single feature point (x,y). As shown by the dashed curve in Figure 3(b), there would be two turning points on the curve; they correspond to a minimum where the estimated FOE coincides with the true FOE, and a maximum whose location depends on the value of (x,y). In particular, it can easily be shown (Xiang, 2001) that the location of the maximum depends on the position of the projection of (x,y) on the bas-relief line relative to the true FOE; if it falls on the left side of the true FOE, the maximum will be on the left, and vice versa. Finally,

it is also clear that as the estimated FOE approaches infinity on either end of the bas-relief line, the curve would approach asymptotically towards a constant value. The total effect of the translational terms would be obtained by summing up all the $\|\dot{\mathbf{p}}_e(\hat{\mathbf{v}}, \mathbf{v}, \mathbf{w}_e)\|^2$ curves for each feature point. Let us denote this summed curve as the TrErr (Translational Error) curve. In this particular example of Figure 3(a) (and for most true FOE not near the image center), all (or most of) the feature points have their projections on the bas-relief line lying on one side of the true FOE. Thus, all the individual curves would have maxima lying on one side of the true FOE. As a result, the TrErr curve would have shape similar to that of the individual curve. That is, it would have an overall maximum on the opposite side of the true FOE with respect to the origin, and when the estimated FOE approaches infinity on either end of the bas-relief line, the residual value would approach asymptotically towards a constant value.

The asymptotic value at infinity will be largely determined by the types of true translation. Predominantly lateral motion causes low asymptotic value (see dashed curves in Figures 3(c) and (d)), with the latter approaching zero as the translational motion approaches that of pure lateral motion.

2. **Rotational error**. How do the rotation parameters enter the picture? If these parameters could be estimated accurately, the SFM problem would be simple. The error profile along the bas-relief valley would be represented by the TrErr curve. In particular, there would be no local minimum within the bas-relief valley. However, *it is precisely the coupling of the rotation with the translation that results in local minima within the bas-relief valley.* By coupling, we mean that the residual error caused by the translational errors can be compensated for by a suitable choice of α_e and β_e. Figures 3 (b) to (d) show this compensating capability of α_e and β_e along the bas-relief line by the dotted curves, where high values on the curves indicate that the compensation is highly effective. We denote these dotted curves as the RotComp (Rotation Compensation) curves.

Referring to Figure 3, as the estimated FOE departs from the true FOE and enters region 1, the RMag constraint is operative and works towards compensating the translation error. However, as $\|(x_{0_e}, y_{0_e})\|$ increases, two factors restrict the applicability of the RMag constraint. Firstly, the corresponding increase of α_e and β_e means that $\|\mathbf{t}_{2,2}\|$ increases, that is, RMag2 constraint comes to the fore, which works against the minimization of J_R. Secondly, as $\mathbf{t}_1$ approaches more and more towards the direction perpendicular to $\mathbf{t}_{2,0}$ and $\mathbf{t}_{2,z}$, there is less advantage to be gained from the cancelation of the $\mathbf{t}_{2,0}$ and $\mathbf{t}_{2,z}$ terms, as long as the directional constraints TDir and RDir are observed. The resulting RotComp curve in region 1 is such that it increases firstly, then decreases asymptotically towards zero. As the estimated FOE departs from the true FOE in the other direction and enters region 2, RotComp would increase first, as in the case of the beginning of region 1. As long as RMag2 is not operative yet, the RotComp is able to follow in tandem the rapid increase of the TrErr curve and therefore to compensate the latter. As the estimated FOE enters region 3, RMag2

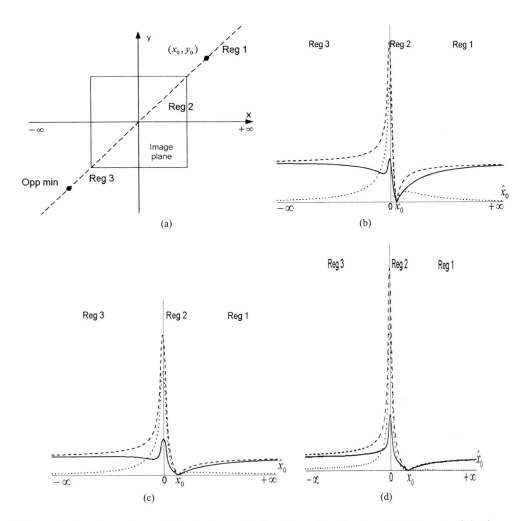

Figure 3: Configuration within the bas-relief valley when the true FOE is out of the image plane. (a) Overall configuration in the image plane; along the bas-relief valley, we divide it into regions 1, 2, and 3 as shown. (b), (c), (d) show the residual errors along the bas-relief valley with increasing amount of lateral translation. The dashed line represents the $\|\dot{\mathbf{p}}_e(\hat{\mathbf{v}}, \mathbf{v}, \mathbf{w}_e)\|^2$ curve for a particular point, the dotted line the RotComp curve, and the solid line the overall J_R curve.

takes effect again. Furthermore t_1 becomes more perpendicular to $t_{2,0}$ and $t_{2,Z}$ thus obviating the need for the RMag constraint. Thus the RotComp curve again decreases asymptotically towards zero.

3. **Formation of local minimum.** The final J_R curve as a result of this coupling between the rotation and the translation would be equal to the subtraction of RotComp curve from the TrErr curve as shown in Figures 3(b) to (c) (the solid curve). Clearly, it is due to the sharp drop-off of the TrErr curve as well as the compensatory effect of the rotational terms that a local minimum forms on the opposite side of the true FOE. We call this minimum the opposite minimum because it always lies on the opposite side of the true FOE. It is located around where the RotComp curve starts to enter region 3, that is, where the RotComp curve is no longer able to follow the TrErr curve. Figure 3(d) illustrates the case where the true FOE is further out (though not at infinity); here the opposite minimum has already been pushed to infinity at the other end of the bas-relief valley.

4. **Factors affecting local minimum.** The exact location and the "depth" of the local minimum depends on various factors discussed below:

   - The type of the true translation affects the shape of the TrErr curve, specifically its maximum location and the asymptotic values. In general, largely lateral translations present a more difficult scenario for most SFM algorithms, because at the opposite end of the bas-relief valley, t_1 and t_2 will be almost entirely perpendicular, resulting in small asymptotic value of J_R. The location of the opposite minimum will approach infinity with residual value approaching zero. A large part of the bas-relief valley becomes very flat, thus presenting a highly ambiguous situation (Figure 4(b)). Furthermore, as far as rotation estimates are concerned, in the limiting case of pure lateral motion, the RMag constraint fails to exert any constraint on the magnitude of α_e and β_e. The reason is as mentioned before: t_1 and t_2 can be in this case made perpendicular to each other at all points through the TDir and RDir constraints; the RMag constraint, which makes t_2 small, becomes redundant.

     Conversely, if the true FOE approaches that of the pure forward motion case, the opposite minimum will merge with the true solution and disappear as shown in Figure 4(a).

   - Field of view determines the effectiveness of the RMag2 constraint. With other conditions fixed, small FOV is a favorable condition for the formation of the opposite minimum. The later the RMag2 constraint sets in, the longer the Rot-Comp curve are able to follow the TrErr curve. Thus the opposite minimum will form further away from the origin, and the residual error J_R will be smaller in value (though the valley around this local minimum may be less steep), making it more likely for the opposite minimum to be picked up as the solution.

- Focal length. A change of focal length brings about several effects. Firstly, the values of α_e and β_e as dictated by the RMag constraint will be smaller. This in turn means that all the $\mathbf{t_{2,2}}$ terms will be reduced in magnitude, both due to the smaller α_e and β_e and the larger f value in the denominator. Lastly, the true FOE $(f\frac{U}{W}, f\frac{V}{W})$ will edge towards infinity with a larger f. All these factors will push the opposite minimum further, as well as making it more conducive for the opposite minimum to be picked up as the solution.

- Unweighted epipolar constraint. Finally, it should be mentioned that if the epipolar constraint is unweighted, it can be shown that there will no maximum in the TrErr curve. Clearly, the resultant coupling with the RotComp curve would yield a minimum in the error surface near the center of the image, a result well-known in the literature.

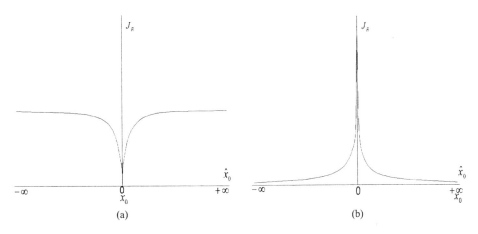

Figure 4: Error profiles of the bas-relief valley in the limiting cases. (a) The true FOE coincides with the origin (b) The true FOE lies at infinity.

3.2.3   Relaxation of the Assumptions on the Distribution of Feature Points and Depth

In the beginning of this section, we made some assumptions on the distribution of feature points and depths, which are necessary for the derivation of the preceding results, especially the formation of the bas-relief valley. It is important to know how the error surface would be affected when these assumptions do not hold.

1. **Uneven feature point distribution**. We first consider the case where the feature points are not evenly distributed (with the assumption on the depth distribution still valid). We can model one aspect of uneven distribution by a shift of the centroid of the feature points from the image center to the point $(\tilde{x}, \tilde{y})$. Since $\mathbf{t_{2,1}}$ is always parallel to $\mathbf{t_{1,1}}$ and $\mathbf{t_{2,2}}$ is always perpendicular to $\mathbf{t_{1,1}}$, the contributions of these two components of $\mathbf{t_2}$ to the error surface will not be affected by the distribution of the

feature points. However, to make the remaining components of $\mathbf{t_2}$, which are $(\mathbf{t_{2,0}} + \mathbf{t_{2,z}})$, perpendicular to $\mathbf{t_1}$, the TDir and RDir constraints need to be modified near the region where the feature points are clustered as follows:

$$\frac{(x_0 - \tilde{x})}{(y_0 - \tilde{y})} = \frac{(\hat{x_0} - \tilde{x})}{(\hat{y_0} - \tilde{y})} \qquad (22)$$

and

$$\frac{\alpha_e}{\beta_e} = -\frac{(\hat{y_0} - \tilde{y})}{(\hat{x_0} - \tilde{x})} \qquad (23)$$

respectively. As a consequence, the bas-relief valley is attracted towards the new centroid when it is in the vicinity of the feature points. However, when the estimated FOE approaches infinity, the direction of $\mathbf{t_1}$ will be mainly determined by $\mathbf{t_{1,0}}$; in other words, the bas-relief valley will be very little affected by the shift of the feature centroid. The resultant bas-relief valley is illustrated in Figure 5.

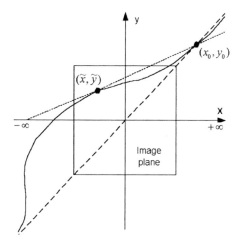

Figure 5: Geometry of the bas-relief valley when the distribution of feature points is uneven, with centroid at $(\tilde{x}, \tilde{y})$. The solid line corresponds to the bas-relief valley; the dashed line corresponds to the bas-relief valley if the feature points are evenly distributed; the dotted line is the line passing through $(\tilde{x}, \tilde{y})$ and (x_0, y_0).

2. **Feature point grouped in local clusters.** Another aspect of unevenly distributed feature points is that the feature points are grouped in local clusters. Here it is well to mention the presence of critical points on the error surface due to the vanishing of the denominator of $\dot{\mathbf{p}}_e(\hat{\mathbf{v}}, \mathbf{v}, \mathbf{w}_e)$. In particular, these critical points are formed when the estimated FOE coincides with (x, y), at which both the numerator and the denominator vanish. These critical points may also be caused by the $\mathbf{n}$ term in the denominator being perpendicular to $(x - \hat{x_0}, y - \hat{y_0})$, the value of $\mathbf{n}$ being dependent on the adopted reconstruction schemes. Such critical points are the possible sources

of local minima or maxima. However, our simulations in the next section show that they do not have a significant effect on the overall error surface as long as the feature points are evenly distributed. However, when feature points are grouped in local clusters, the surrounding error surface can be significantly affected. In particular, each cluster may cause shallow local minimum to form around the cluster.

3. **Uneven depth distribution.** Another factor to consider is that the depth is usually dependent on the feature co-ordinates (x,y) in a complex way. Referring to Figure 2, this depth dependency means that the perturbation terms $t_{1,1} = (x,y)$ and $t_{2,Z} = (-\frac{y_{0_e}}{Z}, \frac{x_{0_e}}{Z})$ are correlated, or equivalently, $(\frac{x}{Z}, \frac{y}{Z})$ and $(-y_{0_e}, x_{0_e})$ are correlated. Thus we can model this dependency as a shift of centroid of the depth-scaled feature points $(\frac{x}{Z}, \frac{y}{Z})$. It can be shown that the effect of this shift on the bas-relief valley is very similar to that caused by a shift of the feature points' centroid. For example, if the depths associated with the feature points at the upper-left corner of the image plane are smaller than those of other feature points, the centroid of the depth-scaled feature points will move toward the upper-left corner. The location of the resulting bas-relief valley will change as if the centroid of the feature points were shifted to the upper-left corner (see Figure 5).

4. **Special case of planar scene.** One special case of depth dependency on the feature co-ordinates is that of a plane, a case well studied by numerous researchers such as Longuet-Higgins (Longuet-Higgins, 1984). It is simple but nevertheless capable of modeling many real scene configurations. Longuet-Higgins showed that if all the object points come from a plane which is expressed as:

$$LX + MY + NZ = 1$$

the motion field caused in the image plane will be identical to the motion field caused by another plane:

$$UX + VY + WZ = 1$$

moving with the translation (L, M, N) and rotation $(\alpha + VN - WM, \beta + WL - UN, \gamma + UM - VL)$. That is, there now exists on the translation error surface another minimum at the location $(f\frac{L}{N}, f\frac{M}{N})$. One can relate this planar case to the case discussed in the preceding paragraph, where depth dependency results in a shift of the centroid of the depth-scaled feature points $(\frac{x}{Z}, \frac{y}{Z})$. Substituting the planar equation for Z into $(\frac{x}{Z}, \frac{y}{Z})$, one can show that the centroid of these depth-scaled feature points is now lying along the direction given by $(f\frac{L}{N}, f\frac{M}{N})$. Thus, we would expect the original bas-relief valley to be pulled towards this effective centroid $(f\frac{L}{N}, f\frac{M}{N})$. Indeed, as we will show through simulation (Figure 8) later, along the direction defined by the origin and the alternative FOE $(f\frac{L}{N}, f\frac{M}{N})$, the residual error surface also forms a valley similar to the bas-relief valley (the bas-relief valley of the alternative motion-scene configuration, as it were). As expected, the two bas-relief valleys will influence each other when they are near each other. The estimates of the rotation will also be affected by the existence of the alternative solution. The RMag2 constraint will be modified since it is now possible to partially cancel the $t_{2,1}$ and $t_{2,2}$ terms with the $t_{2,z}$ term. The result is that while large field of view still improves the estimation of rotation due to the RMag2 constraint, we expect larger errors to remain in the rotation estimates, including that for the rotation around the optical axis.

3.3   Depth Distortion at the Opposite Minimum Solution

In this section, we attempt to investigate how the reconstructed structure would be distorted when the opposite-side minimum is picked up as the solution. In particular, we investigate depth distortion under configurations favourable for the formation of such spurious opposite minimum, namely, the FOV is small and the translation is largely lateral. Rewriting the expression of the distortion factor in equation (11) in terms of its component error terms, we obtain:

$$D \;=\; \frac{(x-\hat{x}_0,y-\hat{y}_0)\cdot \mathbf{n}}{(x-x_0,y-y_0)\cdot \mathbf{n}+Z\left(u_{rot_e},v_{rot_e}\right)\cdot \mathbf{n}} \tag{24}$$

Under the aforementioned favourable conditions, we are able to make two simplifications. Firstly, the condition of small FOV allows us to ignore the second order terms in u_{rot_e} and v_{rot_e}. We also further assume $\gamma_e = 0$ (under noiseless condition, we expect accurate estimation of γ). Secondly, given the large values of $x_0, y_0, \hat{x}_0, \hat{y}_0$ in this configuration, we make these approximations: $(x-\hat{x}_0,y-\hat{y}_0) \approx (-\hat{x}_0,-\hat{y}_0)$ and $(x-x_0,y-y_0) \approx (-x_0,-y_0)$. Equation (24) can then be expressed as:

$$D \;=\; \frac{(-\hat{x}_0,-\hat{y}_0)\cdot \mathbf{n}}{(-x_0,-y_0)\cdot \mathbf{n}+Z\left(-\beta_e f,\alpha_e f\right)\cdot \mathbf{n}} \tag{25}$$

We know that at the opposite minimum solution, the TDir and the RDir constraints hold. These constraints mean that the numerator and the denominator of the above expression are parallel, and thus D is independent of $\mathbf{n}$, the direction of depth reconstruction. We can now write D as

$$D = \frac{1}{\lambda_1 + \lambda_2 Z} \tag{26}$$

where λ_1 and λ_2 are constant, with $\lambda_1 = -\frac{\|(x_0,y_0)\|}{\|(\hat{x}_0,\hat{y}_0)\|}$ (since (x_0,y_0) and $(\hat{x}_0,\hat{y}_0)$ are opposite in direction) and $\lambda_2 = \pm\frac{\|(-\beta_e f,\alpha_e f)\|}{\|(\hat{x}_0,\hat{y}_0)\|}$.

A distortion factor with the form $\frac{1}{\lambda_1 + \lambda_2 Z}$ generates iso-distortion surfaces which are frontal-parallel planes. The resulting distortion transformation is that of a relief transformation which has some nice properties (Koenderink and Van Doorn, 1995). In particular, consider two points in space with depths $Z_1 > Z_2$. Given $\lambda_1 < 0$, it can be shown (Cheong and Ng, 1999) that depth order will be preserved when

$$(\lambda_1 + \lambda_2 Z_1)(\lambda_1 + \lambda_2 Z_2) < 0$$

Equivalently, since $\frac{1}{\lambda_1+\lambda_2 Z}$ is the distortion factor, the above means that depth order will be preserved when $\hat{Z}_1\hat{Z}_2 < 0$, and conversely, depth order will be reversed when $\hat{Z}_1\hat{Z}_2 > 0$. In other words, if $\hat{Z}_1\hat{Z}_2 > 0$, we need to perform a depth order inversion to obtain the correct depth order. Therefore, given the signs of two recovered depths, we can always determine the correct depth order. Of course, it remains open to question if human actually performs the required depth order inversion.

What can we say about the sign of λ_2? If the RMag constraint holds, then

$$\begin{cases} sign(-\beta_e f) &= sign(-\hat{x}_0) \\ sign(\alpha_e f) &= sign(-\hat{y}_0) \end{cases} \tag{27}$$

which means that λ_2 is positive. Under such condition, the iso-distortion surfaces have the following additional properties. The $D = 1$ distortion surface divides the whole space into two parts: the near field in which the space is expanded ($D > 1$) and the far field in which the space is compressed ($D < 1$), with negative distortion factor in the region $0 < Z < -\frac{\lambda_1}{\lambda_2}$. However, the sign of λ_2 may be indeterminate when the true FOE moves towards infinity (as does the opposite minimum). Here the RMag constraint weakens and the RMag2 constraint is ineffective given the small FOV. Under such limiting case, we cannot determine the sign of λ_2.

Before we close this section, a brief remark on the case of translations close to the forward direction is warranted. When the opposite-side minimum is picked up as the solution, the distortion factor can be expressed as:

$$D = \frac{(x-\hat{x}_0, y-\hat{y}_0) \cdot \mathbf{n}}{(x-x_0, y-y_0) \cdot \mathbf{n} + Z(-\beta_e f, \alpha_e f) \cdot \mathbf{n}} \tag{28}$$

In such case, the distortion shows complicated behavior described by the Cremona transformation. This is in accordance with the view presented in (Cheong and Xiang, 2001) that depth recovery is less reliable when forward motion is executed.

3.4 Summary of Results and Discussion

Equation (17) has been critical in our analysis; its simple form renders possible the geometric treatment of the error surface via a consideration of the two vectors $\mathbf{t}_1$ and $\mathbf{t}_2$. The error surface configuration and in particular, the local minima on the surface which are the cause of inherent ambiguity of SFM algorithms, are identified. More importantly, the underlying mechanisms for the formation of such local minima are also investigated in a geometric way which is helpful towards obtaining an intuitive grasp of the problem. The major findings obtained so far are summarized as follows:

1. **Rotation error.** The rotation errors satisfy the following constraints: $\gamma_e = 0$ and $\frac{\alpha_e}{\beta_e} = -\frac{\hat{y}_0}{\hat{x}_0}$ when motion ambiguities arise. The magnitude of the rotation error may be further subject to the constraint of $\alpha_e = \frac{y_{0_e}}{Z_{avg}f}$ and $\beta_e = -\frac{x_{0_e}}{Z_{avg}f}$, but this constraint weakens as the true and the estimated FOE approach infinity. In particular, when the

estimated translation approaches infinity, the RMag constraint is not needed anymore. Only the RMag2 constraint is operative, which tends to make the rotation estimates close to the true solution. Another influential factor is FOV. Under large FOV, accurate rotation estimation is expected. On the other hand, when FOV is small, the rotation parameters are estimated with difficulties.

2. **Translation error**. Bas-relief valley is the major characteristic of the error surface; it is a line defined by the true FOE and the centroid of the feature points. The distribution of the feature points and the depth-scaled feature points will also affect the location of the bas-relief valley. Along the bas-relief valley, there is a local minimum at the opposite side of the global minimum with respect to the origin, which we called the opposite minimum. The residual error along the bas-relief valley also tends to have a local maximum somewhere near the origin and approaches an asymptotic value as the estimated FOE moves towards infinity. The location and the depth of the opposite minimum is determined by several factors. In particular, the opposite minimum will be further away from the image center and its residual value smaller when the FOV is small and the true translation is largely lateral. This opposite minimum in the bas-relief valley poses severe problem to most SFM algorithms.

3. **Depth distortion**. If the SFM algorithms return the opposite minimum as the solution, a distorted structure will be recovered. The behavior of such distortion depends on the location of the opposite minimum. If it is far away from the origin, the distortion transformation between the physical and reconstructed space belongs to relief transformation. Depth order can be preserved if we perform the necessary depth order inversion. In contrast, when the opposite minimum is close to the origin and returned as the solution, the depth distortion is complex and can only be described by a full Cremona Transformation. However it should be noted that if the features and depths are evenly distributed, a SFM algorithm taking proper precaution should be able to avoid this kind of opposite minimum due to its large residual error.

4. **Type of translation**. The type of translation has important influence on the configuration of the residual error surface. Under largely forward translation, the estimation of both translation and rotation is relatively accurate unless the feature points are locally clustered resulting in strong local minima within the image plane. In contrast, the SFM algorithms are more likely to give erroneous motion estimation when the true translation is largely lateral. However, as far as the depth recovery is concerned, translation that is largely lateral results in depth distortion that has nice properties such as preservation of relief.

5. **Type of cost function**. Different cost functions are obtained by setting $\mathbf{n}$ to different directions. Since the bas-relief ambiguity occurs when $\mathbf{t}_1$ and $\mathbf{t}_2$ are roughly perpendicular to each other, making the numerator of $\dot{\mathbf{p}}_e(\hat{\mathbf{v}}, \mathbf{v}, \mathbf{w}_e)$ vanish, different choices of $\mathbf{n}$ has little influence on the formation of the bas-relief valley on the error surface. However the shape of the error surface, and in particular, the error profile along the

bas-relief valley, could be affected by the different choices of $\mathbf{n}$. Indeed, if $\mathbf{n}$ is not set as the "epipolar reconstruction" direction, new local extrema would be introduced on the error surface when $\mathbf{t}_1$ is perpendicular to $\mathbf{n}$ for any feature point, making the denominator of $\dot{\mathbf{p}}_e(\hat{\mathbf{v}}, \mathbf{v}, \mathbf{w}_e)$ vanish. For instance, when $\mathbf{n}$ is set as a constant direction, there will be a large number of local maxima and minima, forming bands running roughly along the $\mathbf{n}^{\perp}$ direction. For the cases of $\mathbf{n}$ equal to constant direction and $\mathbf{n}$ equal to Linear Least Square Reconstruction direction, it can be shown that the opposite minimum on the bas relief valley still persists. Last but not least, it is also clear that the choice of $\mathbf{n}$ would directly affect the properties of the recovered depth.

6. **Minimization strategy**. There are many variants of SFM algorithms based on the differential epipolar constraint: Some estimates translation first and then the rotation (Zhang and Tomasi, 1999; Horn, 1990; Adiv, 1985; Heeger and Jepson, 1992), some estimates rotation first and then the translation (Prazdny, 1980), and others estimate all motion parameters simultaneously (Ma et al., 2000; Kanatani, 1993; Brooks et al., 1998). As long as the various algorithms are purely based on the differential epipolar constraint, the results of our analysis is applicable. For the algorithms that estimate the translation first based on other constraints such as the motion parallax (Rieger and Lawton, 1985), we need to first characterize the error likely to exist in the estimated translation which is beyond the scope of this study. However, assuming an erroneous FOE has been obtained, we know that the corresponding rotational errors will satisfy $\frac{\alpha_e}{\beta_e} = -\frac{\hat{V}}{\hat{U}}$ and $\gamma_e = 0$ if the feature points and the depth-scaled feature points are distributed evenly.

3.5 Experimental Analysis

3.5.1 Visualization of the Cost Functions

In this section, we perform simulations on synthetic images to both visualize and verify the predictions obtained from the preceding theory. We also make additional observations along the way, for instance, regarding the influence of density of feature points on the residual errors. These simulations were carried out based on the "epipolar reconstruction" scheme.

As discussed in the preceding section, we use the translation error surface for visualization purpose. At each point on the plot, the FOE are fixed; then J_R can be expressed as:

$$J_R = \sum_{i=1}^{n} \left(\frac{c_{1_i} - (c_{2_i}\hat{\alpha} + c_{3_i}\hat{\beta} + c_{4_i}\hat{\gamma})}{\delta_i} \right)^2 \qquad (29)$$

where

$$\begin{aligned} c_{1_i} &= u(y - \hat{y}_0) - v(x - \hat{x}_0) \\ c_{2_i} &= \frac{xy}{f}(y - \hat{y}_0) - (\frac{y^2}{f} + f)(x - \hat{x}_0) \end{aligned}$$

$$
\begin{aligned}
c_{3_i} &= \frac{xy}{f}(x - \hat{x}_0) - (\frac{x^2}{f} + f)(y - \hat{y}_0) \\
c_{4_i} &= x(x - \hat{x}_0) + y(y - \hat{y}_0) \\
\delta_i &= \sqrt{(x - \hat{x}_0)^2 + (y - \hat{y}_0)^2}
\end{aligned}
$$

from which the rotation variables can be solved by a typical linear least squares fitting algorithm such as the SVD (singular value decomposition) method. We performed this fitting for each fixed FOE candidate over the whole 2-D search space and obtained the corresponding reprojected flow difference J_R. The residual values were then plotted in such a way that the image intensity encoded the relative value of the residual (bright pixel corresponded to high residual value and vice versa). Furthermore, to illustrate the variation of J_R along the bas-relief valley in details, we also plot the cross-section of the residual error surface along the bas-relief line. Three types of curves were plotted for this purpose, namely, the residual error curve, TrErr, and RotComp as defined before, respectively drawn in solid line, dashed line and dotted line.

The imaging surface was a plane with a dimension of 512×512 pixels; its boundary was delineated by a small rectangle in the center of the plots. The residuals were plotted over the whole FOE search space, subtending the entire hemisphere in front of the image plane. We used visual angle in degree rather than pixel as the FOE search step; thus the co-ordinates in the plots were not linear in the pixel unit. Unless otherwise stated, the synthetic experiments have the following parameters: the focal length was 512 pixels which meant a FOV of approximately 53°; there were 200 object points whose depths ranged randomly from 512 to 1536 pixels; feature points were also distributed randomly over the image plane; true rotational parameters were $(0, 0.001, 0.001)$.

We conducted experiments under the following conditions: 1) varying amount of forward translation, ranging from head-on to lateral motion; 2) small versus large FOV; 3) feature points distributed evenly over the whole image plane versus those clustered at a corner; 4) depth-scaled feature points distributed evenly over the whole image plane versus those clustered at a corner; 5) sparse versus dense flow field; and 6) planar scene versus random scene.

Figure 6 shows the residual error images for different translations. It can be seen that the bas-relief valley becomes more obvious when the translation changed from being purely forward to being purely lateral. Since the feature points distribution were (roughly) even, the TDir constraint was a line passing through the image center and the true FOE. This can be clearly seen from Figures 6(b), (c), and (d) where the translation was not purely forward. Distinct local minima were centered around the true FOE and somewhere on the opposite side of the image center. We also plotted the residual profiles along the bas-relief valleys (note that the residual profiles were plotted in terms of pixels, whereas the residual error surfaces were plotted in terms of visual angles). Apparently, the types of true translation had a significant influence on the formation of the opposite minimum. The opposite minimum disappeared (or merged with the global minimum) for pure forward motion (Figure 6(e)). As the global minimum moved towards the infinity, so did the opposite minimum.

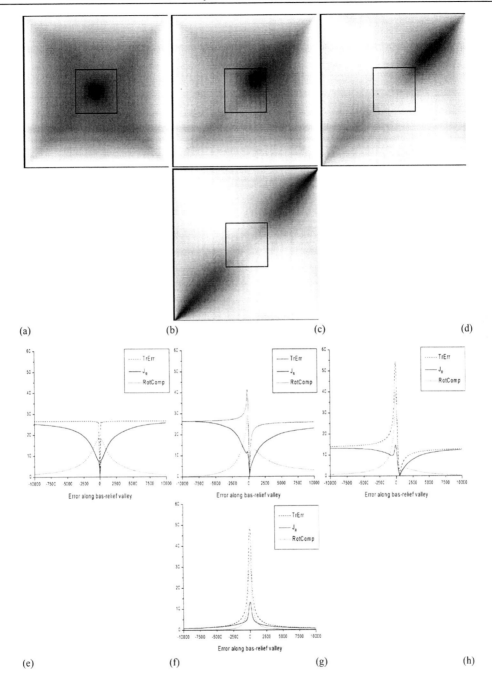

Figure 6: Residual error for different translations. Translational parameters (U, V, W) for (a), (b), (c), and (d) are $(0, 0, 2)$, $(0.5, 0.5, 2)$ $(1, 1, 1)$ and $(1, 1, 0)$ respectively. Figures (e), (f), (g), and (h) correspond to the residual profiles of the bas-relief valleys in (a), (b), (c), and (d) respectively. The dashed, dotted, and solid curves respectively represent the TrErr, RotComp and Jr curves. The residual error surfaces were plotted in terms of visual angles, whereas the residual profiles were plotted in terms of pixels.

The "false" minimum on the opposite side was much shallower than the "true" minimum in the case of non-lateral motion, as can be seen from Figure 6(f), but in Figure 6(h) under lateral motion they are almost equal in depth. The residual profiles also show clearly how the opposite minimum was formed by the coupling of the RotComp curve and the TrErr curve. By looking at the numerical values of the simulation data, we also found that for all the FOE candidates, the errors in the estimated rotational parameters were such that equation (20) held (RDir constraint), while for the rotational estimates around the opposite minimum, we further have their magnitudes satisfying equation (21) (RMag constraint). Under lateral motion (Figure 6(d)), those candidates with the smallest residuals were either the true translation or the translation in the opposite direction, while the estimated rotation satisfied $\frac{\alpha_e}{\beta_e} = -\frac{V}{U}$. Not surprisingly, we found that for these candidates, the magnitudes of $\mathbf{w}_e$ were quite arbitrary (though small) since the RMag constraint is ineffective. It may be noted in passing that from Figure 6 the maxima on the residual error image tended to form a strip perpendicular to the minima strip, and was more prominent for the case of lateral motion.

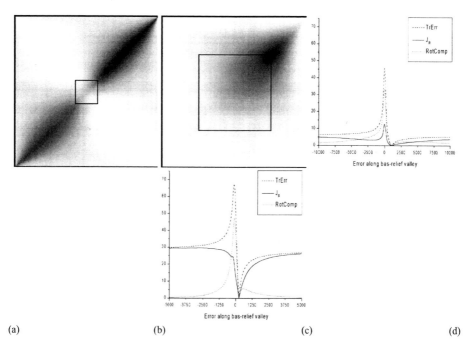

(a)   (b)   (c)   (d)

Figure 7: Influence of the FOV on the residual error images. $\mathbf{v} = (1, 1, 1)$ for both (a) and (b); FOV was 28° for (a) and 90° for (b). Figures (c) and (d) correspond to the residual profiles of the bas-relief valleys in (a) and (b) respectively. Notations as before.

Figure 7 shows the influence of FOV on the residual error images. While examining these plots, it should be kept in mind that in our simulations, larger FOV was obtained by fixing the image size and decreasing the focal length. Thus, under larger FOV the true FOE $(f\frac{U}{W}, f\frac{V}{W})$ would be closer to the image center with the same translational velocity

(U,V,W). The opposite minimum in Figure 7(a) was prominent with small residual value, while in Figure 7(b), the opposite minimum was almost invisible (it was barely visible in Figure 7(d), being located at where the solid curve was just rising towards the asymptotic value in the opposite direction).

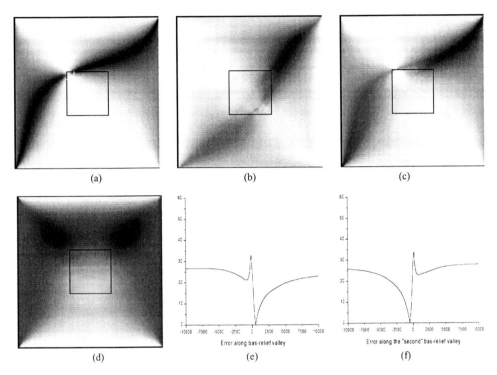

Figure 8: Influence of the distribution of feature points and depth-scaled feature points on the residual error images. (a) Feature points were clustered within a corner highlighted by a gray rectangle; (b) 20 feature points were distributed randomly over the image plane. (c) Feature point distribution was random over the image plane, whereas the depth distribution was such that the depth-scaled feature points were clustered at the upper-left corner of the image plane; (d) All the object points came from a plane with $(L,M,N) = (-0.002, 0.002, 0.002)$. Figures (e) and (f) correspond to the residual profiles of the bas-relief valley and the second bas-relief valley in (d) respectively. $\mathbf{v} = (1,1,1)$ for all the figures.

Figure 8 shows how the distribution of feature points and depth-scaled feature points affect the residual error images. When the feature points were clustered around different parts of the image, it was as if the bas-relief valley were pulled by the centroid of the feature points. This was illustrated in Figure 8(a) where the feature points were clustered in the upper left corner of the image plane. In Figure 8(b), with sparse and randomly distributed feature points, the extrema caused by these feature points can be seen to form around these feature points. Such local extrema always exist but their effect was not significant if the feature points were sufficiently dense, as can be seen from all the other residual error images (with 200 feature points) in this section. Figure 8(c) shows that the location of the centroid of the depth-scaled feature points would also affect the formation of the bas-relief valley. Here, the centroid of the depth-scaled feature points was located at the upper left corner of the image plane, resulting in an error surface as if the feature points were centered in that region. The case of a planar scene was illustrated in Figure 8(d). There were two clear minima, corresponding to the true and the alternative solutions for the planar scene. In addition, as shown by the cross-section of the residual error surface along the bas-relief valley in Figure 8(e), the opposite minimum still exists. Another bas-relief valley was also apparent along the direction defined by the alternative FOE and the origin. Figure 8(f) shows that this second bas-relief valley also has similar profile, that is, it also has a local minimum on the opposite side of the alternative FOE. Finally, the numerical values of the simulation data show that, as predicted, the rotational estimates no longer observed the RMag and the RMag2 constraints.

3.5.2   Inherent Ambiguities and Visual Illusions

Our analyses in the preceding sections show that there are various ambiguities inherent to the SFM problem which may cause erroneous 3-D motion estimation and distorted 3-D space reconstruction. The case of lateral motion was particularly studied because it possesses some unique properties. It is also a motion often found in the biological world and a case heavily studied by visual psychophysicists. In this section, we attempt to explain a well-known human visual illusion, the "rotating cylinder illusion" which is commonly observed in psychophysical experiments. We attribute this illusion to the imprecise estimation of the 3-D motion caused by the inherent ambiguities and the corresponding distorted structure recovered from the erroneous motion estimates.

Rotating Cylinder Illusion

In the psychophysical experiments, the rotating cylinder illusion amounts to the following situation: Dynamic random-dot display representing a rotating cylinder occupies a small portion of the visual field and rotates around a vertical axis passing through the center of the cylinder, as shown in Figure 9. It was found that sometimes the cylinder (as well as other curved objects) was perceived as rotating in a direction opposite to the true one and the correspondingly perceived structure underwent a change in the sign of curvature; that is a convex object was perceived as concave and vice versa (Hoffman, 1998).

Explanations

Some computational models have been proposed to explain the rotating cylinder illusion (Koenderink and Van Doorn, 1991; Soatto and Brockett, 1998). Koenderink and Doorn (Koenderink and Van Doorn, 1991) argued that since the solution of SFM under perspective projection cannot explain the effect (there can be only one solution), it could be that the

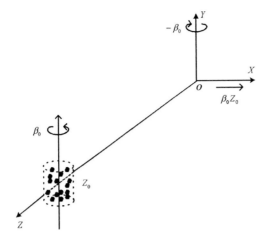

Figure 9: Configuration of the "rotating cylinder illusion". Dynamic random-dot display representing a rotating cylinder rotates about a vertical axis through its centroid with speed β_0. The equivalent egomotion for an observer positioned at o is given by $(\alpha, \beta, \gamma) = (0, -\beta_0, 0)$ and $(U, V, W) = (\beta_0 Z_0, 0, 0)$ where Z_0 is the distance between the optical center and the centroid of the cylinder.

human visual system somehow adopts an affine projection model. Soatto and Brockett (Soatto and Brockett, 1998) attributed these illusions to the presence of noise.

A different view can be inferred from the standpoint of our theory. Given the conditions present in that of the rotating cylinder illusion (small FOV, lateral translation and small depth range), all conducive towards the formation of the opposite minimum, rotating cylinder illusion can be attributed to the erroneous motion estimates caused by the opposite minimum, or more precisely, the error configuration with $\hat{\mathbf{v}} = -\mathbf{v}$, $\frac{\alpha_e}{\beta_e} = -\frac{V}{U}$ and $\gamma_e = 0$. Consider the case where $\beta_0 > 0$, we have under the error configuration $\beta = -\beta_0 < 0, \hat{U} = -\beta_0 Z_0 < 0$. Since the object was perceived as rotating opposite to the veridical direction, we further have: $\hat{\beta} > 0$. Thus it holds that $\beta_e < -\beta_0 < 0$. According to the results in section 3.3, when the relative translation is perceived as opposite to the veridical motion such that $\lambda_1 < 0$, the depth order relationship of any two depths Z_1 and Z_2 would depend on the signs of the perceived depths $\hat{Z}_1$ and $\hat{Z}_2$. We can determine the sign of $\hat{Z}_1$ and $\hat{Z}_2$ as follows. When the rotating cylinder illusion took place, the subject often reported a perceived rotation which had roughly the same speed as the veridical one, that is $\hat{\beta} = \beta_0$ and $\hat{U} = -\beta_0 Z_0$. With all depths under view greater than $\frac{Z_0}{2}$ under the experimental configuration, we immediately obtain the distortion factor $\frac{1}{(-1+\lambda_2 Z)} > 0$ for all the perceived depths. This means that all depths are perceived as positive and thus all depth orders are reversed; it follows that convex object is perceived as concave and vice versa. Since the erroneous 3-D motion at the opposite minimum would be perceived with equal likelihood as the accurate 3-D motion under pure lateral motion, it also explains why the illusion was reported to occur only intermittently.

4   Role of Noise on 3-D Motion Estimation

In practice, optical flow is always estimated with some noise. We express the noise-corrupted flow $\check{\mathbf{p}}$ as:

$$\begin{aligned} \check{\mathbf{p}} &= \dot{\mathbf{p}} + \dot{\mathbf{p}}_n \\ &= (u + u_n, v + v_n, 0)^T \end{aligned} \tag{30}$$

where $\dot{\mathbf{p}}_n$ is the flow component caused by noise. If we replace $\dot{\mathbf{p}}$ by $\check{\mathbf{p}}$ in equation (13), equation (14) still holds, that is, the reprojected flow difference along the $\mathbf{n}$ direction is still zero. It follows that the noise-corrupted cost function, denoted as J_{Rn}, can be obtained as follows:

$$\begin{aligned} J_{Rn} &= J_R \\ &+ 2\sum \left(\begin{array}{c} \frac{(x-\hat{x}_0,y-\hat{y}_0)\cdot(v_{rot_e} - \frac{y_{0e}}{Z}, \frac{x_{0e}}{Z} - u_{rot_e})}{(x-\hat{x}_0,y-\hat{y}_0)\cdot\mathbf{n}} \\ \times \frac{(x-\hat{x}_0,y-\hat{y}_0)\cdot(-v_n,u_n)}{(x-\hat{x}_0,y-\hat{y}_0)\cdot\mathbf{n}} \end{array} \right) \\ &+ \sum \left(\frac{(x-\hat{x}_0,y-\hat{y}_0)\cdot(-v_n,u_n)}{(x-\hat{x}_0,y-\hat{y}_0)\cdot\mathbf{n}} \right)^2 \end{aligned} \tag{31}$$

J_{Rn} consists of three terms. The first term is that of the noise-free case. The second term can be positive or negative, whereas the last term is always zero or positive. These are the most that can be said about the effect of noise without further introducing assumptions about the noise. Next, we investigate the behavior of SFM algorithms under the effects of specific noise types.

4.1   Isotropic Noise Model

Isotropic noise model has been frequently used for noise analysis in the computer vision community. The isotropic noise is defined as an independent Gaussian noise with identical covariance matrix $K = diag\{\sigma^2, \sigma^2, 0\}$. Under this noise model, the effect of noise on the periphery of the search space are small. Referring to the expression $\left(\frac{(x-\hat{x}_0,y-\hat{y}_0)\cdot(-v_n,u_n)}{(x-\hat{x}_0,y-\hat{y}_0)\cdot\mathbf{n}} \right)$ contained in both the second and the third terms of equation (31), it is basically a projection of the random noise on the vectors $(x - \hat{x}_0, y - \hat{y}_0)$ which are approximately constant in the periphery of the search space. The net effect of noise in the periphery is therefore rather benign as shown in Figure 10 where an isotropic noise with a standard deviation fixed at 50% of the average flow speed ($SNR = 7.08dB$) has been added to each component of the flow vector. The same noise has a "stronger" and more complex effect on the topology of the residual in the center of the image. This effect is especially obvious when the features are sparse, which can be seen by comparing Figure 10(b) with Figure 10(c).

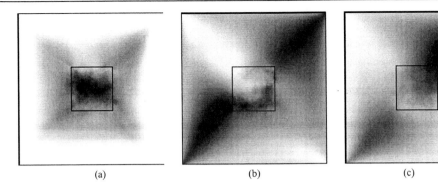

Figure 10: Residual error images for flow fields with isotropic noise. Number of feature points were 200 for (a) and (b) and 2000 for (c). $\mathbf{v} = (0,0,2)$ for (a) and $\mathbf{v} = (1,1,1)$ for both (b) and (c).

Some researchers demonstrated the robustness of their algorithms by conducting experiments using dense flow field with isotropic noise. However such noise model is often unrealistic. We show in the next section what happens when the noise model is non-isotropic.

4.2 Anisotropic Noise Model

A simple anisotropic noise model is one where the noise added to each flow depends on the flow itself. Specifically, for each noiseless flow we add a noise whose horizontal and vertical components are Gaussian with standard deviations proportional to the horizontal and vertical components of the noiseless flow respectively. With this model, the noise tends to point towards the same direction as the noiseless flow. Such a model receives partial theoretical support from (Fermüller et al., 2001) in which a more complicated model was presented compared to the one adopted here.

The effect of noise under this model shows a strong directional anisotropy; this is especially so for the case where the noiseless flow field itself is also predominant in certain direction. Let us denote this direction as $\mathbf{n}_n$. This effect of such anisotropy is most significant at the periphery of the plots where the FOE estimates are far away from image center and have $(x - \hat{x}_0, y - \hat{y}_0)$ approximately pointing in the same direction. When this direction is parallel to $\mathbf{n}_n$, the contribution of the third term in equation (31) would be small, and vice versa. The effect of the second term in equation (31) is also strongly direction-dependent, although the dependence is more complex. Suffice it to say that the resultant residual error images have their local minima being pulled towards the $\mathbf{n}_n$ direction and the periphery of the plots. This is illustrated in Figure 11.

It can be seen from Figure 11 that the influence of a 50% anisotropic noise is quite significant. For the case of $\mathbf{v} = (1,1,1)$ and $\mathbf{w} = (0, 0.0005, 0.001)$ under the scene in view (Figures 11(a) and 11(b)), the noise was biased towards the direction $\mathbf{n}_n = (0.82, 0.56)$. We can see that the bas-relief valley was "pulled" towards the $\mathbf{n}_n$ direction. This effect persisted as we increased the number of feature points, as shown in Figure 11(b). For the case of

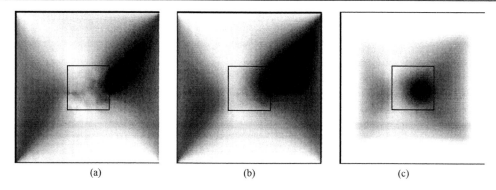

Figure 11: Residual error images for the flow fields with anisotropic noise. Numbers of feature points were 200 for (a), 2000 for (b) and $\mathbf{v} = (1,1,1)$ for (a) and (b). For (c), number of feature points was 800, and $\mathbf{v} = (0,0,2)$. $\mathbf{w} = (0,0.0005,0.001)$ and noise level was 50% for all the images.

forward motion (Figure 11(c)), the value of $\mathbf{n}_n$ is $(0.77,0.64)$. We can see a clear minima strip formed outside the image plane with the global minimum perturbed to $(90,-28)$.

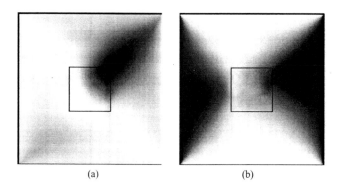

Figure 12: Influence of true rotation. Translational parameters were $(1,1,1)$ for both (a) and (b). Rotational parameters were $(0,0,0)$ for (a) and $(0,0.001,0)$ for (b). Noise level is 50% for all the images.

The implication of the above is that while true rotation parameters do not explicitly appear in the expression of J_{Rn}, their values can influence the performance of SFM algorithms by indirectly affecting the distribution of noise. For instance, a strong rotation around the Y-axis would result in a strong horizontal flow, which will pull the bas-relief valley towards the horizontal direction due to the aforementioned anisotropic noise. Such phenomenon is often observed in practice, the result being that the FOE cannot be reliably estimated when the rotation is dominant. Figure 12 shows the influence of the true rotation on the residual error images under anisotropic noise model.

In real images, features found in one surface patch may have different optical flows from those in another surface patch. According to our anisotropic noise model, the average

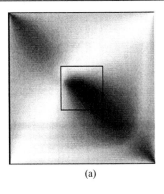 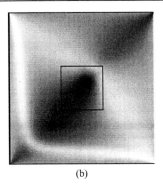 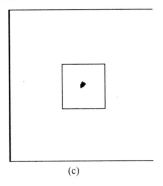

(a)   (b)   (c)

Figure 13: Residual error images using feature points from different patches. All the parameters were the same as Figure 11(c) except in the way we utilized the feature points. The feature points used in (a) and (b) were those found in the top-left and the top-right quarter of the image plane respectively. Figure (c) was the binary image obtained by intersecting the thresholded versions of (a) and (b).

noise directions in these two patches would also be different. If we were to perform SFM separately from these two patches, we would expect the minima strip to be pulled along different directions; however, the region around the true solution would be less affected. This was illustrated in Figures 13(a) and (b). One can capitalize on this characteristic, say, by performing a simple thresholding on the residual values of Figures 13(a) and (b) and intersecting the resulting binary maps. The result was illustrated in Figure 13(c), where the centroid of the "common area" was at $(-12,0)$, closer to the true FOE estimate than the global minimum obtained by using all the feature points in the image simultaneously. This observation can be used to formulate a plausible strategy to improve motion estimation; its feasibility will be further tested on real images in the next section.

4.3   Experiments on Real Images

The aim of this section is to carry out experiments on real images so as to verify the various predictions made, and to study the feasibility of a better algorithm based on the knowledge of the topology of the error surfaces.

Table 1: The parameters of three real image sequences

	Image Size	f	Translation	Rotation
COKE	300×300	439.4	$(x_0,y_0) = (-25,25)$	$(0.0006,0.0006,0.004)$
YosemiteNoCloud	252×316	337.5	$(x_0,y_0) = (0,59)$	$(0.0002,0.0016,-0.0002)$
SOFA1	256×256	309.0	$(U,V,W) = (0.814,0.581,0)$	$(-0.0203,0.0284,0)$

Three familiar real image sequences were used. The parameters of these sequences are listed in Table 1. Among them, only the COKE sequence is genuinely "real", while the other two are computer generated sequences. The YosemiteNoCloud sequence is the well-known Yosemite sequence minus the cloud in the top portion of the images. SOFA1 is the first sequence of the SOFA sequences. SOFA [2] is a package of 9 computer generated image sequences designed for testing research works in motion analysis. SOFA1 describes a simple indoor scene with constant lateral translation and quite significant rotational components. The optical flow was obtained using Lucas's method (Lucas, 1984) with a temporal window of 15 frames. Relatively dense optical flow fields (around 3000 feature points for each sequence) were obtained. Again, the estimated epipolar direction was adopted as the direction for depth reconstruction.

The residual error images were shown in Figure 14, from which several observations can be made.

- Figure 14(c) shows local minima strips along the average optical flow direction, especially outside the image plane. This might be due to the effect of anisotropic noise as discussed above. As for the case of lateral motion in SOFA1, the anisotropic noise also influences the direction of the bas-relief valley. As can be seen in Figure 14(i), the bas-relief valley was pulled towards the average optical flow direction, which was roughly horizontal.

- The effect of the clustered feature points was obvious for each case. Specifically, prominent edges on the image resulted in a clustered feature distribution, as in the case of YosemiteNoCloud. Local minima were formed along the edge, as shown in Figure 14(f).

- Rotation estimates. While the numerical values of the simulation data with synthetic images in Section 3.5 showed that under all configurations except the planar case, γ was invariably estimated with high accuracy for all the FOE candidates, this was not the case for real images. As far as α_e and β_e were concerned, the numerical values also showed that the corresponding RDir and RMag constraints were significantly modified, possibly due to the presence of non-isotropic noise.

Figure 15 demonstrates how we can make use of the knowledge of the topology to design a better algorithm. In the case of anisotropic noise, instead of using all the features simultaneously in an image, we performed SFM separately, each time using features found in different patches. The average optical flows found in different patches will be usually pointing to different directions. The resultant residual error images would thus be pulled differently according to the average flow directions. By combining the separate residual error images, one can obtain a better and more robust FOE estimate. Figure 15 (c) was the binary image obtained by performing an intersection of the thresholded version of Figures 15 (a) and (b) (the residual error images resulting from using different patches). It can be seen that the uncertainty area of the FOE estimate has been much reduced, thereby illustrating the feasibility of the idea. Better strategy for combining the estimation results from different patches can be devised so that factors such as feature number and flow configuration in each patch can be taken into account.

---

[2] courtesy of the Computer Vision Group, Heriot-Watt University (http://www.cee.hw.ac.uk/~mtc/sofa).

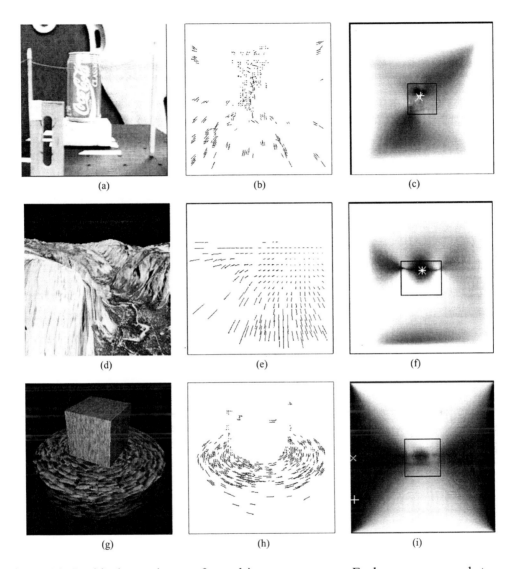

Figure 14: Residual error images for real image sequences. Each row corresponds to one image sequence. From top to bottom: COKE, YosemiteNoCloud and SOFA1. For each row, an actual image of the sequence, the optical flow field and the residual error image are shown from left to right. True FOEs and global minima of the residual error surfaces were highlighted by "+" and "×" on the residual error images respectively.

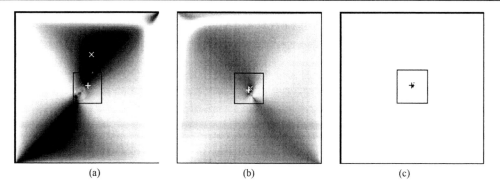

(a)   (b)   (c)

Figure 15: Residual error images resulting from using features in different regions in the COKE image. Figure (a) was obtained by using features from the bottom-left and Figure (b) from the bottom-right quarter of the image plane respectively. Figure (c) was the binary image obtained by performing an intersection of the thresholded version of (a) and (b).

5 Characterizing the Distortion of Shape Recovery from Motion

Given that most existing Structure From Motion (SFM) algorithms cannot recover true motion estimates reliably, it is important to understand the impact such motion errors have on the shape reconstruction. In this section, various robustness issues surrounding different types of second order shape estimates recovered from motion cue are addressed.

5.1 Local Shape Representation

Consider the canonical form of a smooth (quadratic) surface patch:

$$Z = \frac{1}{2}\left(k_{min}X^2 + k_{max}Y^2\right) + d \tag{32}$$

where k_{min}, k_{max} are the two principal curvatures with $k_{min} < k_{max}$. This represents a canonical case whereby the observer is fixating straight ahead at the surface patch located at a distance d unit away, and the $X-$axis and the $Y-$axis are aligned with principal directions (directions of principal curvatures). Rotating the aforementioned surface patch around the $Z-$axis by θ in the anti-clockwise direction, we obtain a more general surface patch whose principal directions do not coincide with the $X - Y$ frame:

$$
\begin{aligned}
Z &= \frac{1}{2}\left(\cos^2\theta(k_{min}X^2 + k_{max}Y^2) + \sin^2\theta(k_{max}X^2 + k_{min}Y^2)\right) \\
&\quad + \cos\theta\sin\theta(k_{min} - k_{max})XY + d
\end{aligned}
\tag{33}
$$

Since the focus of this study is on the recovery of local shape, our analysis will primarily center around the fixation point (X=Y=0). The idea is that if the perceptual tasks require local shape information, the visual system is usually fixated at that point.

Local representations of curved surfaces include the classical differential invariants of

Gaussian and mean curvatures which are computed based on the principal curvatures. A good shape descriptor should correspond to our intuitive idea of shape: shape is invariant under translation and rotation, and more importantly, independent of the scaling operation. Principal curvatures, as well as the Gaussian and mean curvatures satisfy the former but do not satisfy the latter condition because they still contain the information of the amount of curvature. Koenderink (Koenderink and Van Doorn, 1992) proposed two measures of local shape: shape index (S) and curvedness (C) as alternatives to the classical differential shape invariants. S and C are defined respectively as follows:

$$S = \frac{2}{\pi} \arctan \frac{k_{min} + k_{max}}{k_{min} - k_{max}} \tag{34}$$

$$C = \sqrt{\frac{k_{min}^2 + k_{max}^2}{2}} \tag{35}$$

S is a number in the range of $[-1, +1]$ and obviously scale invariant and C is a positive number with the unit m^{-1}. Shape index and curvedness provide us with a description of 3-D quadratic surfaces in terms of their types of shape and amount of curvature. Figure 16 shows the examples of quadratic surfaces with correspondent shape index values.

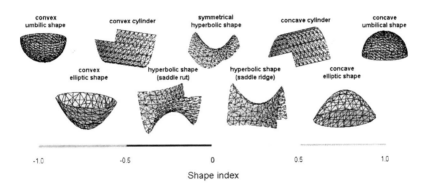

Figure 16: Examples of quadratic surfaces with correspondent shape index values. $S = \pm 1$: umbilic shape (spherical cap and cup); $S = \pm 0.5$: cylindrical shapes; $S \in [-1.0, -0.5]$ or $S \in [0.5, 1.0]$: elliptic shapes; $S \in [-0.5, 0.5]$: hyperbolic shapes.

When shape is reconstructed based on the motion cue, it is more appropriate to use shape index as the local measurement of shape type rather than other differential invariants. Due to the well-known ambiguity of SFM, the scale of the recovered objects and the speed of translation can only be determined up to a common factor. The recovered scale information is thus meaningless unless additional information is available. With shape index, we know that the accuracy with which it can be estimated depends on the accuracy of the 3-D motion estimates.

5.2   Distortion of Shape Recovery under Generic Motions

Many biological organisms often judge structure by making lateral motion, although the lateral translations are usually coupled with certain amount of rotation. For an artificial active vision system, forward and lateral motions are often purposefully executed for various tasks such as depth recovery, which means that the agent executing such motions is at least aware of the generic type of motion being executed. Thus it is reasonable to assume the following:

$$
\begin{cases}
\hat{W} = W = 0 & \text{when lateral motion is executed} \\
\hat{U} = U = \hat{V} = V = 0 & \text{when forward motion is executed}
\end{cases}
\tag{36}
$$

Furthermore, the rotation around the optical axis is usually not executed for an active visual system. Thus we can further assume that $\hat{\gamma} = \gamma = 0$.

5.2.1   Lateral Motion

We consider an agent performing a lateral translational motion $(U,V,0)$, coupled with a rotation (α, β, γ). Under the aforementioned assumption of $\hat{W} = W = \hat{\gamma} = \gamma = 0$, the estimated epipolar direction would be a fixed direction for all the feature points with $\mathbf{n} = \frac{(\hat{U}, \hat{V})}{\sqrt{\hat{U}^2 + \hat{V}^2}}$. Without loss of generality, we can set the estimated epipolar direction to be along the positive X−axis direction, so we have $\hat{V} = 0$ (though V is not necessarily zero). We can easily show that for any other direction, the distortion expression will have identical form with only changes in the values of some parameters. Modifying Equation (11) somewhat to take into account $W = 0$, the distortion factor can readily shown to be:

$$
D = \frac{(\hat{U}^2 + \hat{V}^2)\,Z}{\left(\ (U\hat{U} + V\hat{V})Z + \hat{U}\beta_e\left(X^2 + Z^2\right) - \hat{V}\alpha_e\left(Y^2 + Z^2\right) + (\hat{V}\beta_e - \hat{U}\alpha_e)XY\ \right)}
\tag{37}
$$

from which the mapping from a point (X,Y,Z) in the physical space to a point in the reconstructed space $(\hat{X}, \hat{Y}, \hat{Z})$ can be established.

Consider that a surface $\mathbf{s}(X,Y) = (X,Y,Z(X,Y))$, parameterized[3] by X and Y, is transformed under the mapping to the reconstructed surface $\hat{\mathbf{s}}(\hat{X}, \hat{Y}) = (\hat{X}, \hat{Y}, \hat{Z}(X,Y))$, which is parameterized by $\hat{X}$ and $\hat{Y}$. Using Equation (37), the perceived surface can also be parameterized by X and Y as follows:

$$
\hat{\mathbf{s}}(X,Y) = (DX, DY, DZ)
\tag{38}
$$

where D is defined in (37) and for brevity, $Z(X,Y)$ has been written as Z. Given this parameterization, *Mathematica*[4] can be used to perform algebraic manipulation to obtain

---

[3] Such parametric representation may not be possible for more complex global shapes.

[4] *Mathematica* is a registered trademark of Wolfram Research, Inc.

the expressions for quantities such as $\hat{k}_{min}$ and $\hat{k}_{max}$, as well as to visualize the distortion of the local shape in the following sections.

Curvatures

For the local surface patch given by Equation (33), we obtain the perceived principal curvatures $\hat{k}_{min}$ and $\hat{k}_{max}$ at the fixation point as follows:

$$
\hat{k}_{min} = \frac{1}{2}\left(\frac{(k_{min}+k_{max})U - 2\beta_e}{\hat{U}} - \sqrt{\frac{Q}{\hat{U}^2}} \right)
$$

$$
\hat{k}_{max} = \frac{1}{2}\left(\frac{(k_{min}+k_{max})U - 2\beta_e}{\hat{U}} + \sqrt{\frac{Q}{\hat{U}^2}} \right) \tag{39}
$$

where $Q = U^2(k_{min}-k_{max})^2 + 4(\alpha_e^2+\beta_e^2) - 4U(k_{min}-k_{max})(\beta_e\cos 2\theta + \alpha_e\sin 2\theta)$. Notice that the parameter d which indicates the distance between the surface patch and the observer does not affect the distorted principal curvature (and other shape measures based on principal curvatures).

From (39), it is obvious that Gaussian and mean curvatures are not preserved under the distortion; even the signs of principal curvatures are not necessarily preserved. The principal directions of the perceived surface patch can be obtained by computing the eigenvectors of the Weingarten matrix (Davies and Samuels, 1996) of the parametrical shape representation. It was found that generally the principal directions are not preserved. Only when the principal directions are aligned with the $X-$ and the $Y-$axis and $\alpha_e = 0$, they are preserved, though the directions of the maximum curvature and that of the minimum curvature may swap.

We are also interested in the distortion of the normal curvatures, especially the normal curvatures along the horizontal and vertical directions which are denoted as Z_{XX} and Z_{YY} respectively and given by:

$$
Z_{XX} = k_{min}\cos^2\theta + k_{max}\sin^2\theta
$$

$$
Z_{YY} = k_{min}\sin^2\theta + k_{max}\cos^2\theta \tag{40}
$$

The distorted normal curvatures along the horizontal and vertical directions can be obtained as:

$$
\hat{Z}_{\hat{X}\hat{X}} = \frac{U}{\hat{U}}Z_{XX} - \frac{2\beta_e}{\hat{U}}
$$

$$
\hat{Z}_{\hat{Y}\hat{Y}} = \frac{U}{\hat{U}}Z_{YY} \tag{41}
$$

Interestingly, $\hat{Z}_{\hat{X}\hat{X}}$ and $\hat{Z}_{\hat{Y}\hat{Y}}$ are not affected by α_e. Equation (41) shows that normal curvatures along different directions have different distortion properties.

Shape Index

From the expression of the distorted principal curvatures, the distorted shape index at the fixation point can readily be obtained by substituting (39) into (34):

$$\hat{S} = \frac{2}{\pi} \arctan \left(\frac{(k_{min} + k_{max})U - 2\beta_e}{\hat{U}\sqrt{\frac{Q}{\hat{U}^2}}} \right) \tag{42}$$

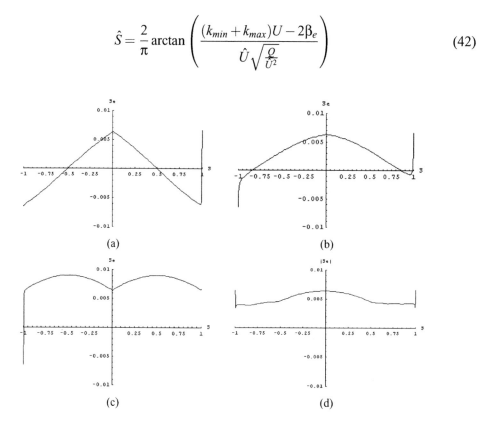

Figure 17: Distortion of shape index with different principal directions, plotted in terms of S_e against S. $\theta = 0$ for (a), $\theta = \frac{\pi}{4}$ for (b) and $\theta = \frac{\pi}{2}$ for (c). (d) was obtained by averaging curves with $\theta = 0 + n \times \frac{\pi}{8}$ where $n = 0, 1, \cdots 8$. All the other parameters are identical: $\alpha_e = 0.005$, $\beta_e = -0.02$, $U = 0.2$ and $\hat{U} = 0.18$.

As the expression for $\hat{S}$ is very complex, we have assorted some diagrams to graphically illustrate the error in the shape index estimate with respect to the true shape index. The curvedness is fixed for each diagram so as to factor out the influence of curvedness. Figure 17 shows the distortion obtained by varying principal directions while fixing all the other parameters. Figures 17(a), (b) and (c) show the sensitivity of shape recovery to the motion estimation errors for three principal directions: $\theta = 0$, $\theta = \frac{\pi}{4}$ and $\theta = \frac{\pi}{2}$. Figure 17(d) show the average sensitivity for different principal directions. Note that the sudden jump of S_e in some of the curves corresponds to the case where the direction of the maximum curvature and that of the minimum curvature have swapped. It can also be seen that the principal direction does affect the robustness of the shape perception. We can also infer from Figure 17 that different shapes have different distortion properties. The average curve shown in Figure 17(d) gives us a rough idea on the overall robustness of the perception of different shapes: the saddle-like shapes (also known as "hyperbolic shapes" with k_{min} and k_{max} having different signs) are more sensitive to the errors in 3-D motion estimation than the concave and convex shapes ("elliptic shapes" or "umbilic shapes" with k_{min} and k_{max} having the same sign).

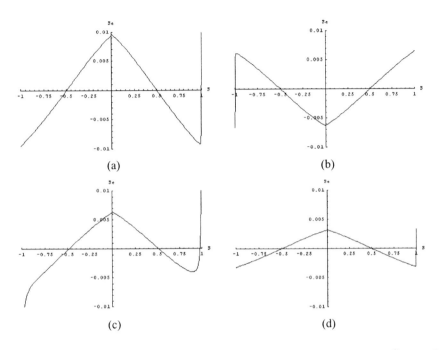

(a)   (b)

(c)   (d)

Figure 18: Other factors that affect the distortion of shape index (see text). $\beta_e = -0.03$ for (a), $\beta_e = 0.02$ for (b) and $\beta_e = -0.02$ for (c) and (d); $\alpha_e = -0.05$ for (c) and $\alpha_e = 0.005$ for (a), (b) and (d); $C = 20$ for (d) and $C = 10$ for (a), (b) and (c). $\theta = 0$, $U = 0.20$ and $\hat{U} = 0.18$ for all the diagrams.

Other factors that may have an impact on the sensitivity of shape perception are studied in Figure 18. β_e has a significant influence on the shape perception. Comparing Figure 18(a) with Figure 17(a), we can see that the error in the estimated shape index increases with increasing value of $|\beta_e|$. The sign of β_e will determine whether the shape index will be under-estimated or over-estimated for each shape type (comparing Figures 18(a) and 18(b)). On the contrary, α_e seems to have little influence on the recovered shape index, as shown in Figure 18(c). This anisotropy between α_e and β_e is related to the directional anisotropy that exists in the distortion of the normal curvatures, derived earlier in Equation (41). This anisotropy is of course a result of the estimated translation being in the horizontal direction. Finally, our analysis seems to support the view that the more curved surfaces will be perceived with high accuracy (see Figure 18(d)).

5.2.2 Distortion under Forward Motion

In this section, we assume pure forward translation is performed. According to Equation 36, we have $\hat{U} = U = \hat{V} = V = 0$. We also assume $\hat{\gamma} = \gamma = 0$ as we did before. Adopting "epipolar reconstruction" approach, the distortion factor can be expressed as:

$$D = \frac{X^2 + Y^2}{X^2 + Y^2 + \alpha_e Z (X^2 Y + YZ^2 + Y^3) - \beta_e Z (XY^2 + XZ^2 + X^3)} \quad (43)$$

Following the same procedure of deriving the distorted curvatures as in the lateral motion case, we can obtain the distorted principal and normal curvature expressions on any point of the distorted surface. The distorted shape index can also be computed thereafter. These expressions are very complex and are thus not presented here. It is noted that at the fixation point, the distorted local shape measurements are undefined due to the undefined distortion factor (see Equation (43)). Our analysis in (Cheong and Xiang, 2001) shows that the distortion factor varies wildly around the fixation point, which implies that the distorted surface patches are also not smooth. Therefore, the distorted local shape measurements at any particular point do not make much sense.

5.2.3 Graphical Illustration

To have an idea on how curved surface patches will be distorted under different generic motions, we use *Mathematica* to obtain the distorted shapes and graphically display them. Figure 19 compares the distortion in the original and recovered surfaces under lateral and forward motions. We consider a vertical cylinder (upper row) and a horizontal cylinder (lower row), given by the equations $Z = \frac{1}{2}X^2 + 4$ and $Z = \frac{1}{2}Y^2 + 4$ respectively. For the lateral motion case and given the parameters stated in the caption of Figure 19, it is easy to show that from Equations (39) and (41) that at the fixation point a vertical cylinder will be perceived as a less curved vertical cylinder and a horizontal cylinder will be perceived as a saddle-like shape. The principal directions remain unchanged. Figures 19(b) and (e) show the full reconstructed surfaces over the entire cylinders; clearly the distorted surface patches are smooth and the distortion behavior at the fixation point seems to be qualitatively representative of the global shape distortion even with a large view angle. The recovered surface patches under forward motion, by contrary, are not smooth and show large distortion. Even when the errors in the rotation estimates are much smaller than those in the case of lateral motion, the perceived surface patches are obviously not quadrics with major distortion occurring around the central region of attention.

5.3 Experiments

In this section, we conducted computational experiments on computer generated and real images to further verify the theoretical predictions.

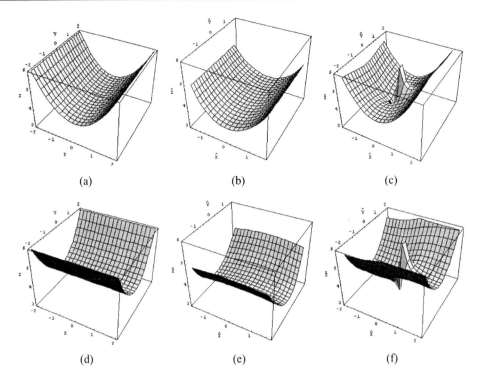

Figure 19: Distortion of cylinders under lateral and forward motion. (a) and (d) are the original vertical and horizontal cylinders respectively. (b) and (e) are the distorted vertical and horizontal cylinders under lateral motion respectively, with $U = 0.9$, $\hat{U} = 1.0$, $V = 0.5$ and $\hat{V} = 0$. (c) and (f) are the distorted vertical and horizontal cylinders under forward motion respectively. The errors in the estimated rotation are: $\alpha_e = 0$ and $\beta_e = 0.05$ for (b) and (e), and $\alpha_e = 0$ and $\beta_e = 0.001$ for (c) and (f). The field of view is 127° for all the diagrams.

5.3.1   Computer Generated Image Sequences

SOFA sequence 1 and 5 (henceforth abbreviated as SOFA1 and SOFA5) were chosen for our experiments, the former depicting a lateral motion and the latter a forward motion. Both of them have an image dimension of 256×256 pixels, a focal length of 309 pixels and a field of view of approximately 45°. The focal length and principal point of the camera were fixed for the whole sequence. Optical flow was obtained using Lucas and Kanade's method (Lucas, 1984), with a temporal window of 15 frames. Depth was recovered for frame 9. We assumed all the intrinsic parameters of the camera were estimated accurately throughout the experiments and concentrated on the impact of errors in the extrinsic parameters.

The 3-D scene for SOFA1 consisted of a cube resting on a cylinder (Figure 20(a)). Clearly the entire feature points on the top face of the cylinder in SOFA1 (which are delineated in Figure 20(a)) lie on a plane. The reconstructed shape of this plane was used to testify our theoretical predictions in the lateral motion case. The camera trajectory for

SOFA1 was a circular route on a plane perpendicular to the world Y-axis, with constant translational parameters $(U,V,W) = (0.8137, 0.5812, 0)$ and constant rotational parameters $(\alpha, \beta, \gamma) = (-0.0203, 0.0284, 0)$. For the reasons we discussed in Section 2, we assume that current SFM algorithms can estimate W accurately (i.e. $\hat{W} = 0$). Given the 3-D motion estimation, depth is estimated as:

$$\hat{Z} = \frac{-\hat{f}(\hat{U}, \hat{V}) \cdot \mathbf{n}}{(u,v) \cdot \mathbf{n} - (\hat{u_{rot}}, \hat{v_{rot}}) \cdot \mathbf{n}} \tag{44}$$

Different SFM algorithms may give different errors in the estimated 3-D motion parameters. Instead of being constraint by one specific SFM algorithm, let us consider all the possible configurations in the 3-D motion estimation errors [5]. The resultant recovered depths are illustrated in Figure 20 using 3-D plot viewed from the side. In particular, Figure 20(b) shows that, with motion parameters accurately estimated, the plane (top of the cylinder) remained as a plane, although the noise in the optical flow estimates made some points 'run away' from the plane. Figure 20(c) depicts the shape recovered when $\beta_e > 0$ and $\hat{U} < 0$. According to Equation (41), we have $\hat{Z}_{\hat{X}\hat{X}} > 0$. It can be seen from Figure 20(c) that the plane was indeed reconstructed as a convex surface. Conversely, when $\beta_e < 0$ and $\hat{U} < 0$, concave surfaces were perceived (Figures 20(d), (e) and (f)). Comparing Figure 20(d) with Figure 20(e), we find that larger $\hat{U}$ in general resulted in smaller curvature distortion, whereas the results of Figure 20(d) with Figure 20(f) show that large β_e resulted in larger curvature distortion. Figures 20(c)–(f) show that the reconstructed surfaces were not curved in the Y direction. All these results confirm to our prediction made in Section 2.

The SOFA5 sequence was used in the next set of experiments to verify predictions in the case of forward motion. The 3-D scene for SOFA5 comprised of a pile of 4 cylinders stacking upon each other and in front of a frontal-parallel background (Figure 21(a)). The camera trajectory for SOFA5 was parallel to the world Z-axis and the corresponding translational and rotational parameters were $(U,V,W) = (0,0,1)$ and $(\alpha, \beta, \gamma) = (0,0,0)$ respectively. Current SFM algorithms have no difficulty in estimating the translation accurately for the SOFA5 sequence (Xiang, 2003). Therefore, the equation for reconstructing depth for each feature point would be:

$$\hat{Z} = \frac{x^2 + y^2}{(u,v) \cdot (x,y) - (\hat{u_{rot}}, \hat{v_{rot}}) \cdot (x,y)}$$

The resultant recovered depths, given different typical errors in the 3-D motion estimates, are shown in Figure 21. Figure 21(b) depicts the case of no errors in the motion parameters. It can be seen that the background plane was preserved roughly. However, when there was small amount of errors in the rotation estimates, a complicated curved surface with significant distortion was reconstructed (shown in Figure 21(c)). It is in accor-

[5] For our experiments, if the subspace algorithm proposed by Heeger and Jepson (Heeger and Jepson, 1992) was used, we obtained the reconstruction shown in Figure 20(c); if the algorithm proposed by Ma et. al. (Ma et al., 2000) was adopted, we obtained the reconstruction shown in Figure 20(d).

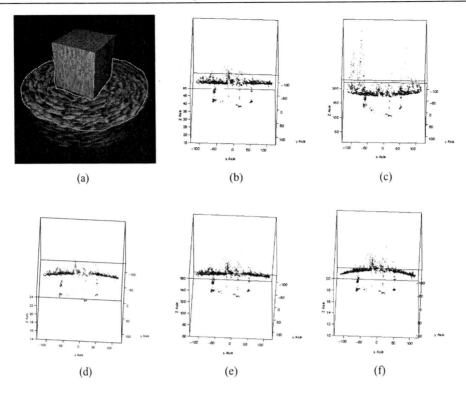

Figure 20: Computer generated lateral motion sequence and shapes recovered when different 3-D motion estimates were used for reconstruction. (a) SOFA1 frame 9 with the top face of the cylinder delineated; (b) Reconstruction with true motion parameters. (c) Reconstruction with $(\hat{U},\hat{V}) = (-1,0)$ and $(\hat{\alpha},\hat{\beta},\hat{\gamma}) = (-0.03,0.01,0)$. (d) Reconstruction with $(\hat{U},\hat{V}) = (-1,0)$ and $(\hat{\alpha},\hat{\beta},\hat{\gamma}) = (-0.01,0.04,0)$. (e) Reconstruction with $(\hat{U},\hat{V} = (-2,0)$ and $(\hat{\alpha},\hat{\beta},\hat{\gamma}) = (-0.01,0.04,0)$. (f) Reconstruction with $(\hat{U},\hat{V} = (-1,0)$ and $(\hat{\alpha},\hat{\beta},\hat{\gamma}) = (-0.01,0.06,0)$.

dance with our prediction that large depth distortion is expected when forward motion is performed.

5.3.2 A Real Image Sequence

Shape recovery was also performed on a real image sequence BASKET[6]. It was taken by a stationary video camera on an upturned basket being rotated on an office chair (Figure 22(a)). The true intrinsic and extrinsic parameters are not available. However, we know that the basket was rotating about a vertical axis which was approximately parallel to the Y-axis of the camera co-ordinate system and located along the Z-axis of the camera co-ordinate system. The equivalent egomotion can thus be expressed as $(U,V,W) = (\beta_0 Z_0,0,0)$ and $(\alpha,\beta,\gamma) = (0,-\beta_0,0)$ where β_0 was the rotational velocity of the basket ($\beta_0 > 0$ in this

---

[6]courtesy of Dr Andrew Calway of Department of Computer Science, University of Bristol.

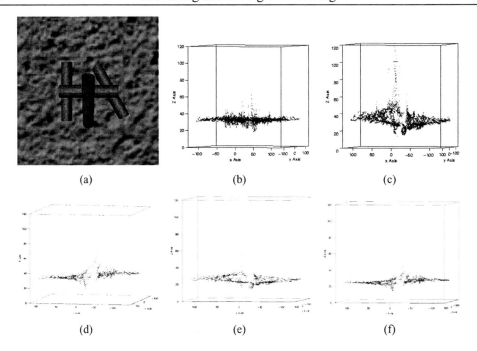

(a)   (b)   (c)

(d)   (e)   (f)

Figure 21: Computer generated forward motion sequence and shape recovered when different 3-D motion estimates were used for reconstruction. (a) SOFA5 frame 9. (b) Reconstruction with true motion parameters. (c) Reconstruction with $(\hat{\alpha}, \hat{\beta}, \hat{\gamma}) = (-0.001, -0.001, 0)$. (d) Reconstruction with $(\hat{\alpha}, \hat{\beta}, \hat{\gamma}) = (0.001, 0.001, 0)$. (e) Reconstruction with $(\hat{\alpha}, \hat{\beta}, \hat{\gamma}) = (-0.001, 0.001, 0)$. (f) Reconstruction with $(\hat{\alpha}, \hat{\beta}, \hat{\gamma}) = (0.001, -0.001, 0)$.

case)and Z_0 the distance between the optical center of the camera and the centroid of the basket. Depth was recovered for frame 9 using Equation (44). Notice that since the camera was not calibrated, we have arbitrarily set the value of the focal length as $\hat{f} = 300$. Fortunately, we have proven in (Cheong and Xiang, 2001) that under lateral motion the errors in the intrinsic parameters would not affect the depth recovery results qualitatively. These recovered shapes, given different errors in the 3-D motion estimates, were shown in Figure 22. Figure 22(b) depicts the shape recovered when $\hat{U} < 0$ and $\beta_e < 0$ ($\beta_e < 0$ since $\beta < 0$ and $\hat{\beta} > 0$). Substituting the signs of these terms into Equation (41), we obtained $\hat{Z}_{\hat{X}\hat{X}} < Z_{XX}$, which means that the recovered surface should be less convex than the true shape. Comparing the results of Figure 22(b) with the true shape measured by us manually, this is indeed the case. Furthermore, according to Equation (41), a smaller β_e or a larger $\hat{U}$ (with the signs of these terms unchanged) would result in an even less convex shape being recovered. Figures 22(c) and (d) clearly support our prediction.

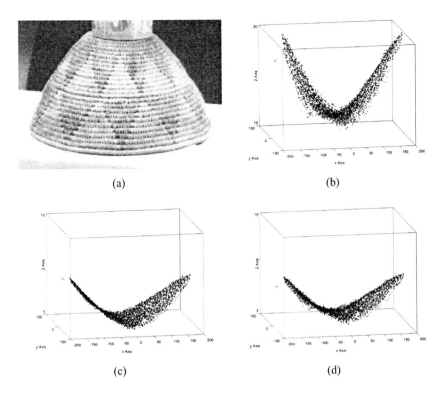

Figure 22: Real lateral motion sequence and shapes recovered when different 3-D motion estimates were used for reconstruction. (a) BASKET frame 9. (b) Reconstruction with $(\hat{U}, \hat{V}) = (-1, 0)$ and $(\hat{\alpha}, \hat{\beta}, \hat{\gamma}) = (0.05, 0.05, 0)$. (c) Reconstruction with $(\hat{U}, \hat{V}) = (-1, 0)$ and $(\hat{\alpha}, \hat{\beta}, \hat{\gamma}) = (0.05, 0.1, 0)$. (d) Reconstruction with $(\hat{U}, \hat{V}) = (-0.5, 0)$ and $(\hat{\alpha}, \hat{\beta}, \hat{\gamma}) = (0.05, 0.05, 0)$.

6 Conclusions and Future Directions

Understanding the inherent motion ambiguities is critical for addressing the SFM problem. To this end, we have developed a geometrically motivated motion error analysis method which is capable of depicting the topological structures of the various optimization cost functions. The motion error configurations likely to cause ambiguities were made clear, under noiseless and noisy conditions and under different motion types. Other conditions that may affect the location of the ambiguities such as feature distribution and density were also considered. This in turn was followed by an investigation on the reliability of shape recovery from motion cue given errors in the estimates for motion parameters. Experiments on both synthetic and real image sequences were carried out. Results obtained on real image sequences seemed to confirm the anisotropic noise model. Results also showed that factors such as sparseness of features, coupled with anisotropic noise distribution, may have great impact on the residual error distribution. Our findings suggest that second order shapes are recovered with varying degrees of uncertainty depending on the motion-scene configuration, given the inherent motion ambiguities.

This work represents part of the ongoing study towards fully understanding the behavior of SFM algorithms. More work needs to be done to extend our understanding in areas such as uncalibrated motion ambiguities. Though we focus on the calibrated case throughout the study, our approach can be readily adopted to analyzing the behavior of SFM algorithms under uncalibrated case. More details and some preliminary results can be found in (Xiang, 2001) and (Cheong and Xiang, 2001). Some partial analyses with a view towards such understanding has been carried out in (Cheong and Peh, 2004) using the depth-is-positive constraint. Similarly, the preceding depth distortion analysis should lead to the more important question of what can be done with such distorted depth. Finally, by understanding how the topology changes, it opens up the possibility of a more robust algorithm. The important conclusion of this work is that the SFM algorithms assume different behaviors under different motion-scene configurations, corroborating the view that current SFM algorithms can perform well only in restricted domains (Oliensis, 2000a). It follows that if we can characterize and identify the different behaviors and domains, it then becomes possible to propose better algorithms or to fuse the results of several existing algorithms.

References

Adiv, G. 1985. Determining 3-D motion and structure from optical flow generated by several moving objects. *IEEE Trans. PAMI*, 7:384–401.

Adiv, G. 1989. Inherent ambiguities in recovering 3-D motion and structure from a noisy flow field. *IEEE Trans. PAMI*, 11:477–489.

Brooks, M.J., Chojnacki, W. and Baumela, L. 1997. Determining the egomotion of an uncalibrated camera from instantaneous optical flow. *Journal of the Optical Society of America A*, 14(10):2670–2677.

Brooks, M.J., Chojnacki, W., Hengel, A.V.D. and Baumela, L. 1998. Robust Techniques for the Estimation of Structure from Motion in the Uncalibrated Case. In *Proc. Conf. ECCV*, pp. 283–295.

Cheong, L-F., Fermüller, C. and Aloimonos, Y. 1998. Effects of errors in the viewing geometry on shape estimation. *Computer Vision and Image Understanding*, 71(3):356–372.

Cheong, L-F. and Ng, K. 1999. Geometry of Distorted Visual Space and Cremona Transformation. *International Journal of Computer Vision*, 32(2):195–212.

Cheong, L-F. and Peh, C.H. 2004. Depth distortion under calibration uncertainty. *Computer Vision and Image Understanding*, 93(3):221–244.

Cheong, L-F. and Xiang, T. 2001. Characterizing Depth Distortion under Different Generic Motions. *International Journal of Computer Vision*, 44(3):199–217.

Chiuso, A., Brockett, R. and Soatto, S. 2000. Optimal Structure From Motion: Local Ambiguities and Global Estimates. *International Journal of Computer Vision*, **39**(3):195–228.

Cutting J.E. and Vishton P.M. 1995. Perceiving layout and knowing distances: The integration, relative potency, and contextual use of different information about depth. In W. Epstein and S. Rogers (Eds.), Perception of Space and Motion. San Diego: Academic Press.

Daniilidis, K. and Spetsakis, M.E. 1997. Understanding Noise Sensitivity in Structure from Motion. In *Visual Navigation: From Biological Systems to Unmanned Ground Vehicles*, Y. Aloimonos (Ed.), Lawrence Erlbaum Assoc. Pub..

Davies A. and Samuels P. 1996. An introduction to computational Geometry for curves and surfaces, Clarendon Press.

Dutta, R. and Snyder, M.A. 1990. Robustness of correspondence-based structure from motion. In *Proc. Int. Conf. on Computer Vision*, Osaka, Japan, PP. 106–110.

Fermüller, C., Shulman, D. and Aloimonos, Y. 2001. The statistics of Optical Flow. *Computer Vision and Image Understanding*, **82**:1–32.

Fermüller, C. and Aloimonos, Y. 2000. Observability of 3-D Motion. *International Journal of Computer Vision* **37**(1):43–63.

Grossmann, E. and Victor, J.S. 2000. Uncertainty analysis of 3-D reconstruction from uncalibrated views. *Image and Vision Computing*, **18**(9): 686–696.

Hoffman, D.D. 1998. Visual Intelligence, W.W. Norton and Company, INC.

Heeger, D.J. and Jepson, A.D. 1992. Subspace methods for recovering rigid motion I: Algorithm and implementation. *International Journal of Computer Vision* **7**:95–117.

Horn, B.K.P. 1987. Motion fields are hardly ever ambiguous. *International Journal of Computer Vision*, **1**:259–274.

Horn, B.K.P. 1990. Relative orientation. *International Journal of Computer Vision* **4**:59–78.

Kahl, F. 1995. Critical motions and ambiguous Euclidean reconstructions in auto-calibration. In *Proc. Int. Conf. on Computer Vision*, pp. 469–475.

Kanatani, K. 1993. 3-D Interpretation of Optical Flow by Renormalization. *International Journal of Computer Vision* **11**(3):267–282.

Koenderink, J.J. and van Doorn, A.J. 1991. Affine Structure from Motion. *J. Optic. Soc. Am.*, **8**(2):377–385.

Koenderink, J.J. and van Doorn, A.J. 1992. Surface shape and curvature scales. *Image and Vision Computing*, **10**(8):557–564.

Koenderink, J.J. and van Doorn, A.J. 1995. Relief: Pictorial and Otherwise. *Image and Vision Computing,* **13**(5):321–334.

Longuet-Higgins, H.C. 1981. A computer algorithm for reconstruction of a scene from two projections. *Nature,* **293**:133-135.

Longuet-Higgins, H.C. 1984. The visual ambiguity of a moving plane. *Proc. R. Soc. Lond.,* **B233**:165-175.

Lucas, B.D. 1984. Generalized Image Matching by the Method of Differences. PhD Dissertation, Carnegie-Mellon University.

Ma, Y., Košecká, J. and Sastry, S. 2000. Linear Differential Algorithm for Motion Recovery: A Geometric Approach. *International Journal of Computer Vision* **36**(1):71–89.

Ma, Y., Košecká, J. and Sastry, S. 2001. Optimization Criteria and Geometric Algorithms for Motion and Structure Estimation. *International Journal of Computer Vision,* **44**(3):219–249.

Maybank, S.J. 1993. Theory of Reconstruction from Image Motion. Springer, Berlin.

Negahdaripour, S. Critical surface pairs and triplets. *International Journal of Computer Vision,* **3**:293–312.

Oliensis, J. 2000. A critique of structure-from-motion algorithms. *Computer Vision and Image Understanding,* **80**:172–214.

Oliensis, J. 2000. A New Structure From Motion Ambiguity. *IEEE Trans. PAMI ,* **22**(7):685–700.

Oliensis, J. 2004. The Least-Squares Error for Structure from Infinitesimal Motion. *ECCV,* 531–545.

Prazdny, K. 1980. Egomotion and relative depth map from optical flow. *Biological Cybernetics,* **36**:87-102.

Rieger, J.H. and Lawton, D.T. 1985. Processing differential image motion. *J. Optic. Soc. Am. A,* **2**:354–359.

Soatto, S. and Brockett, R. 1998. Optimal structure from motion: local ambiguities and global estimates. In *Proc. IEEE Conf. on Computer Vision and Pattern Recognition,* pp. 282–288.

Spetsakis, M.E. 1994. Models of Statistical Visual Motion Estimation. *CVGIP* **60**:300–312.

Szeliski, R. and Kang, S.B. 1997. Shape ambiguities in structure from motion. *IEEE Trans. PAMI* **19**(5):506–512.

Weng, J., Huang, T.S and Ahuja, N. 1991. *Motion and Structure from Image Sequences,* Springer-Verlag.

Xiang, T. 2001. *Understanding the Behavior of Structure From Motion Algorithms: A Geometric Approach.* Ph. D. dissertation, National University of Singapore.

Young, G.S. and Chellapa, R. 1992. Statistical analysis of inherent ambiguities in recovering 3-D motion from a noisy flow field. *IEEE Trans. PAMI* , **14**:995–1013.

Zhang, Z. 1998. Understanding the Relationship Between the Optimization Criteria in Two-View Motion Analysis. In *Proc. Conf. ICCV*, pp. 772–777.

Zhang, T. and Tomasi, C. 1999. Fast, Robust, and Consistent Camera Motion Estimation. In *Proc. IEEE Conf. on Computer Vision and Pattern Recognition*, pp. 164–170.

Xiang, T. and Cheong, L-F. 2003. Understanding the behavior of structure from motion algorithms: a geometric approach. *International Journal of Computer Vision*, **51**(2): 111-137.

In: Computer Vision and Robotics
Editor: John X. Liu, pp. 245-296

ISBN 1-59454-357-7
© 2006 Nova Science Publishers, Inc.

Chapter 7

ALGEBRAIC CURVES IN STRUCTURE FROM MOTION

Jeremy Yirmeyahu Kaminski and Mina Teicher[*†]
Bar-Ilan University, Department of Mathematics and Statistics,
Ramat-Gan, Israel.

Abstract

We introduce multiple-view geometry for algebraic curves, with applications in both static and dynamic scenes. More precisely, we show when and how the epipolar geometry can be recovered from algebraic curves. For that purpose, we introduce a generalization of Kruppa's equations, which express the epipolar constraint for algebraic curves. For planar curves, we show that the homography through the plane of the curve in space can be computed. We investigate the question of three-dimensional reconstruction of an algebraic curve from two or more views. In the case of two views, we show that for a generic situation, there are two solutions for the reconstruction, which allows extracting the right solution, provided the degree of the curve is greater or equal to 3. When more than two views are available, we show that the reconstruction can be done by linear computations, using either the dual curve or the variety of intersecting lines. In both cases, no curve fitting is necessary in the image space.

For dynamic scenes, we address the question of recovering the trajectory of a moving point, also called trajectory triangulation, from moving, non-synchronized cameras. Two cases are considered. First we address the case where the moving point itself is tracked in the images. Secondly, we focus on the case where the tangents to the motion are detected in the images. Both cases yield linear computations, using the dual curve or the variety of intersecting lines.

Eventually, we present several experiments on both synthetic and real data, which demonstrate that our results can be used in practical situations.

[*]This work is partially supported partially supported by EU-network HPRN-CT-2009-00099(EAGER), the Emmy Noether Research Institute for Mathematics and the Minerva Foundation of Germany, the Israel Science Foundation grant 8008/02-3 (Excellency Center "Group Theoretic Methods in the Study of Algebraic Varieties").

[†]E-mail address: {kaminsj,teicher}@macs.biu.ac.il

1 Introduction

Visual servoing is a central engineering issue. In that context, geometric computer vision is an important device, that usually focuses on points and lines in static environments or linear motions. A panorama of the past decade work, in that framework, can be found in [14, 25] and references to earlier work in [13].

In contrast, we introduce multiple-view geometry for algebraic curves, that leads to applications in both static and dynamic scenes. For clarity, we shall separate these two cases.

1.1 Static Scenes

We first consider static configurations. When considering curves in static scenes, the litterature, until recently, was rather sparse, especially for non-planar algebraic curves. Given known projection matrices [44, 39, 40] show how to recover the 3D position of a conic section from two and three views, and [46] show how to recover the homography matrix of the conic plane, and [10, 50] show how to recover a quadric surface from projections of its occluding conics.

Reconstruction of higher-order curves were addressed in [31, 30, 5, 42, 43]. In [5] the matching curves are represented parametrically where the goal is to find a re-parameterization of each matching curve such that in the new parameterization the points traced on each curve are matching points. The optimization is over a discrete parameterization, thus, for a planar algebraic curve of degree n, which is represented by $\frac{1}{2}n(n+3)$ points, one would need $n(n+3)$ minimal number of parameters to solve for in a non-linear bundle adjustment machinery — with some prior knowledge of a good initial guess. In [42, 43] the reconstruction is done under infinitesimal motion assumption with the computation of spatio-temporal derivatives that minimize a set of non-linear equations at many different points along the curve. In [30] only planar algebraic curves were considered, whereas in [31], the plane of non-planar algebraic curve 3D reconstruction is addressed.

On the problem of recovering the camera geometry (projection matrices, epipolar geometry, multi-view tensors) from matching projections of algebraic curves, the literature is sparse. [28, 30] show how to recover the fundamental matrix from matching conics with the result that 4 matching conics are minimally necessary for a unique solution. [30] generalize this result to higher order curves, but consider only planar curves and [31] generalizes this result to non-planar curves.

In this paper we address the general issue of multi-view geometry of algebraic curves from both angles: (i) recovering camera geometry (fundamental matrix or homography in the case of planar curve), and (ii) reconstruction of the curve from its projections across two or more views. Short versions of this work were published in [30, 31].

We start with the recovery of camera parameters from matching curves. We show how one can compute, without any knowledge on the camera, the homography induced by a planar curve. Then we introduce the generalized Kruppa equation, which embodies the epipolar constraint of two projections of an algebraic curve. In this context, we establish a necessary and sufficient condition on the degree and genus of the algebraic curves, for the epipolar geometry to be determined up to a finite-fold ambiguity.

We then shift to the reconstruction of algebraic curves from multiple views. We address three representations of curves:

1. The standard representation, by the set of points lying on the curve, in which we show that the reconstruction from two views of a curve of degree d admits two solutions, one of degree d and the other of degree $d(d-1)$.

2. The dual curve representation, made of the tangent planes to the curve, for which we derive the minimal number of views necessary for reconstruction as a function of the curve degree and genus.

3. The representation by the set of lines intersecting the curve in space, for which we also derive the minimal number of views necessary for reconstruction as a function of curve degree alone.

The latter two cases do not require any curve fitting in the image space.

1.2 Dynamic Scenes

Next we turn our attention to dynamic scenes, where both scene points and cameras may move simultaneously. Recently a new body of research has appeared which considers configurations of independently moving points, first with the pioneering work of Avidan and Shashua [3], and then in other contributions [9, 18, 22, 36, 51, 54, 55]. A common assumption of these works is that the motion must occur along a straight line or a conic section. When the motion is linear and at constant velocity, the recovery of the trajectory is done linearly. However for quadratic trajectories, the computations are non linear. Some authors have also considered the case where the motion is captured by tangential measurements within the images [47], but remained in the case of linear or quadratic trajectories.

We present a complete generalization and address the problem of general trajectory triangulation of moving points from non-synchronized cameras [32, 33]. Two cases are considered:

1. The motion is captured in the images by tracking the moving point itself.

2. The tangents of the motion only are extracted from the images.

In both problems, since the cameras are allowed to move and are not synchronized (the pictures from two different cameras are made at different time instants), the conventional "triangulation" based photogrammetry approach will not apply here. Therefore we looked for more sophisticated mathematical tools to solve these problems.

The motion is recovered by triangulation of the trajectory of the moving point, which is regarded as a piecewise algebraic curve. For this purpose, we use the second and third representations, as introduced above.

In both cases these representations of curves allow:

1. The recovery of a more standard representation of the trajectory.

2. The computation of the set of positions of the moving point at each time instant an image was made.

Furthermore, a theorem is given on the number of independent constraints, a camera provides on the motion of the point, as a function of the camera motion.

1.3   Paper Organization

The paper is organized as follows. First in a background section 2, we present the mathematical material necessary for an overall understandings of this work. In section 3, we present of full multiple-view theory for curves in static environments. We start by showing how the epipolar geometry can be recovered from curves. Then we address the issue of three-dimensional reconstruction of curves, from still images. The case of dynamic scenes is the subject of section 4, where we show how trajectory triangulation can be done, from moving non-synchronized cameras, from point-based and tangential measurements. Eventually, we present experments for both static and dynamic scenes, which demonstrate that the theoretical results, developped in the core of the paper, can also be applied in practice, for any situation where structure from motion must be used.

2   Background in Algebraic Geometry

This section provides a brief introduction to algebraic geometry, and algebraic curves in particular, necessary for the overall understanding of our work. For more technical and comprehensive introduction, we refer to [23, 24, 26].

2.1   Algebraic Variety

The central concept to be used in this paper is the definition of an *affine variety*. This is the locus of common zeros of a family of polynomials. More precisely, consider a family of polynomials in n variables. The locus of their common zeros defines an algebraic variety in the n-dimensional affine space. When the polynomials are all homogeneous, this variety can also be regarded as being embedded in the $(n-1)$-dimensional projective space, and is called a *projective variety*. Let us go through a more formal formulation.

2.1.1   Affine Variety

The affine space of dimension n, meaning the set of $n-$tuples of numbers $(x_1, ..., x_n)$, will be denoted by $\mathbb{A}^n$. When dealing with applications, the points have real coordinates and the varieties are defined by equations with real coefficients. However algebraic varieties are well behaved when regarded as complex varieties. So we shall consider all of them as varieties defined by equations with complex coefficients. This allows using all the power of algebraic geometry. For the computations, however, after the result is obtained over the complex numbers, we eventually consider only the real solutions.

Definition 1 *Given a polynomial F in the polynomial ring $\mathbb{C}[X_1,....,X_n]$, we say that a point P in the $n-$dimensional affine space $\mathbb{A}^n$, regarded as a $n-$tuple $(P_1,.....,P_n)$, is a zero of F if $F(P) = 0$.*

Definition 2 *An algebraic variety in the affine space $\mathbb{A}^n$ is a subset of $\mathbb{A}^n$ being the common zeros of a family of polynomial $\{F_i\}$:*

$$Z(\{F_i\}) = \{P \in \mathbb{A}^n | F_i(P) = 0 \text{ for all } i\}$$

It is clear that the variety will not be affected if we add to the family all the combinations: $\sum_i G_i F_i$, where all but a finite number of G_i are zeros. Hence a variety of $\mathbb{A}^n$ is always defined by the ideal ([34]) generated by its defining family. We immediately deduce that:

Proposition 1 *The union of two varieties is a variety. The intersection of any family of varieties is a variety. The empty set and the whole space are varieties.*

Then the varieties are the closed sets of a topology called the *Zariski topology*. This topology is extremely different from the usual topology, based on Euclidean distance. Indeed the closed sets in the Zariski topology are very small, since they are made only of points, which are solutions of a system of polynomial equations. On the contrary the open sets are very large.

Given an ideal I in $\mathbb{C}[X_1,...,X_n]$, we shall denote by $Z(I)$ the variety it defines. Since the polynomial ring is noetherian, we get:

Proposition 2 *A variety is always defined by a finite set of equations, namely the generators of the ideal, which defines the variety.*

As each ideal defines a variety, any subset S of $\mathbb{A}^n$ defines an ideal as follows:

$$I(S) = \{F \in \mathbb{C}[X_1,...,X_n] | \forall P \in S, F(P) = 0\}$$

There is a relationship between ideals of $\mathbb{C}[X_1,...,X_n]$ and varieties of $\mathbb{A}^n$. The following properties can be proven:

Proposition 3   *1. If $S_1 \subset S_2$ are two subsets of $\mathbb{A}^n$, then $I(S_2) \subset I(S_1)$.*

2. If $I_1 \subset I_2$ are two ideals of $\mathbb{C}[X_1,...,X_n]$, then $Z(I_2) \subset Z(I_1)$.

3. For any two subsets S_1 and S_2 of $\mathbb{A}^n$, we have: $I(S_1 \cup S_2) = I(S_1) \cap I(S_2)$.

4. For any ideal $I \subset \mathbb{C}[X_1,...,X_n]$, $I(Z(I)) = \sqrt{I}$, the radical of I, defined by $\sqrt{I} = \{F \in \mathbb{C}[x_1,...,x_n] | \exists r, F^r \in I\}$.

5. *For any subset $S \in \mathbb{A}^n$, $Z(I(S)) = \bar{S}$, the closure of S.*

Note that the fourth point is a direct consequence of the Hilbert's Nullstellensatz:

Theorem 1 *When the ground field is algebraically closed (which is the case for complex varieties), the following holds. Let I be an ideal of $\mathbb{C}[X_1, ..., X_n]$, and let $F \in \mathbb{C}[X_1, ..., X_n]$ which vanishes at all points of $Z(I)$. Then $F^r \in I$ for some integer $r > 0$.*

We end this very short introduction by defining irreducibity and dimension.

Definition 3 *A variety is said to be irreducible, if it cannot be expressed as the union of two non-empty proper sub-varieties.*

Any variety can be decomposed as the union of irreducible subvarieties. Given such a decomposition, we can discard all the irreducible subvarieties which are included in larger irreducible subvarieties. Then the decomposition, obtained by this process, is minimal. The subvarieties appearing in the minimal decomposition, are called the *irreducible components* of the variety.

Definition 4 *The dimension of a variety X if the supremum of all integers k such that there exists a chain $Z_0 \subset Z_1 \subset .... \subset Z_k$ of distinct irreducible sub-varieties of X.*

The definition of the dimension is based on the Zariski topology and is very easy to write. In contrast, defining the dimension by the number of independant parameters necessary to describe the variety, is more challenging. However it can been proven that these approaches are equivalent.

2.1.2 Projective Variety

As we defined an affine variety to be a subset of an affine space $\mathbb{A}^n$, defined by polynomials equations, we shall define a projective variety to be a subset of a projective space $\mathbb{P}^n$ defined by **homogeneous** polynomials.

The properties of projective varieties are very similar to those of affine varieties.

2.2 Relation between Affine and Projective Varieties

As mentioned above, an affine varieties is defined by a set of polynomials equations in n variables. Such a set defines an affine variety in the $n-$dimensional affine space. On the other hand to define projective varieties, we need homogeneous polynomials. Consider a set of homogeneous polynomials in $n+1$ variables. This set defines a projective variety in the $n-$dimensional projective space. However one can also consider this set defines an affine variety in the $(n+1)-$dimensional affine space. This latter variety is called the *affine cone* over the former projective variety, since it is a cone whose section with a generic hyperplane gives a model of the related projective variety.

There exists also another relation, of major importance, between affine and projective varieties. Consider a projective variety, say $V_p \subset \mathbb{P}^n$, defined by homogeneous equations $f_i(X_1,...,X_{n+1}) = 0$. If the hyperplane at infinity in $\mathbb{P}^n$ is defined by $x_{n+1} = 0$, then the variety defined by $f_i(X_1,...,X_n,1) = 0$ is an affine variety, say V_a, included in n−dimensional affine space. V_p is called the *projective closure* of V_a. V_a can be regarded as the affine piece of V_p and the varieties defined by $f_i(X_1,...,X_n,0) = 0$ is made of the points infinity of V_p.

2.3 Algebraic Planar Curves

An algebraic curve is an algebraic variety of dimension 1. Since we shall make an extensive use of algebraic curves in the sequel, we present here material more specifically related to algebraic curves. We shall first focus on planar algebraic curves, meaning curves embedded in a plane. Such a curve is defined by a single polynomial.

2.3.1 Definitions

Definition 5 *A polynomial f is said to be square-free if it cannot be written as a product like: $f = g^2 h$, where g and h are non constant polynomials.*

Definition 6 *A planar algebraic curve C is a subset of points, whose projective coordinates satisfy an homogeneous square-free polynomial equation: $f(x,y,z) = 0$. The degree of f is called the order or degree of C. The curve is said to be irreducible, when the polynomial f cannot be divided by a non-constant polynomial.*

Note that when two polynomials define the same curve, they must be equal up to a scalar. For convenience and shorter formulation, we define *a form $f \in \mathbb{C}[x,y,z]$ of degree n to be an homogeneous polynomial in x,y,z of total degree n.*

As mentionned before, the degree of a planar curve is simply the degree of its defining polynomial. To better understand the geometric meaning of this notion, consider a curve C embedded in the projective plane defined by $f(x,y,z) = 0$. Let L be a line generated by two points $\mathbf{a}$ and $\mathbf{b}$. A point $\mathbf{p} = \mathbf{a} + \lambda\mathbf{b}$ on L is located on the curve if $J(\lambda) = f(\mathbf{a} + \lambda\mathbf{b})$ has a root. Since the degree of J, as function of λ, is the degree of f, the degree of C is the number of points a general line in the plane meets the curve.

2.3.2 Tangency and Singularities

Let (a_x, a_y, a_z) and (b_x, b_y, b_z) be the projective coordinates of $\mathbf{a}$ and $\mathbf{b}$ respectively. The intersection of L and C is made of the points $\{\mathbf{p}_\lambda\}$, such that the parameters λ satisfy the equation:

$$J(\lambda) \equiv f(a_x + \lambda b_x, a_y + \lambda b_y, a_z + \lambda b_z) = 0$$

Taking the first-order term of the a Taylor-Lagrange expansion:

$$
\begin{aligned}
J(\lambda) &= f(\mathbf{a}) + \lambda(\tfrac{\partial f}{\partial x}(\mathbf{a})b_x + \tfrac{\partial f}{\partial y}(\mathbf{a})b_y + \tfrac{\partial f}{\partial z}(\mathbf{a})b_z) \\
&= f(\mathbf{a}) + \lambda \operatorname{grad}_{\mathbf{a}}(f)^T \mathbf{b} \\
&= 0
\end{aligned}
$$

If $f(\mathbf{a}) = 0$, $\mathbf{a}$ is located on the curve. Furthermore let assume that $\operatorname{grad}_{\mathbf{a}}(f)^T \mathbf{b} = 0$, then the line L and the curve C meet at $\mathbf{a}$ in two coincident points. A point is said to be *regular* is $\operatorname{grad}_{\mathbf{a}}(f) \neq \mathbf{0}$. Otherwise it is a *singular* (or *multiple*) point. The set of singular points or singularities is denoted by $sing(C)$. When the point $\mathbf{a}$ is regular, *the line L is said to be tangent to the curve C at $\mathbf{a}$.*

Proposition 4 *The set of singularities $sing(C)$ is finite.*

Proof: A singular point $\mathbf{a}$ is such that: $f(\mathbf{a}) = 0$ and $\operatorname{grad}_{\mathbf{a}}(f) = 0$. Thus it is located as the intersection of four distinct curves. Hence there are at most a finite number of such points. ∎

Definition 7 *Let C be a curve, defined by a polynomial f. Let $\mathbf{p}$ be a point on C. The multiplicity of $\mathbf{p}$ is the smallest m such as there exists $(i, j, k) \in \mathbb{N}^3$ with $i + j + k = m$ and:*

$$
\frac{\partial^m f}{\partial^i x \partial^j y \partial^k z}(\mathbf{p}) \neq 0.
$$

The multiplicity of $\mathbf{p}$ is denoted by $m(\mathbf{p}, C)$ or simply $m(\mathbf{p})$ if there is no ambiguity.

Note that $m(\mathbf{p}, C) = 1$ if $\mathbf{p}$ is a regular point of C. Otherwise $m(\mathbf{p}, C) > 1$. If $m(\mathbf{p}, C) = 2$, the point is called a *double point*, if $m(\mathbf{p}, C) = 3$, a *triple point*, etc. If the point $\mathbf{p}$ is translated to the origin, then the affine part of the curve (obtained by pluging $z = 1$ in f) is given by the following polynomial:

$$
f_{affine} = f_m + f_{m+1} + \ldots + f_d,
$$

where f_i is a form of degree i and m the multiplicity of $\mathbf{p}$. Since f_m is a form in two variables, we can write it as a product of linear factors $f_m = \prod l_i^{r_i}$, where l_i are distinct lines. The lines l_i are tangent to the curve at the singular point $\mathbf{p}$. A singular point is said to be *ordinary* if all its tangents are distinct. A double point is a *node* if it is ordinary and a *cups* otherwise.

Definition 8 *Given a planar algebraic curve C, the dual curve is defined in the dual plane, as the closure of the set of all lines tangent to C at simple points. The dual curve is algebraic and thus can be described as the set of lines (u, v, w), that are the zeros of a form $\phi(u, v, w) = 0$.*

Proposition 5 *Let C be a curve of degree d, which only singularities are nodes. The degree of the dual curve D is:*

$$
m = d(d - 1) - 2(\#nodes),
$$

where $(\#nodes)$ is the number of nodes.

2.3.3 Inflexions Points and Hessian Curves

We will also need to consider the notion of inflexion point:

Definition 9 *An inflexion point **a** of a curve C is a simple point of it whose tangent intersects the curve in at least three coincident points. This means that the third order term of the Taylor-Lagrange development of $J(\lambda)$ must vanish too.*

It will be useful to compute the inflexion points. For this purpose we define the Hessian curve $H(C)$ of C, which is given by the determinantal equation:

$$\mid \frac{\partial^2 f}{\partial x_i \partial x_j} \mid = 0$$

It can be proven (see [48]) that the points where a curve C meets its Hessian curve $H(C)$ are exactly the inflexion points and the singular points. Since the degree of $H(C)$ is $3(d-2)$, there are $3d(d-2)$ inflexion and singular points counting with the corresponding intersection multiplicities (Bezout's theorem, see [48]).

2.3.4 Genus and Rational Curves

The genus of the algebraic curve can be defined in numerous manners. Some definitions are topological, some are analytic, and some are algebraic. For further details, the reader should consult [45]. Here it is sufficient to provide a partial definition of it.

Definition 10 *For a planar algebraic curve, which degree is d and which only singularities are nodes, the genus is defined as being the following number:*

$$g = \frac{(d-1)(d-2)}{2} - (\#nodes),$$

where #nodes is the number of nodes.

The genus is zero if and only if the curve is globally given by a parametric representation by homogeneous polynomials of same degree in the projective case, or by rational functions in the affine case. Such a curve is called *rational*.

2.4 Algebraic Spatial Curves

An algebraic spatial curve is defined as being the intersection of two or more algebraic surfaces. In a more formal way, it is defined by a set of homogeneous equations:

$$\mathbf{F}_i(X, Y, Z, T) = 0$$

In the body of this article, this representation is referred as the point-based representation. A point $\mathbf{P} = [X, Y, Z, T]^T$ on the curve is said to be singular if: $\text{grad}_\mathbf{P}(F_i) = 0$ for all i.

2.4.1 Dual Curve

Let X be an irreducible curve in $\mathbb{P}^3$, defined by the following family of polynomials: $\{F_i\}_i$. As in the case of planar curves, there is a natural concept of duality. The *dual curve* is the set of planes tangent to the curve at a simple point. It turns out that the dual curve is also an algebraic variety of the dual three-dimension projective space, $\mathbb{P}^{3*}$ (Figure 1). Moreover in general the dual curve is simply a surface of $\mathbb{P}^{3*}$, meaning a variety of dimension 2 (see [23]). The relation between a curve and its dual curve is bijective in the sense that a curve is completely determined by its dual curve. Therefore a spatial algebraic curve can be represented either as the solution of a family of equations or by its dual curve.

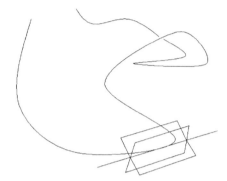

Figure 1: The dual curve is the set of planes tangent to the curve at simple points.

Given a curve $X \subset \mathbb{P}^3$, let $\Upsilon \in \mathbb{C}[A,B,C,D]$ be the polynomial defining the dual curve $X^* \subset \mathbb{P}^{3*}$. The computation of Υ from X is done by the following elimination problem:

Eliminate X,Y,Z,λ_i from the following system:

$$
\begin{aligned}
F_i(X,Y,Z,1) &= 0 \\
\begin{bmatrix} A \\ B \\ C \end{bmatrix} - \sum \lambda_i \mathrm{grad}_{\mathbf{P}}(F_i(X,Y,Z,1)) &= 0 \\
AX + BY + CZ + D &= 0,
\end{aligned}
$$

where $\mathbf{P} = [X,Y,Z]^T$. This system simply expresses the fact that the tangent plane to the curve at a point $[X,Y,Z,1]$ must be a linear combination of the gradients (that is the normals) of the surfaces defining the curve. The practical computation is done by an elimination engine, say Gröbner basis engine for instance [7, 16, 17].

The conversion from the dual curve to the original curve is done by a similar elimination problem since the duality is an involution, that is the dual curve of the dual curve is the curve itself. Hence if $\Upsilon \in \mathbb{C}[A,B,C,D]$ is the polynomial defining the dual curve $X^* \subset \mathbb{P}^{3*}$, the original curve X is recovered as follows:

Eliminate A,B,C,λ from the following system:

$$\Upsilon(A,B,C,1) = 0$$

$$\begin{bmatrix} X \\ Y \\ Z \end{bmatrix} - \lambda \text{grad}_{\blacksquare}(\Upsilon(A,B,C,1)) = 0$$

$$AX + BY + CZ + T = 0,$$

where $\blacksquare = [A,B,C]^T$.

2.4.2   Curve Representation in $\mathbb{G}(1,3)$

There exists a third and very useful representation of spatial algebraic curves.

It is well known that a line in $\mathbb{P}^3$ can be represented by its Plücker coordinates as point of $\mathbb{P}^5$ lying on special quadric, called the Grassmannian of lines of $\mathbb{P}^3$ and denoted by $\mathbb{G}(1,3)$ [4, 14, 23]. Therefore we shall denote by $\mathbf{L}$ a line in $\mathbb{P}^3$ and by $\widehat{\mathbf{L}}$ its Plücker coordinates which makes it a point of $\mathbb{P}^5$. We proceed to show that a curve in $\mathbb{P}^3$ can be represented as a subvariety of $\mathbb{G}(1,3)$ which leads to very useful applications.

A smooth irreducible curve X which degree is d and embedded in $\mathbb{P}^3$ is entirely determined by the set of lines meeting it [23] (Figure 2).

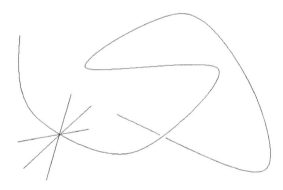

Figure 2: The set of lines intersecting a spatial curve completely determines the curve itself.

We define the following set of lines:

$$\Lambda = \{\mathbf{L} \subset \mathbb{P}^3 | \mathbf{L} \cap X \neq \emptyset\}$$

The following facts are well known [23]:

1. Λ is an irreducible subvariety of $\mathbb{G}(1,3)$.

2. There exists a homogeneous polynomial Γ, which degree is d, such that:

$$Z(\Gamma) \cap \mathbb{G}(1,3) = \Lambda,$$

where $Z(\Gamma) = \{\widehat{\mathbf{L}} \in \mathbb{P}^5 | \Gamma(\widehat{\mathbf{L}}) = 0\}$.

3. Γ is defined modulo the $d-$th graded piece of the ideal defining $\mathbb{G}(1,3)$, that is modulo $I(\mathbb{G}(1,3))_d$.

4. The dimension of the $d-$th graded piece of the homogeneous coordinate ring of $\mathbb{G}(1,3)$, that is of $S(\mathbb{G}(1,3))_d$ is for $d \geq 2$:

$$N_d = \binom{d+5}{d} - \binom{d-2+5}{d-2}.$$

5. It is sufficient to pick N_d generic points on Λ to find Γ modulo $I(\mathbb{G}(1,3))_d$. Each such point $\widehat{\mathbf{L}}$ yields one linear equation on Γ:

$$\Gamma(\widehat{\mathbf{L}}) = 0.$$

f we put this high-flown terminology down to earth, it says this. The subvariety Λ is the intersection of $\mathbb{G}(1,3)$ with some hypersurface defined by a polynomial Γ. There exists a whole family of hypersurfaces that intersect in $\mathbb{G}(1,3)$ exactly over Λ. This means that the polynomial Γ is not uniquely defined and is computable only modulo the quadratic equation defining $\mathbb{G}(1,3)$. That is, two polynomials Γ_1 and Γ_2 such that $\Lambda = Z(\Gamma_1) \cap \mathbb{G}(1,3) = Z(\Gamma_2) \cap \mathbb{G}(1,3)$ satisfy an equation as follows:

$$\Gamma_2(L_1, ..., L_6) = \Gamma_1(L_1, ..., L_6) + R(L_1, ..., L_6)(L_1L_6 - L_2L_5 + L_3L_4),$$

for some polynomial R. Therefore the space of all possible Γ is the quotient of the space of all homogeneous polynomials of degree d in 6 variables, denoted S_5, by $I \cap S_5$, where I is the ideal generated by the polynomial defining $\mathbb{G}(1,3)$. Therefore its dimension is $N_d = \binom{d+5}{d} - \binom{d-2+5}{d-2}$. Eventually, in order to recover the original curve X in $\mathbb{P}^3$ it is sufficient to compute one of the polynomials Γ. This justifies the following definition.

Definition 11 *Any element of the equivalence class of Γ is said to be the Chow polynomial of the curve X.*

Moreover, it is sufficient to find N_d linearly independent equations to find the Chow polynomial of X.

The previous properties provides us with a way to compute Γ from a set of discrete measurements extracted from image sequences. However we proceed to show, how one can recover Γ from the usual point based representation of the curve. Let $\{F_i\}$ the family of polynomials defining the curve. The computation of Γ is done as follows:

Eliminate X,Y,Z from the following system:

$$F_i(X,Y,Z,1) \; = \; 0$$

$$\widehat{\mathbf{L}} \vee \begin{bmatrix} X \\ Y \\ Z \\ 1 \end{bmatrix} \; = \; 0,$$

where $\widehat{\mathbf{L}} = [L_1, L_2, L_3, L_4, L_5, L_6] \in \mathbb{P}^5$ represents a line meeting the curve and $\vee$ is the join operator (see [4, 14]). The join of $\widehat{\mathbf{L}}$ and the point $\mathbf{P} = [X,Y,Z,1]^T$ vanishes to express the fact that $\mathbf{P}$ lies on the line represented by $\widehat{\mathbf{L}}$.

When given Γ as a result of the previous elimination, we shall compute its normal form (to get a canonical representation of Γ) with respect to the equation defining $\mathbb{G}(1,3)$:

$$L_1 L_6 - L_2 L_5 + L_3 L_4 = 0$$

Given the Chow polynomial, it is easy to obtain the point-based representation. Let Γ be the Chow polynomial of the curve. Follow the following procedure:

1. Pick three generic points, Q_1, Q_2, Q_3 on the plane at infinity (last coordinate zero).

2. Consider a point $P = [X,Y,Z,1]^T$ in the affine piece of $\mathbb{P}^3$. The point P is lying on X if any linear combination of the three lines $(PQ_i)_{i=1,2,3}$ is a zero of the Chow polynomial. This yields $\binom{d+2}{d}$ equations defining completely the curve X.

This simple algorithm makes use of the fact that a point $\mathbf{P}$ of $\mathbb{P}^3$ is represented as a plane included in $\mathbb{G}(1,3)$ generated by three lines passing through $\mathbf{P}$. Hence a point is lying on the curve if the plane that represents it in $\mathbb{G}(1,3)$ is included into the hypersurface defined by the Chow polynomial of the curve.

In many applications, the sequences of locations of a moving point at each time instant an image was made is much more useful than the standard representation of the trajectory. Therefore we show how the Chow polynomial can be directly used for computing the intersection of the trajectory with a line. This line will be an optical ray in practice.

As before we shall denote the Chow polynomial by Γ. Each 2D point $\mathbf{p}$ generates an optical ray $\mathbf{L_p}$ which is a zero of Γ. $\mathbf{L_p}$ is given by $\widehat{\mathbf{M}}\mathbf{p}$ where $\widehat{\mathbf{M}}$ is a 6×3 matrix, which is a polynomial function of the camera matrix $\mathbf{M}$. $\widehat{\mathbf{M}}$ maps an image point to the optical ray it generates with the optical center [14]. In order to compute where this optical ray meets the curve, we use the following procedure:

1. Pick three random and arbitrary points $\mathbf{R}_1, \mathbf{R}_2, \mathbf{R}_3$ in $\mathbb{P}^3$.

2. Give a parametric representation of $\mathbf{L_p}$: $\mathbf{P}(\lambda) = \mathbf{A} + \lambda \mathbf{B}$ ($\mathbf{A}$ and $\mathbf{B}$ are two distinct points of the line, arbitrary chosen).

3. Build the following parametric lines $\mathbf{L}_i(\lambda) = \mathbf{P}(\lambda) \vee \mathbf{R}_i$, for $i = 1,2,3$.

4. For every triplet α, β, γ, we have $\Gamma(\alpha \mathbf{L}_1(\lambda) + \beta \mathbf{L}_2(\lambda) + \gamma \mathbf{L}_3(\lambda)) = 0$, for $\mathbf{P}(\lambda)$ to be located on the curve. This yields $\frac{1}{2}(d+1)(d+2)$ equations on λ.

5. Find the solution of the previous system. Let λ_0 be that solution. Note that for solving the previous system it is enough to solve the first uni-variate polynomial equation and then to check which root is common to all the equations.

6. The point $\mathbf{P}(\lambda_0) = \mathbf{A} + \lambda_0 \mathbf{B}$ is the 3D point that we are looking for.

This algorithm, as the previous one, is also based on the fact that to a point in $\mathbb{P}^3$ corresponds a unique plane in $\mathbb{G}(1,3)$. This plane is spanned by three lines passing through the point in $\mathbb{P}^3$. Therefore the points $\mathbf{R}, i = 1, 2, 3$ are used to generate these three lines.

Note that in this algorithm in some very special cases, there might be more than one solution to the system in λ. This will be the case if the optical ray meets the curve at more than one point. This happens if an image was made when the moving point passed through either a point of the trajectory that occludes another point or a point of tangency of the curve with a line passing through the camera center. Those two cases are singular so they appear with probability zero. However in case of a singular point, the ambiguity can solved by looking at neighboring points.

3   Static Configurations

Recovering epipolar geometry from curve correspondences requires the establishment of an algebraic relation between the two image curves, involving the fundamental matrix. Hence such an algebraic relation may be regarded as an extension of Kruppa's equations. In their original form, these equations have been introduced to compute the camera-intrinsic parameters from the projection of the absolute conic onto the two image planes [38]. However it is obvious that they still hold if one replaces the absolute conic by any conic that lies on a plane that does not meet any of the camera centers. In this form they can be used to recover the epipolar geometry from conic correspondences [28, 30]. Furthermore it is possible to extend them to any planar algebraic curve [30]. Moreover a generalization for arbitrary algebraic spatial curves is possible and is a step toward the recovery of epipolar geometry from matching curves [31].

We start by the case of planar curves which is much more simple than the general case.

We shall use the following notations in this section. X will denote a spatial curve, either planar or not, whereas the image curves will be denoted by $Y_i, i = 1, 2$ and defined by polynomials $f_i, i = 1, 2$. The dual image curves Y_i^* are defined by the polynomials ϕ_i. The camera matrices will always be denoted by $\mathbf{M}_i$. $\mathbf{F}, \mathbf{e}_1$ and $\mathbf{e}_2$ are respectively the fundamental matrix, the first and the second epipole. We will also need to consider the two following mappings, that we call in the sequel the epipolar mappings, $\gamma : \mathbf{p} \to \mathbf{e}_1 \wedge \mathbf{p}$ and $\xi : \mathbf{p} \to \mathbf{F}\mathbf{p}$. Both are defined on the first image plane; γ associates a point to its epipolar line in the first image, while ξ sends it to its epipolar line in the second image.

3.1 Epipolar Geometry from Planar Curves

The spatial curve X is assumed to be planar and might have singular points.

3.1.1 Introductory Properties

Let $\mathbf{H}$ be the homography induced by the plane of curve in space. Given a point $\mathbf{p}$ in the first image lying on Y_1, $\mathbf{Hp}$, also denoted $h(\mathbf{p})$, must lie in the second image curve Y_2. Since these curves are irreducible, there exists a simple relation between Y_1, Y_2 and h:

$$\exists \lambda, \ \forall x,y,z, \ f_2(h(x,y,z)) = \lambda f_1(x,y,z) \tag{1}$$

Now we proceed to prove some further elementary properties about two views of a planar curve. Those properties will be necessary in the sequel. We shall denote by ε_i the set of epipolar lines, in image i, tangent to Y_i at regular points.

Proposition 6 *The two sets ε_1 and ε_2 are projectively isomorphic. Furthermore the elements of ε_1 and ε_2 are in correspondence through the homography $\mathbf{H}$.*

Proof: The line joining $\mathbf{e}_1$ and $\mathbf{p}$, is tangent to Y_1 at $\mathbf{p}$ if $\lambda = 0$ is a double root of the equation: $f_1(\mathbf{p} + \lambda \mathbf{e}_1) = 0$. This is equivalent to say that $\mathrm{grad}_{\mathbf{p}}(f_1)^T \mathbf{e}_1 = 0$. Moreover if $\mathbf{p}' \cong \mathbf{Hp}$ (where $\cong$ means equality up a to scale factor), $\mathrm{grad}_{\mathbf{p}'}(f_2)^T \mathbf{e}_2 = \mathrm{grad}_{\mathbf{Hp}}(f_2)^T \mathbf{He}_1 = \mathrm{d}f_2(h(\mathbf{p})) \circ \mathrm{d}h(\mathbf{p}).\mathbf{e}_1 = \mathrm{d}(f_2 \circ h)(\mathbf{p}).\mathbf{e}_1 = \lambda \mathrm{d}f_1(\mathbf{p}).\mathbf{e}_1 = \lambda \mathrm{grad}_{\mathbf{p}}(f_1)^T \mathbf{e}_1$.

Therefore the line generated by $\mathbf{e}_1$ and $\mathbf{p}$ is tangent to Y_1 if and only if the line given by $\mathbf{e}_2 \wedge \mathbf{Hp}$ is tangent to Y_2. Given a line $\mathbf{l} \in \varepsilon_1$, its corresponding line $\mathbf{l}' \in \varepsilon_2$ is given by: $\mathbf{H}^{-T}\mathbf{l} = \mathbf{l}'$. [1]   ■

Note that since epipolar lines are transformed in the same way through any regular homography, the two sets ε_1 and ε_2 are in fact projectively related by any homography.

Proposition 7 *The inflexions (respectively the singularities) of the two image curves are projectively related by the homography through the plane of the curve in space.*

Proof: The simple relation (1) implies this double property:

$$\begin{bmatrix} \frac{\partial f_1}{\partial x}(\mathbf{p}) \\ \frac{\partial f_1}{\partial y}(\mathbf{p}) \\ \frac{\partial f_1}{\partial y}(\mathbf{p}) \end{bmatrix} = \mathbf{H}^T \begin{bmatrix} \frac{\partial f_2}{\partial x}(\mathbf{Hp}) \\ \frac{\partial f_2}{\partial y}(\mathbf{Hp}) \\ \frac{\partial f_2}{\partial y}(\mathbf{Hp}) \end{bmatrix}.$$

$$[\tfrac{\partial^2 f_1}{\partial x_i \partial x_j}(\mathbf{p})] = \mathbf{H}^T [\tfrac{\partial^2 f_2}{\partial x_i \partial x_j}(\mathbf{Hp})]\mathbf{H}$$

The first relations implies the conservation of the singularities by homography, whereas the second relation implies the conservation of the whole Hessian curve by homography.   ■
  Using further derivation, we immediately get:

---

[1] By duality $\mathbf{H}^T$ sends the lines of the second image plane into the lines of the first image plane. Here we have showed that $\mathbf{H}^T$ induces to one-to-one correspondence between ε_2 and ε_1.

Proposition 8 *The multiplicity and the number of distinct tangents at a singularity is the same in both images.*

3.1.2 Extended Kruppa's Equations

Roughly speaking, the extended Kruppa's equations state that the sets of epipolar lines tangent to the curve in each image are projectively related. A similar observation has been made in [2] for epipolar lines tangent to apparent contours of objects, but it was used within an optimization scheme. Here we are looking for closed-form solutions, where no initial guesses are required. In order to develop such a closed-form solution for the computation of the epipolar geometry, we need a more quantitative approach, which is given by the following theorem:

Theorem 2 *For a generic position of the camera centers, the dual image curves and the epipolar mappings are related as follows. There exists a non-zero scalar* $\lambda \in \mathbb{C}$, *such as for all* **p** *in the first image plane, the following equality holds:*

$$\phi_2(\xi(\mathbf{p})) = \lambda \phi_1(\gamma(\mathbf{p})) \tag{2}$$

Proof: First it is clear, by proposition 6, that both sides of the this equations define the same algebraic set, that is the union of the tangent lines to Y_1 passing through the first epipole $\mathbf{e}_1$. This set has been denoted by ε_1. Moreover by propositions 7 and 8 the two dual curves have same degree, which is a necessary condition for the equation to hold. It is left to show that each tangent appears with the same multiplicity in each side. It is easily checked by a short computation (where $\wedge$ is the cross product, which embodies the same information than the join operator $\vee$): $\phi_2(\xi(\mathbf{p})) = \phi_2(\mathbf{e}_2 \wedge \mathbf{H}\mathbf{p}) = \phi_2(h(\mathbf{e}_2) \wedge h(\mathbf{p})) \cong \phi_2 \circ ({}^t h)^{-1}(\mathbf{p})$. [2] Then it is sufficient to see that the dual formulation of equation (1) is given by $\phi_2 \circ ({}^t h)^{-1} \cong \phi_1$. ∎

3.1.3 Recovering Epipolar Geometry from Matching Conics

Let $\mathbf{C}_1$ (respectively $\mathbf{C}_2$) be the full rank (symmetric) matrix of the conic in the first (respectively second) image. The equations of the dual curves are $\phi_1(u,v,w) = \mathbf{l}^T \mathbf{C}_1^* \mathbf{l} = 0$ and $\phi_2(u,v,w) = \mathbf{l}^T \mathbf{C}_2^* \mathbf{l} = 0$ where $\mathbf{l} = [u,v,w]^T$, $\mathbf{C}_1^*$ and $\mathbf{C}_2^*$ are the adjoint matrices of $\mathbf{C}_1$ and $\mathbf{C}_2$ (see [48]).

Hence the extended Kruppa's equations reduce in that case to the classical ones:

$$\mathbf{F}^T \mathbf{C}_2^* \mathbf{F} \cong [\mathbf{e}_1]_\times \mathbf{C}_1^* [\mathbf{e}_1]_\times. \tag{3}$$

From equation (3), one can extract a set, denoted E_λ, of six equations on $\mathbf{F}$, $\mathbf{e}_1$ and an auxiliary unknown λ. By eliminating λ it is possible to get five bi-homogeneous equations on $\mathbf{F}$ and $\mathbf{e}_1$.

---

[2] Indeed for a regular 3×3 matrix $\mathbf{H}$: $\mathbf{H}\mathbf{x} \wedge \mathbf{H}\mathbf{y} = det(H)\mathbf{H}^{-T}(\mathbf{x} \wedge \mathbf{y})$. Then since ϕ_2 is a homogeneous polynomial, the last equality is true up to the scale factor, $det(H)^{deg(\phi_2)}$.

Theorem 3 *The six equations, E_λ, are algebraically independent.*

Proof: Using the following regular isomorphism: $(\mathbf{F}, \mathbf{e}_1, \lambda) \longmapsto (\mathbf{D}_2^\star \mathbf{F} \mathbf{D}_1^{-1}, \mathbf{D}_1 \mathbf{e}_1, \lambda) = (\mathbf{X}, \mathbf{y}, \lambda)$, where $\mathbf{D}_1 = \sqrt{\mathbf{C}_1}$ and $\mathbf{D}_2^\star = \sqrt{\mathbf{C}_2^\star}$, the original equations are mapped into the upper-triangle of $\mathbf{X}^T \mathbf{X} = \lambda[\mathbf{y}]_\times^2$. Given this simplified form, it is possible of compute a Gröbner basis [7]. Then we can compute the dimension of the affine variety in the variables $(\mathbf{X}, \mathbf{y}, \lambda)$, defined by these six equations. The dimension is 7, which shows that the equations are algebraically independent. ■

Note that the equations E_λ imply that $\mathbf{F}\mathbf{e}_1 = \mathbf{0}$ (one can easily deduce it from the equation 3 [3]). In order to count the number of matching conics, in generic positions, that are necessary and sufficient to recover the epipolar geometry, we eliminate λ from E_λ and we get a set E that defines a variety V of dimension 7 in a 12-dimensional affine space, whose points are $(\mathbf{e}_1, \mathbf{F})$. The equations in E are bi-homogeneous in $\mathbf{F}$ and $\mathbf{e}_1$ and V can also be regarded as a variety of dimension 5 into the bi-projective space $\mathbb{P}^2 \times \mathbb{P}^8$, where $(\mathbf{e}_1, \mathbf{F})$ lie. Now we project V into $\mathbb{P}^8$, by eliminating $\mathbf{e}_1$ from the equations, we get a new variety V_f which is still of dimension 5 and which is contained into the variety defined by $det(\mathbf{F}) = 0$, whose dimension is 7 [4]. Therefore two pairs of matching conics in generic positions define two varieties isomorphic to V_f which intersect in a three-dimensional variety $(5 + 5 - 7 = 3)$. A third conic in generic position will reduce the intersection to a one-dimensional variety $(5 + 3 - 7 = 1)$. A fourth conic will reduce the system to a zero-dimensional variety. These results can be compiled into the following theorem:

Theorem 4 *{Four conics} or {three conics and a point}*
or {two conics and three points} or {one conic and five points}
in generic positions are sufficient to compute the epipolar geometry.

A similar result has been formulated independently in [28].

3.1.4 Recovering Epipolar Geometry from Higher Order Planar Curves

Now we proceed to analysis the case of planar algebraic curves which degree $d \geq 3$. Since one cannot get a three-dimensional reconstruction from a single planar algebraic curve, a further analysis of the dimension of the set of solutions of the extended Kruppa's equations, in that case, is not really relevant. Therefore we concentrate on recovering the epipolar geometry through the computation of the homography induced by the plane of the curve in space.

We will show next that a single matching pair of planar curves, which genus $g \geq$ 1(which implies that the degree $d \geq 3$), is sufficient for uniquely recovering the homography matrix induced by the plane of the curve in space, whereas two pairs of matching curves (residing on distinct planes) are sufficient for recovering the fundamental matrix.

---

[3]It is clear that we have: $\mathbf{F}^T \mathbf{C}_2^\star \mathbf{F} \mathbf{e}_1 = \mathbf{0}$. For any matrix $\mathbf{M}$, we have: $\ker(\mathbf{M}^T) = im(\mathbf{M})^T$. In addition, $\mathbf{C}_2$ is invertible. Hence $\mathbf{F}\mathbf{e}_1 = \mathbf{0}$.

[4]Since it must be contained into the projection to $\mathbb{P}^8$ of the hypersurface defined by $det(\mathbf{F}\mathbf{e}_1) = 0$

From equation (1), we get $\binom{d+2}{d} - 1$ equations on the entries of the homography matrix. Let V the variety in $\mathbb{P}^8$ defined by these equations. We give a sufficient conditions for V to be discrete.

Proposition 9 *For V being a finite set, it is sufficient that $g \geq 1$.*

Proof: The homography matrix must leave the Weierstrass points ([20, 24]) globally invariant. The number of Weierstrass point is always finite and is at least $2g + 2$. Then for $g \geq 1$, there are enough Weierstrass points to constraint the homography matrix up to a finite-fold ambiguity. ∎

Hence follows immediately:

Theorem 5 *Two planar algebraic curves which genus are greater or equal to 1 and that are lying on two generic planes are sufficient to recover the epipolar geometry.*

However solving equations (1) requires big machinery either symbolic or numerical. Moreover the computation of Weierstrass points, introduced in the proof of the previous proposition, is quite heavy. Hence we propose a simpler algorithm that works for a large category of planar curves. This simpler algorithm is true for non-oversingular curves, e.g. when a technical condition about the singularities of the curve holds. A curve of degree d, whose only singular points are either nodes or cusps, satisfy the Plücker's formula (see [53]):

$$3d(d-2) = i + 6\delta + 8\kappa,$$

where i is the number of inflexion points, δ is the number of nodes, and κ is the number of cusps. For our purpose, a curve is said to be *non-oversingular* when its only singularities are nodes and cusps and when $i + s \geq 4$, where s is the number of all singular points.

Since the inflexion and singular points in both images are projectively related through the homography matrix (proposition 7), one can compute the homography through the plane of the curve in space as follows:

1. Compute the Hessian curves in both images.

2. Compute the intersection of the curve with its Hessian in both images. The output is the set of inflexion and singular points.

3. Discriminate between inflexion and singular points by the additional constraint for each singular point $\mathbf{a}$: $\mathrm{grad}_{\mathbf{a}}(f) = \mathbf{0}$.

At first sight, there are $i! \times s!$ possible correspondences between the sets of inflexion and singular points in the two images. But it is possible to further reduce the combinatory by separating the points into two categories. The points are normalized such that the last coordinates is 1 or 0. Then separate real points from complex points. It is clear that real points are mapped to real points. Now by genericity (the plane of the curve in space does not pass through the camera centers), complex points are mapped to complex points. Indeed consider the point $\mathbf{p} = (\alpha, \beta, \varepsilon) + i(\gamma, \delta, 0)$, where $\varepsilon \in \{0, 1\}$. Let $\mathbf{A} = (a_{ij})_{ij}$ be the homography matrix. Then $\mathbf{Ap} = (a_{11}(\alpha + i\gamma) + a_{12}(\beta + i\delta) + a_{13}\varepsilon, a_{21}(\alpha + i\gamma) + a_{22}(\beta + i\delta) + a_{23}\varepsilon, a_{31}(\alpha + i\gamma) + a_{32}(\beta + i\delta) + a_{33}\varepsilon)$. In order $\mathbf{Ap}$ to be a real point, two algebraic conditions on $\mathbf{A}$ and $\mathbf{p}$ must be satisfied. This cannot hold in general. Therefore we can conclude that each category of the first image must be matched with the same category in the second image. This should be used to reduce the combinatory. Then the right solution can be selected as it should be the one that makes the system of equations extracted from (1) the closest to zero or the one that minimizes the Hausdorff distance (see [27]) between the set of points from the second image curve and the reprojection of the set of points from the first image curve into the second image. For better results, one can compute the Hausdorff distance on inflexion and singular points separately, within each category.

3.2 Epipolar Geometry from Non-planar Curves

Here X be a smooth irreducible curve in $\mathbb{P}^3$, whose degree is $d \geq 2$. Since the case of planar curve has been treated above, X is assumed to be a *non-planar curve*. Before defining and proving the extended Kruppa's equations for arbitrary curve, we need to establish a number of facts about the projection of a spatial curve onto a plane by a pinhole camera.

3.2.1 Single View of a Spatial Curve

Let $\mathbf{M}$ be the camera matrix, $\mathbf{O}$ the camera center, R the retinal plane and Y the image curve. Our first concern is the study of the singularities of the Y.

Proposition 10 *The curve Y will always contain singularities.*

Proof: Singularities correspond to optical rays that meet the curve X in at least two distinct points or are tangent to X. Let $\mathbb{G}(1,3)$ denote the Grassmannian of lines in $\mathbb{P}^3$. Consider the subvariety S_X of $\mathbb{G}(1,3)$ generated by the set of chords and tangents of X, a chord being a line meeting X at least twice. Let $S(X) = \cup_{L \in S_X} L$ be the union of lines that are elements of S_X. It is well known that $S(X)$ is an irreducible subvariety of $\mathbb{P}^3$ and that $\dim(S(X)) = 3$, unless X is planar, which has been excluded (see [23]). Hence $S(X) = \mathbb{P}^3$ and $\mathbf{O} \in S(X)$, that is, at least one chord is passing through $\mathbf{O}$. ∎

The process of singularity formation with projection is illustrated in the figure 3. In the proposition below we will investigate the nature of those singularities.

Proposition 11 *For a generic position of the camera center, the only singularities of Y will be nodes.*

Proof: This is a well known result. See [26] for further details. ∎

We define the *class* of a planar curve to be the degree of its dual curve. Let m be the class of Y. We prove that m is constant for a generic position of the camera center.

Proposition 12 *For a generic position of $\mathbf{O}$, the class of Y is constant.*

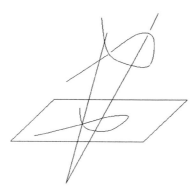

Figure 3: A singularity of the image curve corresponds to an optical ray meeting the curve in space at least twice.

Proof: For each position of $\mathbf{O} \notin R$, we get a particular curve Y or equivalently Y^*, which is the dual curve of Y. Let $\mathcal{Y}$ be this family of dual curves parameterized by $\mathbf{O}$. Then $\mathcal{Y} \subset \mathbb{P}^{2\star} \times \mathbb{P}^3$. Then it is known that $\mathcal{Y}$ is flat [26, 23, 12, 11] over an open set of $\mathbb{P}^3$. Since $\mathbb{P}^3$ is irreducible, this open set must be dense. On the other hand, the degree is constant for a flat family of varieties. Hence the degree of $Y^\star$ is constant for a generic position of $\mathbf{O}$. ∎

Note that using Proposition 11, it is possible to give a formula for m as a function of the degree d and the genus g of X. For a generic position of $\mathbf{O}$, Y will have same degree and genus as X. As mentionned above, we have:

$$m = d(d-1) - 2(\sharp\text{nodes}),$$
$$g = \frac{(d-1)(d-2)}{2} - (\sharp\text{nodes}),$$

where $\sharp$nodes denotes the number of nodes of Y (note that this includes complex nodes). Hence the genus, the degree and the class are related by:

$$m = 2d + 2g - 2.$$

3.2.2 Extended Kruppa's Equations

We are ready now to investigate the recovery of the epipolar geometry from matching curves. The major result here is that the *Extended Kruppa's Equations* still hold in the general case of non-planar algebraic curves.

Theorem 6 Extended Kruppa's equations

For a generic position of the camera centers with respect to the curve in space, there exists a non-zero scalar λ, *such that for all points* $\mathbf{p}$ *in the first image, the following equality holds:*

$$\phi_2(\xi(\mathbf{p})) = \lambda\phi_1(\gamma(\mathbf{p})) \tag{4}$$

Once again observe that if X is a conic and $\mathbf{C}_1$ and $\mathbf{C}_2$ the matrices that respectively represent Y_1 and Y_2, the extended Kruppa's equations reduce to the classical Kruppa's equations, that is: $[\mathbf{e}_1]_x^T \mathbf{C}_1^\star [\mathbf{e}_1]_x \cong \mathbf{F}^T \mathbf{C}_2^\star \mathbf{F}$, where $\mathbf{C}_1^\star$ and $\mathbf{C}_2^\star$ are the adjoint matrices of $\mathbf{C}_1$ and $\mathbf{C}_2$.

Proving these extended Kruppa's equations requires the establishment of these three steps:

1. ϕ_1 and ϕ_2 have equal degrees, that is, the image curves have the same *class*.

2. If ε_i is the set of epipolar lines tangent to the curve in image i, then ε_1 and ε_2 must be projectively isomorphic.

3. Moreover in each corresponding pair of epipolar lines, $(\mathbf{l}, \mathbf{l}') \in \varepsilon_1 \times \varepsilon_2$, the multiplicity of $\mathbf{l}$ and $\mathbf{l}'$ as points of the dual curves $Y_1^\star$ and $Y_2^\star$ must be the same.

The first step is obviously necessary for the equation (4) to hold. It is true, as stated in Proposition 12. As previously, let m be the common degree of ϕ_1 and ϕ_2. The second point is necessary to prove that both sides of equation (4) represent the same geometric set, that is here the union of lines in ε_1. This point holds thanks to the following lemma:

Lemma 1 *The sets of epipolar lines tangent to the image curve in each image are projectively related.*

Proof: This is due to the fact that a pair of corresponding epipolar lines is the trace in the images of a plane containing the baseline joining the two camera centers. Hence the two lines are tangent to the image curves if and only if the plane they define (which contains the baseline) is tangent to the curve in space. Therefore the two sets ε_1 and ε_2 are composed of the traces in the images of the planes containing the baseline and tangent to the curve in space. Hence these two sets are projectively related. ∎

Finally we can prove Theorem 6.

Proof: Since each side of equation (4) represents the union of lines in ε_1, it can be factorized into linear factors, satisfying the following:

$$\phi_1(\gamma(x,y,z)) = \prod_i (\alpha_{1i}x + \alpha_{2i}y + \alpha_{3i}z)^{a_i}$$
$$\phi_2(\xi(x,y,z)) = \prod_i \lambda_i(\alpha_{1i}x + \alpha_{2i}y + \alpha_{3i}z)^{b_i},$$

where $\sum_i a_i = \sum_j b_j = m$. It is left to prove that for all i, $a_i = b_i$. Observe that $a_i > 1$, when the epipolar line is tangent to the image curve in at least two distinct points. Hence one must prove that the number of tangency points for each element of ε_1 is the same for its corresponding element in ε_2. This number must also be the number of tangency points of the plane, defined by these two corresponding elements in ε_1 and ε_2. Hence this third assertion holds using the same argument as in Lemma 1.   ∎

By eliminating the scalar λ from the extended Kruppa's equations (4) we obtain a set of bi-homogeneous equations in $\mathbf{F}$ and $\mathbf{e}_1$. Hence they define a variety in $\mathbb{P}^2 \times \mathbb{P}^8$. This gives rise to an important question. How many of those equations are algebraically independent, or in other words what is the dimension of the set of solutions? This is the issue of the next section.

3.2.3   Dimension of the Set of Solutions

Let $\{E_i(\mathbf{F}, \mathbf{e}_1)\}_i$ be the set of bi-homogeneous equations on $\mathbf{F}$ and $\mathbf{e}_1$, extracted from the extended Kruppa's equations (4). Our first concern is to determine whether all solutions of equation (4) are admissible, that is whether they satisfy the usual constraint $\mathbf{Fe}_1 = 0$. Indeed we prove the following statement:

Proposition 13 *As long as there are at least 2 distinct lines through $\mathbf{e}_1$ tangent to Y_1, equation (4) implies that rank$\mathbf{F} = 2$ and $\mathbf{Fe}_1 = 0$.*

Proof: The variety defined by $\phi_1(\gamma(\mathbf{p}))$ is then a union of at least 2 distinct lines through $\mathbf{e}_1$. If equation (4) holds, $\phi_2(\xi(\mathbf{p}))$ must define the same variety.

There are 2 cases to exclude: If rank$\mathbf{F} = 3$, then the curve defined by $\phi_2(\xi(\mathbf{p}))$ is projectively equivalent to the curve defined by ϕ_2, which is $Y_1^\star$. In particular, it is irreducible.

If rank$\mathbf{F} < 2$ or rank$\mathbf{F} = 2$ and $\mathbf{Fe}_1 \neq 0$, then there is some $\mathbf{a}$, not a multiple of $\mathbf{e}_1$, such that $\mathbf{Fa} = 0$. Then the variety defined by $\phi_2(\xi(\mathbf{p}))$ is a union of lines through $\mathbf{a}$. In neither case can this variety contain two distinct lines through $\mathbf{e}_1$, so we must have rank$\mathbf{F} = 2$ and $\mathbf{Fe}_1 = 0$.   ∎

As a result, in a generic situation every solution of $\{E_i(\mathbf{F}, \mathbf{e}_1)\}_i$ is admissible. Let V be the subvariety of $\mathbb{P}^2 \times \mathbb{P}^8 \times \mathbb{P}^2$ defined by the equations $\{E_i(\mathbf{F}, \mathbf{e}_1)\}_i$ together with $\mathbf{Fe}_1 = 0$ and $\mathbf{e}_2^T \mathbf{F} = \mathbf{0}^T$, where $\mathbf{e}_2$ is the second epipole. We next compute the lower bound on the dimension of V, after which we would be ready for the calculation itself.

Proposition 14 *If V is non-empty, the dimension of V is at least $7 - m$.*

Proof: Choose any line $\mathbf{l}$ in $\mathbb{P}^2$ and restrict $\mathbf{e}_1$ to the affine piece $\mathbb{P}^2 \setminus \mathbf{l}$. Let (x, y) be homogeneous coordinates on $\mathbf{l}$. If $\mathbf{Fe}_1 = 0$, the two sides of equation (4) are both unchanged by replacing $\mathbf{p}$ by $\mathbf{p} + \alpha \mathbf{e}_1$. So equation (4) will hold for all $\mathbf{p}$ if it holds for all $\mathbf{p} \in \mathbf{l}$. Therefore equation (4) is equivalent to the equality of 2 homogeneous polynomials of degree m in x and y, which in turn is equivalent to the equality of $(m + 1)$ coefficients. After eliminating λ, we have m algebraic conditions on $(\mathbf{e}_1, \mathbf{F}, \mathbf{e}_2)$ in addition to $\mathbf{Fe}_1 = 0$, $\mathbf{e}_2^T \mathbf{F} = \mathbf{0}^T$.

The space of all epipolar geometries, that is, solutions to $\mathbf{Fe}_1 = 0$, $\mathbf{e}_2^T \mathbf{F} = \mathbf{0}^T$, is irreducible of dimension 7. Therefore, V is at least $(7 - m)$-dimensional.   ∎

For the calculation of the dimension of V we introduce some additional notations. Given a triplet $(\mathbf{e}_1, \mathbf{F}, \mathbf{e}_2) \in \mathbb{P}^2 \times \mathbb{P}^8 \times \mathbb{P}^2$, let $\{\mathbf{q}_{1\alpha}(\mathbf{e}_1)\}$ (respectively $\{\mathbf{q}_{2\alpha}(\mathbf{e}_2)\}$) be the tangency points of the epipolar lines through $\mathbf{e}_1$ (respectively $\mathbf{e}_2$) to the first (respectively second) image curve. Let $\mathbf{Q}_\alpha(\mathbf{e}_1, \mathbf{e}_2)$ be the 3d points projected onto $\{\mathbf{q}_{1\alpha}(\mathbf{e}_1)\}$ and $\{\mathbf{q}_{2\alpha}(\mathbf{e}_2)\}$. Let $\mathbf{L}$ be the baseline joining the two camera centers. We next provide a sufficient condition for V to be discrete.

Proposition 15 *For a generic position of the camera centers, the variety V will be discrete if, for any point $(\mathbf{e}_1, \mathbf{F}, \mathbf{e}_2) \in V$, the union of $\mathbf{L}$ and the points $\mathbf{Q}_\alpha(\mathbf{e}_1, \mathbf{e}_2)$ is not contained in any quadric surface.*

Proof: For generic camera positions, there will be m distinct points $\{\mathbf{q}_{1\alpha}(\mathbf{e}_1)\}$ and $\{\mathbf{q}_{2\alpha}(\mathbf{e}_2)\}$, and we can regard $\mathbf{q}_{1\alpha}$, $\mathbf{q}_{2\alpha}$ locally as smooth functions of $\mathbf{e}_1$, $\mathbf{e}_2$.

We let W be the affine variety in $\mathbb{C}^3 \times \mathbb{C}^9 \times \mathbb{C}^3$ defined by the same equations as V. Let $\Theta = (\mathbf{e}_1, \mathbf{F}, \mathbf{e}_2)$ be a point of W corresponding to a non-isolated point of V. Then there is a tangent vector $\vartheta = (\mathbf{v}, \Phi, \mathbf{v}')$ to W at Θ with Φ not a multiple of $\mathbf{F}$.

If χ is a function on W, $\nabla_{\Theta, \vartheta}(\chi)$ will denote the derivative of χ in the direction defined by ϑ at Θ. For

$$\chi_\alpha(\mathbf{e}_1, \mathbf{F}, \mathbf{e}_2) = \mathbf{q}_{2\alpha}(\mathbf{e}_2)^T \mathbf{F} \mathbf{q}_{1\alpha}(\mathbf{e}_1),$$

the extended Kruppa's equations imply that χ_α vanishes identically on W, so its derivative must also vanish. This yields

$$\begin{aligned}
\nabla_{\Theta, \vartheta}(\chi_\alpha) &= (\nabla_{\Theta, \vartheta}(\mathbf{q}_{2\alpha}))^T \mathbf{F} \mathbf{q}_{1\alpha} \\
&\quad + \mathbf{q}_{2\alpha}^T \Phi \mathbf{q}_{1\alpha} + \mathbf{q}_{2\alpha}^T \mathbf{F}(\nabla_{\Theta, \vartheta}(\mathbf{q}_{1\alpha})) \\
&= 0.
\end{aligned} \tag{5}$$

We shall prove that $\nabla_{\Theta, \vartheta}(\mathbf{q}_{1\alpha})$ is in the linear span of $\mathbf{q}_{1\alpha}$ and $\mathbf{e}_1$. (This means that when the epipole moves slightly, $\mathbf{q}_{1\alpha}$ moves along the epipolar line, see figure 4.)

Consider $\kappa(t) = f(\mathbf{q}_{1\alpha}(\mathbf{e}_1 + t\mathbf{v}))$, where f is the polynomial defining the image curve Y_1. Since $\mathbf{q}_{1\alpha}(\mathbf{e}_1 + t\mathbf{v}) \in Y_1$, $\kappa \equiv 0$, so the derivative $\kappa'(0) = 0$. On the other hand, $\kappa'(0) = \nabla_{\Theta, \vartheta}(f(\mathbf{q}_{1\alpha})) = \mathrm{grad}_{\mathbf{q}_{1\alpha}}(f)^T \nabla_{\Theta, \vartheta}(\mathbf{q}_{1\alpha})$.

Thus we have $\mathrm{grad}_{\mathbf{q}_{1\alpha}}(f)^T \nabla_{\Theta, \vartheta}(\mathbf{q}_{1\alpha}) = 0$. But also $\mathrm{grad}_{\mathbf{q}_{1\alpha}}(f)^T \mathbf{q}_{1\alpha} = 0$ and $\mathrm{grad}_{\mathbf{q}_{1\alpha}}(f)^T \mathbf{e}_1 = 0$. Since $\mathrm{grad}_{\mathbf{q}_{1\alpha}}(f) \neq \mathbf{O}$ ($\mathbf{q}_{1\alpha}$ is not a singular point of the curve), this shows that $\nabla_{\Theta, \vartheta}(\mathbf{q}_{1\alpha})$, $\mathbf{q}_{1\alpha}$, and $\mathbf{e}_1$ are linearly dependent. $\mathbf{q}_{1\alpha}$ and $\mathbf{e}_1$ are linearly independent, so $\nabla_{\Theta, \vartheta}(\mathbf{q}_{1\alpha})$ must be in their linear span.

We have that $\mathbf{q}_{2\alpha}^T \mathbf{F} \mathbf{e}_1 = \mathbf{q}_{2\alpha}^T \mathbf{F} \mathbf{q}_{1\alpha} = 0$, so $\mathbf{q}_{2\alpha}^T \mathbf{F} \nabla_{\Theta, \vartheta}(\mathbf{q}_{1\alpha}) = 0$: the third term of equation (5) vanishes. In a similar way, the first term of equation (5) vanishes, leaving

$$\mathbf{q}_{2\alpha}^T \Phi \mathbf{q}_{1\alpha} = 0.$$

The derivative of $\chi(\mathbf{e}_1, \mathbf{F}, \mathbf{e}_2) = \mathbf{F} \mathbf{e}_1$ must also vanish, which yields:

$$\mathbf{e}_2{}^T \Phi \mathbf{e}_1 = 0.$$

From the first equality, we deduce that for every $\mathbf{Q}_\alpha$, we have:

$$\mathbf{Q}_\alpha^T \mathbf{M}_2^T \Phi \mathbf{M}_1 \mathbf{Q}_\alpha = 0.$$

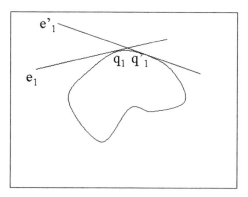

Figure 4: The point $\mathbf{q}_{1\alpha}(\mathbf{e}_1)$ is a smooth function of $\mathbf{e}_1$. When $\mathbf{e}_1$ moves slightly, $\mathbf{q}_{1\alpha}(\mathbf{e}_1)$ moves along the epipolar line.

From the second equality, we deduce that every point $\mathbf{P}$ lying on the baseline must satisfy:

$$\mathbf{P}^T \mathbf{M}_2^T \Phi \mathbf{M}_1 \mathbf{P} = 0.$$

The fact that Φ is not a multiple of F implies that $\mathbf{M}_2^T \Phi \mathbf{M}_1 \neq 0$, so together these two last equations mean that the union $\mathbf{L} \cup \{\mathbf{Q}_\alpha\}$ lies on a quadric surface. Thus if there is no such quadric surface, every point in V must be isolated. ∎

Observe that this result is consistent with the previous proposition, since there always exist a quadric surface containing a given line and six given points. However in general there is no quadric containing a given line and seven given points. Therefore we can conclude with the following theorem.

Proposition 16 *For a generic position of the centers of projections, there is no quadric containing the line L and the tangency points $\mathbf{Q}_\alpha(\mathbf{e}_1, \mathbf{e}_2)_{\alpha=1,\ldots,m}$.*

Proof: First, one can observe that when the genus g of X is equal or greater to 2, then the result is obvious. Indeed the intersection of a quadric and X has at most $2d$ points (by Bezout's theorem) and there are $m > 2d$ points of tangency, which are distinct for a generic line L.

To handle, the general case, let us introduce some notations. Consider the product of m instances of $\mathbb{P}^3$: $H = \mathbb{P}^3 \times \ldots \times \mathbb{P}^3$. Then a $m-$tuple $(Q_1, \ldots, Q_m) \in H$ is such that the points

Q_i lie on a quadric if the matrix W has rank at most 9, where:

$$\mathbf{W} = \begin{bmatrix} x_1^2 & \cdots & x_m^2 \\ x_1 y_1 & \cdots & x_m y_m \\ x_1 z_1 & \cdots & x_m z_m \\ x_1 t_1 & \cdots & x_m t_m \\ y_1^2 & \cdots & y_m^2 \\ y_1 z_1 & \cdots & y_m z_m \\ y_1 t_1 & \cdots & y_m t_m \\ z_1^2 & \cdots & z_m^2 \\ z_1 t_1 & \cdots & z_m t_m \\ t_1^2 & \cdots & t_m^2 \end{bmatrix}$$

and $[x_i, y_i, z_i, t_i]$ are the homogeneous coordinates of Q_i. This defines a closed subvariety of H, that we shall denote S.

Consider now the following set:

$$\Sigma = \{(L, \mathbf{Q}_1(\mathbf{e}_1, \mathbf{e}_2), ..., \mathbf{Q}_m(\mathbf{e}_1, \mathbf{e}_2)) \in \mathbb{G}(1,3) \times H\},$$

where $\mathbb{G}(1,3)$ is the Grassmannian of lines of $\mathbb{P}^3$. The set Σ of course depends on both L (that is on the center of projections) and X. Let us show that Σ is an algebraic variety.

A point $Q \in \mathbb{P}^3$ is a tangency point of a plane containing L with X, if and only if the two following conditions are satisfied: (i) $Q \in X$ and (ii) the tangent to X at Q intersects L (in projective space).

The Plücker coordinates of the tangent T_Q to X at Q are homogenous polynomial functions of the coordinates of Q (given by the Gauss map). T_Q intersects L if and only if the join $T_Q \vee L$ vanishes [4], which yields a bi-homogeneous equation on the coordinates of Q and of those of L. We shall denote this equation $\phi(Q, L) = 0$.

For a polynomial $F \in R[X, Y, Z, T]$, where R is some ring, we write F_i for the polynomial in $R[X_i, Y_i, Z_i, T_i]$, obtained from F by substituting the variables X, Y, Z, T by the the variables X_i, Y_i, Z_i, T_i. Therefore the set Σ is made of the common zeros of the polynomials $F_{11}, ..., F_{r1}, ..., F_{1m}, ..., F_{rm}, \phi_1, ..., \phi_m$, where $F_1, ..., F_r$ are the polynomials defining X. Thus Σ can be viewed as an closed subvariety of $\mathbb{G}(1,3) \times H$.

Let π_1 and π_2 be the canonical projections: $\pi_1 : \Sigma \longrightarrow \mathbb{G}(1,3)$ and $\pi_2 : \Sigma \longrightarrow H$. Therefore for a line L, if there exists a quadric containing L and the tangency points $\mathbf{Q}_1(\mathbf{e}_1, \mathbf{e}_2), ..., \mathbf{Q}_m(\mathbf{e}_1, \mathbf{e}_2)$ then $\pi_2(\pi_1^{-1}(L))$ is included into the closed subvariety S defined above.

Thus a line L, for which there exists a quadric containing L and the tangency points, must lie in in $\pi_1(\pi_2^{-1}(S))$. This is a subset of a proper closed subvariety of $\mathbb{G}(1,3)$. This completes the proof. ∎

Eventually, we conclude this section by the following corollary.

Corollary 1 *For a generic position of the centers of projection, the generalized Kruppa equation defines the epipolar geometry up to a finite-fold ambiguity if and only if $m \geq 7$.*

Since different curves in generic position give rise to independent equations, this result means that the sum of the classes of the image curves must be at least 7 for V to be a finite set. Observe that this result is consistent with the fact that four conics ($m = 2$ for each conic) in general position are sufficient to compute the fundamental matrix, as shown in [28, 30]. Now we proceed to translate the result in terms of the geometric properties of X directly using the degree and the genus of X, related to m by the following relation: $m = 2d + 2g - 2$. Here are some examples for sets of curves that allow the recovery of the fundamental matrix:

1. Four conics ($d = 2, g = 0$)in general position.

2. Two rational cubics ($d = 3, g = 0$) in general position.

3. A rational cubic and two conics in general position.

4. Two elliptic cubics ($d = 3, g = 1$) in general position (see also [30]).

5. A general rational quartic ($d = 4, g = 0$), and a general elliptic quartic ($d = 4, g = 1$).

3.3 Three-dimensional Reconstruction

We turn our attention to the problem of reconstructing an algebraic curve from two or more views, given known camera matrices (epipolar geometries are known). The basic idea is to intersect together the cones defined by the camera centers and the image curves. However this intersection can be computed in three different spaces, giving rise to different algorithms and applications. Given the representation in one of those spaces, it is possible to compute the two other representations as shown in section 2 and in [29].

We shall mention that in [19] a scheme is proposed to reconstruct the outline of an algebraic surface from a single view by blowing-up the projection. This approach results in a spatial curve defined up to an unknown projective transformation. In fact the only computation this reconstruction allows is the recovery of the projective properties of the curve. Moreover this reconstruction is valid for irreducible curves only. However reconstructing from two projections not only gives the projective properties of the curve, but also the relative depth of it with respect to others objects in the scene and furthermore the relative position between irreducible components.

3.3.1 Homography Recovery for Planar Curve

We begin by restricting our attention to planar curves. We proceed to the recovery of the homography matrix induced by the plane of the curve in space. This approach reduces the reconstruction problem to finding the roots of a uni-variate polynomial. Let $\mathbf{e}_i, i = 1, 2$ be the first and second epipole. Let $\mathbf{S}$ be any homography, which can be extracted from the epipolar geometry by $\mathbf{S} = [\mathbf{e}_2]_\times \mathbf{F}$, where $[\mathbf{e}_2]_\times$ is the matrix representing the cross-product by $\mathbf{e}_2$. Then the homography through the plane of the curve in space can be written as: $\mathbf{H} = \mathbf{S} + \mathbf{e}_2\mathbf{h}^T$ (see [49, 35]). We define μ to be such that: $\mathbf{He}_1 = \mu\mathbf{e}_2$. [5] As seen in equation (1), the two image curves and the homography $\mathbf{H}$ are related as follows:

$$\exists\lambda, \forall\mathbf{p}, f_1(\mathbf{p}) = \lambda f_2(\mathbf{Hp})$$

This implies:

$$\lambda\mathrm{grad}_\mathbf{p}(f_1) = \mathbf{H}^T\mathrm{grad}_{\mathbf{Hp}}(f_2),$$

for any $\mathbf{p}$ in the first image. Let $\mathbf{g}_i = \mathrm{grad}_{\mathbf{e}_i}(f_i), i = 1, 2$. Thus the previous expression, applied to $\mathbf{e}_1$, can be re-written as follows:

$$\lambda\mathbf{g}_1 = \mu^{d-1}(\mathbf{S}^T + \mathbf{he}_2^T)\mathbf{g}_2.$$

We define β to be $\beta = \mathbf{e}_2^T\mathbf{g}_2 = df_2(\mathbf{e}_2)$ and η to be $\eta = \frac{\lambda}{\mu^{d-1}}$. Hence $\mathbf{h} = \frac{1}{\beta}(\eta\mathbf{g}_1 - \mathbf{S}^T\mathbf{g}_2)$. Thus $\mathbf{H} = \mathbf{S} + \frac{1}{\beta}\mathbf{e}_2(\eta\mathbf{g}_1^T - \mathbf{g}_2\mathbf{S}) = (\mathbf{I} - \frac{1}{\beta}\mathbf{e}_2\mathbf{g}_2^T)\mathbf{S} + \frac{\eta}{\beta}\mathbf{e}_2\mathbf{g}_1^T$. Substituting this expression of $\mathbf{H}$ into equation (1) yields:

$$\lambda f_1(\mathbf{p}) = f_2((\mathbf{I} + \frac{1}{\beta}\mathbf{e}_2\mathbf{g}_2^T)\mathbf{Sp} + \frac{\eta}{\beta}\mathbf{e}_2\mathbf{g}_1^T\mathbf{p}).$$

Then after elimination of λ, we get a set of $\binom{d+2}{d} - 1$ equations of degree d in η. Hence the problem is equivalent to find the common solutions to this set of equations. This can be achieved by picking one equation, solving it and keeping only solutions that are solutions of the whole system.

Now we turn back our attention to the general case. We shall investigate three different types of representation, each leading to a particular reconstruction algorithm.

3.3.2   Reconstruction in Point Space

Let the camera projection matrices be $\mathbf{M}_1$ and $\mathbf{M}_2$. They are assumed to be known at any level of ambiguity: projective, affine or Euclidean. Hence the two cones defined by the image curves and the camera centers are given by: $\Delta_1(\mathbf{P}) = f_1(\mathbf{M}_1\mathbf{P})$ and $\Delta_2(\mathbf{P}) = f_2(\mathbf{M}_2\mathbf{P})$. The reconstruction is defined as the curve whose equations are $\Delta_1 = 0$ and $\Delta_2 = 0$. It is clear that the original space curves is contained within the intersection on these two viewing cones. However since each cone has degree d (the same than the space curve), by Bezout theorem, the intersection must have degree d^2. This implies that the intersection contains more than just the original space curve. It turns out, by the following theorem, that this intersection contains, in general, only two irreducible components (separated curves). One has degree d and the other has degree $d(d - 1)$. Therefore if $d \geq 3$, since the original space has degree d, we can extract the right component which is the answer of the reconstruction problem. When $d = 2$ (the curve is a conic), however, the reconstruction problem admits two solutions, which are both conics, and a third view is necessary to select the right conic.

---

[5]Note that it is not possible to normalize $\mathbf{H}$ such that $\mathbf{He}_1 = \mathbf{e}_2$, because $\mathbf{H}$ is given as a function of $\mathbf{S}$ by $\mathbf{H} = \mathbf{S} + \mathbf{e}_2\mathbf{h}^T$, which constraints its norm.

Theorem 7 *For a generic position of the camera centers, that is when no epipolar plane is tangent twice to the curve X, the curve defined by $\{\Delta_1 = 0, \Delta_2 = 0\}$ has two irreducible components. One has degree d and is the actual solution of the reconstruction. The other one has degree $d(d-1)$.*

Proof: For a line $\mathbf{l} \subset \mathbb{P}^3$, we write $\sigma(\mathbf{l})$ for the pencil of planes containing $\mathbf{l}$. For a point $\mathbf{p} \in \mathbb{P}^2$, we write $\sigma(\mathbf{p})$ for the pencil of lines through $\mathbf{p}$. There is a natural isomorphism between $\sigma(\mathbf{e}_i)$, the epipolar lines in image i, and $\sigma(\mathbf{L})$, the planes containing both camera centers. Consider the following covers of $\mathbb{P}^1$:

1. $X \xrightarrow{\eta} \sigma(\mathbf{L}) \cong \mathbb{P}^1$, taking a point $x \in X$ to the epipolar plane that it defines with the camera centers.

2. $Y_1 \xrightarrow{\eta_1} \sigma(\mathbf{e}_1) \cong \sigma(\mathbf{L}) \cong \mathbb{P}^1$, taking a point $y \in Y_1$ to its epipolar line in the first image.

3. $Y_2 \xrightarrow{\eta_2} \sigma(\mathbf{e}_2) \cong \sigma(\mathbf{L}) \cong \mathbb{P}^1$, taking a point $y \in Y_2$ to its epipolar line in the second image.

If ρ_i is the projection $X \to Y_i$, then $\eta = \eta_i \rho_i$. Let B the union set of branch points of η_1 and η_2. It is clear that the branch points of η are included in B. Let $S = \mathbb{P}^1 \setminus B$, pick $t \in S$, and write $X_S = \eta^{-1}(S)$, $X_t = \eta^{-1}(t)$. Let μ_{X_S} be the monodromy: $\pi_1(S, t) \longrightarrow \mathrm{Perm}(X_t)$, where $\mathrm{Perm}(Z)$ is the group of permutations of a finite set Z. It is well known that the path-connected components of X are in one-to-one correspondence with the orbits of the action of $\mathrm{im}(\mu_{X_S})$ on X_t. Since X is assumed to be irreducible, it has only one component and $\mathrm{im}(\mu_{X_S})$ acts transitively on X_t. Then if $\mathrm{im}(\mu_{X_S})$ is generated by transpositions, this will imply that $\mathrm{im}(\mu_{X_S}) = \mathrm{Perm}(X_t)$. In order to show that $\mathrm{im}(\mu_{X_S})$ is actually generated by transpositions, consider a loop in $\mathbb{P}^1$ centered at t, say l_t. If l_t does not go round any branch point, then l_t is homotopic to the constant path in S and then $\mu_{X_S}([l_t]) = 1$. Now in B, there are three types of branch points:

1. branch points that come from nodes of Y_1: these are not branch points of η,

2. branch points that come from nodes of Y_2: these are not branch points of η,

3. branch points that come from epipolar lines tangent either to Y_1 or to Y_2: these are genuine branch points of η.

If the loop l_t goes round a point of the first two types, then it is still true that $\mu_{X_S}([l_t]) = 1$. Now suppose that l_t goes round a genuine branch point of η, say b (and goes round no other points in B). By genericity, b is a simple two-fold branch point, hence $\mu_{X_S}([l_t])$ is a transposition. This shows that $\mathrm{im}(\mu_{X_S})$ is actually generated by transpositions and so $\mathrm{im}(\mu_{X_S}) = \mathrm{Perm}(X_t)$.

Now consider $\tilde{X}$, the curve defined by $\{\Delta_1 = 0, \Delta_2 = 0\}$. By Bezout's Theorem $\tilde{X}$ has degree d^2. Let $\tilde{x} \in \tilde{X}$. It is projected onto a point y_i in Y_i, such that $\eta_1(y_1) = \eta_2(y_2)$. Hence $\tilde{X} \cong Y_1 \times_{\mathbb{P}^1} Y_2$; restricting to the inverse image of the set S, we have $\tilde{X}_S \cong X_S \times_S X_S$. We can therefore identify $\tilde{X}_t$ with $X_t \times X_t$. The monodromy $\mu_{\tilde{X}_S}$ can then be given by $\mu_{\tilde{X}_S}(x,y) = (\mu_{X_S}(x), \mu_{X_S}(y))$. Since $\mathrm{im}(\mu_{X_S}) = \mathrm{Perm}(X_t)$, the action of $\mathrm{im}(\mu_{\tilde{X}_S})$ on $X_t \times X_t$ has two orbits, namely $\{(x,x)\} \cong X_t$ and $\{(x,y)|x \neq y\}$. Hence $\tilde{X}$ has two irreducible components. One has degree d and is X, the other has degree $d^2 - d = d(d-1)$. ■

Solving the system defined by $\Delta_1(\mathbf{P}) = 0$ and $\Delta_2(\mathbf{P}) = 0$ can be done by Gröbner basis computation. Then as mentioned above, for $d \geq 3$, the actual solution can be extracted. However the case of planar curves can be treated more easily.

3.3.3 Reconstruction in the Dual Space

As above, let X be the curve in space, that we want to reconstruct. Let $X^\star$ be the dual variety of X, that is, the set of planes tangent to X. Since X is supposed not to be a line, the dual variety $X^\star$ must be a hypersurface of the dual space [23]. Hence let Υ be a minimal degree polynomial that represents $X^\star$. Our first concern is to determine the degree of Υ.

Proposition 17 *The degree of Υ is m, that is, the common degree of the dual image curves.*

Proof: Since $X^\star$ is a hypersurface of $\mathbb{P}^{3\star}$, its degree is the number of points where a generic line in $\mathbb{P}^{3\star}$ meets $X^\star$. By duality it is the number of planes in a generic pencil that are tangent to X. Hence it is the degree of the dual image curve. Another way to express the same fact is the observation that the dual image curve is the intersection of $X^\star$ with a generic plane in $\mathbb{P}^{3\star}$. Note that this provides a new proof that the degree of the dual image curve is constant for a generic position of the camera center. ■

For the reconstruction of $X^\star$ from multiple view, we will need to consider the mapping from a line $\mathbf{l}$ of the image plane to the plane that it defines with the camera center. Let $\mu : \mathbf{l} \mapsto \mathbf{M}^T \mathbf{l}$ denote this mapping [15]. There exists a link involving Υ, μ and ϕ, the polynomial of the dual image curve: $\Upsilon(\mu(\mathbf{l})) = 0$ whenever $\phi(\mathbf{l}) = 0$. Since these two polynomials have the same degree (because μ is linear) and ϕ is irreducible, there exist a scalar λ such that

$$\Upsilon(\mu(\mathbf{l})) = \lambda\phi(\mathbf{l}),$$

for all lines $\mathbf{l} \in \mathbb{P}^{2\star}$. Eliminating λ, we get $\frac{(m+2)(m+1)}{2} - 1$ linear equations on Υ. Since the number of coefficients in Υ is $\frac{(m+3)(m+2)(m+1)}{6}$, we can state the following result:

Proposition 18 *The reconstruction in the dual space can be done linearly using at least* $k \geq \frac{m^2+6m+11}{3(m+3)}$ *views.*

Proof: The least number of views must satisfy $k(\frac{(m+2)(m+1)}{2} - 1) \geq \frac{(m+3)(m+2)(m+1)}{6} - 1$. ■

The lower bounds on the number of views k for few examples are given below:

1. for a conic locus, $k \geq 2$,

2. for a rational cubic, $k \geq 3$,

3. for an elliptic cubic, $k \geq 4$,

4. for a rational quartic, $k \geq 4$,

5. for a elliptic quartic, $k \geq 4$.

Moreover it is worth noting that the fitting of the dual image curve is not necessary. It is sufficient to extract tangents to the image curves at distinct points. Each tangent $\mathbf{l}$ contributes to one linear equation on Υ: $\Upsilon(\mu(\mathbf{l})) = 0$. However one cannot obtain more than $\frac{(m+2)(m+1)}{2} - 1$ linearly independent equations per view.

3.3.4   Reconstruction in $\mathbb{G}(1,3)$

The spatial curve X admits Γ as a Chow polynomial. Let d be the common degree of X and Γ. Let f be the polynomial defining the image curve, Y. Consider the mapping that associates to an image point its optical ray: $\nu : \mathbf{p} \mapsto \widehat{\mathbf{M}}\mathbf{p}$, where $\widehat{\mathbf{M}}$ is a 3×6 matrix, which entries are polynomials functions of $\mathbf{M}$ [15]. Hence the polynomial $\Gamma(\nu(\mathbf{p}))$ vanishes whenever $f(\mathbf{p})$ does. Since they have same degree and f is irreducible, there exists a scalar λ such as for every point $\mathbf{p} \in \mathbb{P}^2$, we have:

$$\Gamma(\nu(\mathbf{p})) = \lambda f(\mathbf{p}).$$

This yields $\binom{d+2}{d} - 1$ linear equations on Γ.

Hence a similar statement to that in Proposition 18 can be made:

Proposition 19 *The reconstruction in $\mathbb{G}(1,3)$ can be done linearly using at least $k \geq \frac{1}{6}\frac{d^3+5d^2+8d+4}{d}$ views.*

For some examples, below are the minimal number of views for a linear reconstruction of the curve in $\mathbb{G}(1,3)$:

1. for a conic locus, $k \geq 4$,

2. for a cubic, $k \geq 6$,

3. for a quartic, $k \geq 8$.

As in the case of reconstruction in the dual space, it is not necessary to explicitly compute f. It is enough to pick points on the image curve. Each point yields a linear equation on Γ: $\Gamma(\nu(\mathbf{p})) = 0$. However for each view, one cannot extract more than $\frac{1}{2}d^2 + \frac{3}{2}d$ independent linear equations.

4   Dynamic Configurations

4.1   Trajectory Triangulation from Point Measurements

Now we turn to consider the case of dynamic scenes. A moving point is viewed in one or several sequences. The camera matrices are assumed to be known over each sequence.

However the cameras are *not assumed to be synchronized*. We want to recover the trajectory of the moving point. The trajectory recovery is then naturally done in the Grassmannian of lines $\mathbb{G}(1,3)$. Let Γ be a Chow polynomial of the curve X generated by the motion of the point and let be d its degree. For now we will assume that d is known. We shall show in section 4.1.3 how this assumption can be addressed.

We first generate the number of independent measurements necessary to recover the point trajectory. Then we analyze carefully to which extent the measurements extracted from a particular sequence are independent and provide enough constraints to recover the trajectory. Finally we generalize our results to a general framework for trajectory recovery.

4.1.1 How Many Measurements are Necessary?

Each 2D point extracted from the images contributes one linear equation in Γ:

$$\Gamma(\widehat{\mathbf{M}}\mathbf{p}) = 0, \tag{6}$$

where $\mathbf{M}$ is the camera matrix and $\widehat{\mathbf{M}}$ is the 6×3 matrix mapping each image point to its optical ray, as before. Thus the following result is immediate:

Proposition 20 *The recovery of the trajectory of a moving point can be done linearly using at least $k \geq \frac{1}{12}d^4 + \frac{2}{3}d^3 + \frac{23}{12}d^2 + \frac{7}{3}d$ independent measurements.*

Proof: The number of degrees of freedom of the Chow polynomial that must be constrained is $N_d - 1 = \binom{d+5}{d} - \binom{d-2+5}{d-2} = \frac{1}{12}d^4 + \frac{2}{3}d^3 + \frac{23}{12}d^2 + \frac{7}{3}d$. ∎

The lower bounds on k for few examples:

1. for a moving point on a conic locus, $k \geq 19$,

2. for a moving point on a cubic, $k \geq 49$,

3. for a moving point on a quartic, $k \geq 104$.

A natural question is to know how many independent equations of type 6 each camera provides. This is the issue described in the next section.

4.1.2 Which Measurements are Actually Independent?

We start by the simple case of a static camera.

Proposition 21 *A static camera provides $\binom{d+2}{d} - 1$ constraints on Γ.*

Proof: All the optical rays generated by a static camera belong to a plane included into the Grassmannian $\mathbb{G}(1,3)$. Therefore a static camera allows to recover the intersection with a plane of the subvariety of $\mathbb{G}(1,3)$ we are looking for. This intersection is a curve of degree d. Therefore a static camera yields $\binom{d+2}{d} - 1$ constraints on the Chow polynomials of the curve. ∎

Note that when the camera is static, all the optical rays pass through the same point, the camera center. Hence the space of curves of a given degree constrained to pass through the optical rays contains the space of curves of the same degree passing through the camera center. This has a practical consequence. If several static cameras are viewing a moving point, the recovery of the trajectory might lead to a parasite solution, i.e. a curve passing through the camera centers. If such a parasite solution exists it must be eliminated.

Let us now consider a moving camera over a trajectory modeled by an algebraic curve of degree k. The question is to know how many independent measurements this camera provides on a point moving over a trajectory of degree d. When the number of measurements is large enough, this can be viewed as computing the number of degrees of freedom of the family of curves of degree d over the surface generated by the optical rays generated by the tracked point and the camera center. This is a question of algebraic geometry.

Theorem 8 Fundamental Theorem

A camera moving along a curve of degree $k \geq 1$ provides $H(k,d)$ constraints on the Chow Polynomial Γ of the trajectory of degree $d \geq 1$ a moving point , where

$$
\begin{aligned}
H(1,1) &= & 4 \\
H(2,1) &= & 5 \\
H(1,2) &= & 12 \\
H(2,2) &= & 17 \\
H(k,2) &= & 18 \text{ for } k \geq 3 \\
H(k,d) &= & \left\{ \begin{array}{ccc} N_d - \binom{d-k+5}{5} + \binom{d-k+3}{5} - 1 & \text{if } k \leq d-2 \\ N_d - 7 & \text{if } k = d-1 \\ N_d - 1 & \text{if } k \geq d \end{array} \right\} \text{ for } d \geq 3
\end{aligned}
$$

Proof: The proof is based on cohomological computation [26]. Let X (resp. Y) be the point (resp. camera center) trajectory. Each observation generates an optical ray joining the camera center and the point. Let $\mathbf{L}_1, ..., \mathbf{L}_n$ be these n lines joining X and Y. Let Γ_X and Γ_Y be the Chow polynomial of X and Y respectively. We shall denote by $Z(\Gamma_X)$ and $Z(\Gamma_Y)$ the sets where they vanish. Let $V = Z(\Gamma_X) \cap Z(\Gamma_Y) \cap \mathbb{G}(1,3)$. For $n >> 1$, we have

$$
\{ \Gamma \in H^0(\mathbb{P}^5, O_{\mathbb{P}^5}(d)) : \Gamma(L_i) = 0, i = 1, ..., n \} =
$$
$$
\{ \Gamma \in H^0(\mathbb{P}^5, O_{\mathbb{P}^5}(d)) : \Gamma_{|V} \equiv 0 \} = I_{V,\mathbb{P}^5}(d).
$$

So, we want to compute $dim(I_{V,\mathbb{P}^5}(d))$, which is the dimension of the space of Γ_X, or, equivalently, $h^0(V, O_V(d)) = h^0(O_{\mathbb{P}^5}(d)) - dim(I_{V,\mathbb{P}^5}(d))$. Since V is a complete intersection of degree $(d,k,2)$ in $\mathbb{P}^5$, the dimension of $I_{V,\mathbb{P}^5}(d)$ should be equal to

$$
h^0(O_{\mathbb{P}^5}(d-2)) + h^0(O_{\mathbb{P}^5}(d-k)) - h^0(O_{\mathbb{P}^5}(d-k-2)) + 1.
$$

As a consequence

$$
h^0(V, O_V(d)) = N_d - \left(h^0(O_{\mathbb{P}^5}(d-k)) - h^0(O_{\mathbb{P}^5}(d-k-2)) + 1 \right).
$$

■

If several independently moving cameras are viewing a moving point, then the number of constraints this whole camera rig provides is the sum of each $H(k,d)$ for each camera until the $N_d - 1$ constraints are obtained.

4.1.3 A General Framework for Trajectory Recovery from Known Cameras

At this point we are in a position to propose a general framework for trajectory recovery. A set of *non-synchronized* cameras $\mathbf{M}_i, i = 1, ..., m$ which are either static or moving is viewing at a set of points $\mathbf{P}_j, j = 1, ..., n$ either static or moving. Since each camera is regarded as a dynamic system, the camera matrices are time dependent. Hence the camera matrix i at time k_i will be denoted by $\mathbf{M}_{ik_i}$. Note that the cameras are independent and in particular they are not supposed to be synchronized. Therefore the time samples are different between every two cameras. This means that the indices k_i are independent between two cameras. Let $\mathbf{p}_{ijk_i}$ be the projection of the point $\mathbf{P}_j$ onto the camera i at time k_i. All the $\mathbf{M}_{ik_i}$ are known for all i and all k_i. This can be achieved during a preprocess by tracking static points over each sequence.

For a given point $\mathbf{P}_j$, the optical rays $\mathbf{L}_{ijk_i} = \widehat{\mathbf{M}}_i \mathbf{p}_{ijk_i}$, for all i and k_i, meet the trajectory of $\mathbf{P}_j$. Then according to the geometric entity generated for all i and k_i by those rays, the motion of $\mathbf{P}_j$ can be recovered. Here we provide a table that gives the correspondence between the motion of the point and the geometry of the optical rays.

Motion of $\mathbf{P}_j$	Geometric entity generated by $\{\mathbf{L}_{ijk}\}$
Static point	Plane in $\mathbb{P}^5$ included in $\mathbb{G}(1,3)$
Point moving on a line	Hyperplane section of $\mathbb{G}(1,3)$
Point moving on a conic	Intersection of a quadric of $\mathbb{P}^5$ with $\mathbb{G}(1,3)$
...	...
Point moving on a curve of degree d	Intersection of a hypersurface of degree d with $\mathbb{G}(1,3)$

Therefore this framework provides us with a way of segmenting static points from moving points and then to reconstruct the location of the former and the trajectory of the latter.

This framework can be seen as a complete generalization of [3], where only the case of moving points on a line was presented using the formalism of the linear line complex.

4.2 Trajectory Triangulation from Tangential Measurements

This second part is devoted to trajectory triangulation from tangential measurements of the motion. For this purpose, we shall use the dual curve of the trajectory. A theoretical result, similar to theorem 8, is given. Finally the interesting case of a synchronized stereo rig receives a special attention.

Let $\mathbf{P}$ be a moving point in $\mathbb{P}^3$. At each time instant, only the tangent of the motion is extracted from the images. Hence the natural reconstruction scheme to be used is based on the dual space representation. Let X be the curve generated by the motion of the point and X^* be its dual curve. Let d and m be the degree of X and X^* respectively. Then X^* is given by a polynomial Υ of degree m.

4.2.1 How Many Measurements are Necessary ?

Consider that a moving point is viewed by either static or dynamic *non-synchronized* cameras. Assume that in each image, the tangent to the trajectory is extracted. Each such tangent $\mathbf{l}$ yields a linear constraint on Υ:

$$\Upsilon(\mathbf{M}^T\mathbf{l}) = 0,$$

where $\mathbf{M}$ is the camera matrix [6]. This leads immediately to the following results:

Proposition 22 *The reconstruction of the trajectory of a moving point can be done by tangential measurements using at least* $k \geq \frac{(m+3)(m+2)(m+1)}{6} - 1$ *independent measurements.*

Note that the case of conics were presented in [47]. Here we summarize the minimal value of k in few cases:

1. for a moving point on a conic locus, $k \geq 9$,

2. for a moving point on a rational cubic, $k \geq 34$,

3. for a moving point on an elliptic cubic, $k \geq 83$,

4. for a moving point on a rational quartic, $k \geq 83$.

4.2.2 Which Measurements are Actually Independent?

We shall proceed in a very similar way than for the case of the Chow polynomial. Therefore we will start by analyzing the case where the center of projection is static.

Proposition 23 *A static camera provides* $\binom{m+2}{m} - 1$ *constraints on* Υ.

---

[6]If $\mathbf{M}$ is the camera matrix, by duality $\mathbf{M}^T$ maps a line of the image to a plane. It is easy to see that this plane passes trough the camera center $\mathbf{O}$. Therefore it is the plane generated by $\mathbf{O}$ and the image line

Proof: The set of all the planes passing through the camera center is a plane of the dual projective space $\mathbb{P}^{3*}$. Therefore a static camera allows to recover the intersection in $\mathbb{P}^{3*}$ of the dual curve with a plane . This intersection is a curve of degree m. Therefore a static camera yields $\binom{m+2}{m} - 1$ constraints on the polynomial Υ defining the dual curve.   ∎

Now we shall consider the general case of a moving camera. Here the result is much simpler than in the case of the Chow polynomial.

Theorem 9 Fundamental Theorem *When the camera is moving, independently of the point, along any trajectory, we have enough constraints to recover the dual trajectory of the moving point.*

Proof: In order to compute the polynomial Υ defining the dual curve X^*, we need $N = \binom{m+3}{m} - 1$ tangent planes to X in a generic situation. Let Y be the curve along which the camera center is moving. In the projective space $\mathbb{P}^3$, any plane is meeting any curve. Hence the set of all the planes meeting Y is the dual space itself. Therefore it is enough to pick N point over Y and N tangents of X. Since the motions of the moving point and of the camera centers are independent, N images are in general sufficient.   ∎

4.2.3 Synchronized Stereo Rig

The case of a synchronized stereo rig has not been considered in the context of point measurements, because in that case such a synchronized stereo rig would allow a classical triangulation technique. However in the case of tangential measurements, using a synchronized stereo rig can reduce the number of necessary measurements. Note that since the information is made from the tangents to the motion, not from the moving point itself, the reconstruction cannot be done point wise despite the synchronization between the two cameras. However this synchronization can be used to reduce the number of views necessary for the reconstruction as follows. At each time t, there are now two measurements l_1 and l_2 which are the tangents of the motion in the two sequences. Each tangent contributes one linear equation on Υ. Moreover the pencil of planes defined by $\mathbf{M}_1^T l_1$ and $\mathbf{M}_2^T l_2$ (where $\mathbf{M}_i$ are the camera matrices) is included into the variety X^* and then Υ must vanish over all points of this pencil. This can be expressed more algebraically. For all λ_1 and λ_2, we have:

$$\Upsilon(\lambda_1 \mathbf{M}_1 l_1 + \lambda_2 \mathbf{M}_2 l_2) = 0.$$

This yields $\binom{m+1}{m} = m + 1$ linear equations on Υ. Therefore the following lower bound is easily obtained:

Proposition 24 *The reconstruction of the trajectory of a moving point from a synchronized stereo rig using tangential measurements, can be done using at least $k \geq \frac{1}{6} \frac{m(m^2+6m+11)}{m+1}$ independent pairs of images.*

Some values of k:

1. for a moving point on a conic locus, $k \geq 3$,

2. for a moving point on a rational cubic, $k \geq 7$,

3. for a moving point on an elliptic cubic, $k \geq 12$,

4. for a moving point on a rational quartic, $k \geq 19$.

5 Experiments

5.1 Epipolar Geometry

5.1.1 Homography Recovery from a Single Planar Curve by Point Extraction

In the first experiment, we consider the problem of recovering the homography matrix induced by a planar cubic across two images (see Figure 5) using the method described in 3.1.4 (i.e. without prior knowledge of the epipolar geometry). The cubic equations of the image curves were recovered by least-squares fitting. The recovered homography was then used to re-project the curve from one image onto the other. The reprojection error was at subpixel values (see Figure 6).

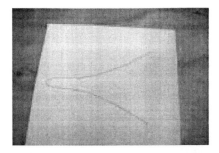

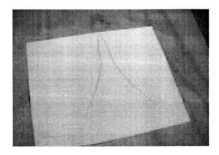

Figure 5: The two images of a cubic curve.

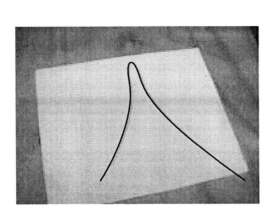

Figure 6: The reprojected curve is overlaid on the second image cubic. The bottom display shows an enlarged section of the curve and the overlaid reprojected curve — the error is at subpixel values.

5.1.2 Recovering Homography from Planar Curves Correspondences

Given two images of the same curve of order 4 (see Figure 7) and the epipolar geometry, we compute the plane and the homography matrix, using the algebraic approach described in 3.3.1. To demonstrate the accuracy of the algorithm, the reprojection of the curve in the second image is shown in the figure 8. The 3D rendering of the correct solution is shown in the figure 9.

Finally, the equation of the correct solution on its plane is given by:

$$f(x,y,z) = \frac{9006922504387547}{9007199254740992}z^4 - \frac{4947731105035649}{1152921504606846976}yz^3 +$$
$$\frac{1070847909255857}{14757395258967641292}y^2z^2 - \frac{5458927196207623}{12089258196146291747061760}y^3z +$$
$$\frac{3969428158337415}{24758800785707605497982484480}y^4 - \frac{7563069091264439}{1152921504606846976}xz^3 +$$
$$\frac{5911661048544087}{295147905179352825856}xyz^2 - \frac{7447102119819593}{302231454903657293676544}xy^2z +$$
$$\frac{3625625302714855}{6189700196426901374495621112}xy^3 + \frac{4936178943362411}{295147905179352825856}x^2z^2 -$$
$$\frac{8944822903795571}{302231454903657293676544}x^2yz + \frac{7158022235457567}{309485009821345068724781056}x^2y^2 -$$
$$\frac{6146225343803339}{302231454903657293676544}zx^3 + \frac{7423176283805271}{6189700196426901374495621112}x^3y +$$
$$\frac{6539339092801811}{6189700196426901374495621112}x^4$$

The curve is drawn on figure 10.

5.1.3 Epipolar Geometry from Points and Conic Correspondences

We proceeded to the recovery of the epipolar geometry from conics and points correspondences extracted from real images. The extraction has been done manually and the conics were fitted by classical least square optimization.

The recovery of the epipolar geometry has been done using four conics and 1 point. First the fundamental matrix is computed using three conics and 1 point, which leads to a

Figure 7: The curves of order 4 as an input of the reconstruction algorithm.

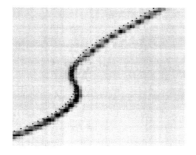

Figure 8: Reprojection of the curve onto the second image.

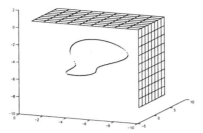

Figure 9: The curve of order 4 as an output of the reconstruction algorithm.

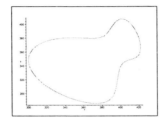

Figure 10: The original curve.

finite number of solutions (see theorem 4). The computation is too intense for the standard computer algebra packages. We have found that **Fast Gb** [7], a powerful software tool for Gröbner basis computation, introduced by J.C. Faugere [16, 17] is one of the few packages that can handle this kind of computation. Then the additional conic is used to select the right solution.

The images used for the experiments together with results and comments are presented in figure 11

5.1.4   Recovering Epipolar Geometry from Spatial Curves Correspondences

We proceed to the computation of the epipolar geometry from a rational cubic and two conics. The curves are randomly chosen, as well as the camera.

---

[7]Logiciel conçu et réalisé au laboratoire LIP6 de l'université Pierre et Marie CURIE.

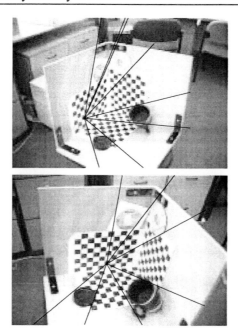

Figure 11: The two images that were used. The epipoles and the corresponding epipolar lines tangents to the conics are drawn on the images.

Hence the cubic is defined by the following system:

$$226566665956452626ZX - 19148549932360861699ZT -$$
$$791130248041963297YZ - 1198609868087508022Z^2 +$$
$$893468169675527814XT + 2859405018489194422T^2 -$$
$$179632615056970090YT + 277960038226472656Y^2 = 0$$

$$555920076452945312XY + 656494420457765614ZX -$$
$$1755155973545148735YZ - 17491544508000074954Z^2 +$$
$$984240461094724954XT - 61309565864179510YT -$$
$$1802588912007356295ZT + 2913197457767954747T^2 = 0$$

$$1111840152905890624X^2 - 2905335341664005486ZX -$$
$$793850352563738017YZ + 12868901614348436582Z^2 +$$
$$1713207647519936006XT - 248798847306328202YT -$$
$$2942349361064284313ZT + 398814386951585134T^2 = 0$$

The first and the second conic are respectively defined by:

$$25X + 9Y + 40Z + 61T = 0$$
$$40X^2 - 78XY + 62ZX + 11XT + 88Y^2 +$$
$$YZ + 30YT + 81Z^2 - 5ZT - 28T^2 = 0$$

and

$$4X - 11Y + 10Z + 57T = 0$$
$$-82X^2 - 48XY - 11ZX + 38XT - 7Y^2 +$$
$$58YZ - 94YT - 68Z^2 + 14ZT - 35T^2 = 0$$

The camera matrices are given by:

$$\mathbf{M}_1 = \begin{bmatrix} -87 & 79 & 43 & -66 \\ -53 & -61 & -23 & -37 \\ 31 & -34 & -42 & 88 \end{bmatrix}$$

$$\mathbf{M}_2 = \begin{bmatrix} -76 & -65 & 25 & 28 \\ -61 & -60 & 9 & 29 \\ -66 & -32 & 78 & 39 \end{bmatrix}$$

Then we form the Extended Kruppa's Equations for each curve. From a computational point of view, it is crucial to enforce the constraint that each λ is different from zero. Mathematically this means that the computation is done in the localization with respect to each λ.

As expected, we get a zero-dimension variety which degree is one. Thus there is a single solution to the epipolar geometry given by the following fundamental matrix:

$$\mathbf{F} = \begin{bmatrix} -\dfrac{511443}{13426} & -\dfrac{2669337}{13426} & -\dfrac{998290}{6713} \\ \dfrac{84845}{2329} & \dfrac{23737631}{114121} & \dfrac{14061396}{114121} \\ \dfrac{1691905}{228242} & \dfrac{3426650}{114121} & \dfrac{8707255}{228242} \end{bmatrix}$$

5.2 Three-Dimensional Reconstruction

5.2.1 Reconstruction of Spatial Curves

We start with a synthetic experiment followed later by a real image one. Consider the curve X, drawn in figure 12, defined by the following equations:

$$F_1(x,y,z,t) = x^2 + y^2 - t^2$$
$$F_2(x,y,z,t) = xt - (z - 10t)^2$$

The curve X is smooth and irreducible, and has degree 4 and genus 1. We define two camera matrices:

$$\mathbf{M}_1 = \begin{bmatrix} 1 & 0 & 0 & 5 \\ 0 & 0 & 1 & -2 \\ 0 & -1 & 0 & -10 \end{bmatrix}$$

$$\mathbf{M}_2 = \begin{bmatrix} 1 & 0 & 0 & -10 \\ 0 & 0 & -1 & 0 \\ 0 & 1 & 0 & -10 \end{bmatrix}$$

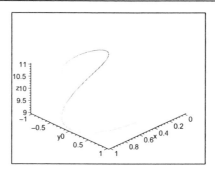

Figure 12: A spatial quartic

The reconstruction of the curve from the two projections has been made in the point space, using **Fast Gb**. As expected there are two irreducible components. One has degree 4 and is the original curve, while the second has degree 12.

5.2.2   Heteroscedastic Estimation

We shall see below that both reconstruction in $\mathbb{G}(1,3)$ and trajectory triangulation lead to estimate a parameter θ in some projective space $\mathbb{P}^n$, where θ is constrained by a set of equations $\mathbf{Z}_i^T \theta = 0$. The vector $\mathbf{Z}_i$ which represents an hyperplane in the space where θ lies come from measurements $\mathbf{x}_i$. Let ϕ be the function that sends a measurement $\mathbf{x}_i$ to the vector $\mathbf{Z}_i$. In practice $\mathbf{x}_i$ is some feature extracted from the image, namely a point in our cases.

The function ϕ is not linear. This causes that even if the basic measurements $\mathbf{x}_i$ are all corrupted by the same additive noise, the vector $\mathbf{Z}_i$ will be distorted by different noises. Therefore, assume that $\mathbf{x}_i = \tilde{\mathbf{x}}_i + \varepsilon_i$, where $\tilde{\mathbf{x}}_i$ is the true value and ε_i an additive noise. Assume that $\varepsilon_i \sim N(0, \mathbf{C}_{\mathbf{x}i})$. Then at the first order we have:

$$\mathbf{Z}_i = \phi(\mathbf{x}_i) = \phi(\tilde{\mathbf{x}}_i) + d\phi(\tilde{\mathbf{x}}_i).\varepsilon_i,$$

so the covariance matrix of $\mathbf{Z}_i$, say $\mathbf{C}_{\mathbf{z}i}$ is given by:

$$\mathbf{C}_{\mathbf{z}i} = \mathbf{J}_{\mathbf{x}i} \mathbf{C}_{\mathbf{x}i} \mathbf{J}_{\mathbf{x}i}^{-T},$$

where $\mathbf{J}_{\mathbf{x}i}$ is the Jacobian matrix of ϕ at $\tilde{\mathbf{x}}_i$. This shows that even if the matrices $\mathbf{C}_{\mathbf{x}i}$ are all equal, the matrices $\mathbf{C}_{\mathbf{z}i}$ are still different. Note that at the first order the mean of $\mathbf{Z}_i$ is still zero. Since each vector $\mathbf{Z}_i$ is distorted by a different noise, we will refer to this situation as a *heteroscedastic noise* estimation problem.

It is well known that in case of heteroscedastic noise, the standard least square algorithm even in its normalized form has large instability [37]. In order to remedy to this problem, three major methods exist:

    1. renormalization method of Kanatani, see [6],

2. approximated likelihood estimation, see [6],

3. general error in variable estimation, see [37].

Please refer to these articles for further details. Those three methods give excellent results when the dimensionality of the problem is small. In our case, the number of parameters is rather high (more than 20). Therefore none of these method is completely robust. Their performance depend on the initial guess of the solution.

5.2.3 3D reconstruction Using the Grassmannian $\mathbb{G}(1,3)$

For the next experiment, we consider seven images of an electric wire — one of the views is shown in figure 13 and the image curve after segmentation and thinning is shown in figure 14. Hence for each of the images, we extracted a set of points lying on the thread. No fitting is performed in the image space. For each image, the camera matrix is calculated using the calibration pattern. Then we proceeded to compute the Chow polynomial Γ of the curve in space. The curve X has degree 3. Once Γ is computed, a reprojection is easily performed, as shown in figure 15.

Figure 13: An electric thread.

Figure 14: An electric thread after segmentation and thinning.

5.3 Dynamic Scenes

As mentioned above, the Chow polynomial is not uniquely defined. In order to get a unique solution, we have to add some constraints to the estimation problem which do not distort the

Figure 15: Reprojection on a new image.

geometric meaning of the Chow polynomial. This is simply done by imposing the Chow polynomial to vanish over W_d additional arbitrary points of $\mathbb{P}^5$ which do not lie on $\mathbb{G}(1,3)$. The number of additional points necessary to get a unique solution is $W_d = \binom{d+5}{d} - N_d$, where d is the degree of the Chow polynomial.

5.3.1 Synthetic Trajectory Triangulation

Let $\mathbf{P} \in \mathbb{P}^3$ be a point moving on a cubic, as follows:

$$\mathbf{P}(t) = \begin{bmatrix} t^3 \\ 2t^3 + 3t^2 \\ t^3 + t^2 + t + 1 \\ t^3 + t^2 + t + 2 \end{bmatrix}$$

It is viewed by a moving camera. At each time instant a picture is made ,we get a 2D point $\mathbf{p}(t) = [x(t), y(t)]^T = [\frac{\mathbf{m}_1^T(t)\mathbf{P}(t)}{\mathbf{m}_3^T(t)\mathbf{P}(t)}, \frac{\mathbf{m}_2^T(t)\mathbf{P}(t)}{\mathbf{m}_3^T(t)\mathbf{P}(t)}]^T$, where $\mathbf{M}^T(t) = [\mathbf{m}_1(t), \mathbf{m}_2(t), \mathbf{m}_3(t)]$ is the transpose of the camera matrix at time t.

Then we build the set of optical rays generated by the sequence. The Chow polynomial is then computed and given below:

$$\begin{aligned}
\Gamma(L_1,...,L_6) = & -72L_2^2L_3 + L_1^3 - 5L_1L_4L_5 - \\
& 18L_1L_3L_6 + 57L_2L_3L_5 + 48L_2L_4L_5 - 43L_1L_2L_4 - \\
& 10L_1L_3L_5 + 21L_1L_5L_6 - 30L_1L_4L_6 - 108L_2L_3L_6 + \\
& 41L_1L_2L_5 + 69L_1L_2L_6 - 26L_1L_2L_3 - 36L_2L_4^2 - \\
& 21L_2L_5^2 + 3L_3L_5^2 - 9L_3^2L_5 - 12L_4^2L_5 + 6L_4L_5^2 + \\
& 4L_1^2L_4 + 20L_2^3 - 13L_3^3 + 8L_4^3 - L_5^3 + 108L_2^2L_6 - \\
& 120L_2^2L_5 + 27L_3^2L_6 - 25L_1^2L_6 + 57L_2L_3^2 + \\
& 84L_2^2L_4 + 7L_1L_3^2 - L_1^2L_5 + 31L_1L_2^2 + \\
& 5L_1^2L_3 + L_1L_5^2 - 11L_1^2L_2 + 7L_1L_4^2
\end{aligned}$$

At this point we perform the algorithm described in section 2 and get exactly the sequence of locations of the moving point $\mathbf{P}(t)$. We show in figure 16 the recovered discrete locations of the point in 3D.

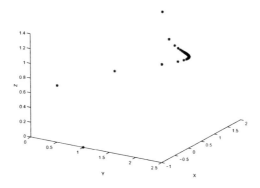

Figure 16: The 3D locations of the point

5.3.2 Trajectory Triangulation from Real Images

A point is moving over a conic section. Four static non-synchronized cameras are looking at it. We show in figure 17 one image of one sequence.

Figure 17: A moving point over a conic section

The camera matrices are computed using the calibration pattern. Every 2D measurement $\mathbf{p}(t)$ is corrupted by additive noise, which we consider as an isotropic Gaussian noise $N(0, \sigma)$. The variance is estimated to be about 2 pixels.

For each 2D point $\mathbf{p}(t)$, we form the optical ray it generates $\mathbf{L}(t) = \widehat{\mathbf{M}}\mathbf{p}(t)$. Then the estimation of the Chow polynomial is made using the optical rays $\mathbf{L}(t)$. In order to avoid the problem of scale, the Plücker coordinates of each line are normalized such that the last coordinate is equal to one. Hence the lines are represented by vectors in a five-dimensional affine space, denoted by $\mathbf{L}_a(t)$. Hence if θ is a vector containing the coefficient of the Chow polynomial Γ, θ is the solution of the following problem:

$$Z(\mathbf{L}_a(t))^T\theta = 0, \text{ for all } t,$$

with $\| \theta \| = 1$. As mentioned above, this estimation problem is characterized by a heteroscedastic noise. More precisely, each $Z(\mathbf{L}_a(t))$ has the following covariance matrix:

$$\mathbf{C}_{\mathbf{L}(t)} = \mathbf{J}\widehat{\mathbf{M}} \begin{bmatrix} \sigma & 0 & 0 \\ 0 & \sigma & 0 \\ 0 & 0 & 0 \end{bmatrix} \widehat{\mathbf{M}}^T\mathbf{J}^T,$$

where $\mathbf{M}$ is the camera matrix and $\mathbf{J}$ is the Jacobian matrix of the normalization of $\mathbf{L}(t)$. That is for $\mathbf{L}(t) = [L_1, L_2, L_3, L_4, L_5, L_6]^T$, we have:

$$\mathbf{J} = \begin{bmatrix} \frac{1}{L_6} & 0 & 0 & 0 & 0 & -\frac{L_1}{L_6^2} \\ 0 & \frac{1}{L_6} & 0 & 0 & 0 & -\frac{L_2}{L_6^2} \\ 0 & 0 & \frac{1}{L_6} & 0 & 0 & -\frac{L_3}{L_6^2} \\ 0 & 0 & 0 & \frac{1}{L_6} & 0 & -\frac{L_4}{L_6^2} \\ 0 & 0 & 0 & 0 & \frac{1}{L_6} & -\frac{L_5}{L_6^2} \end{bmatrix}$$

The result to be presented has been computed using the method introduced in [6]. However the two other methods have similar performance. The result is stable where starting with a good initial guess. In order to handle more general situations we further stabilize it by incorporating some extra constraints that come some our *a-priori* knowledge of the form of the solution. Then the final result is very robust and is presented in figure 18.

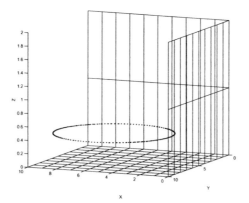

Figure 18: The trajectory rendered in the calibration pattern

5.3.3 Trajectory Triangulation and Dual Curve

The experiments using the dual curve are more complex. These experiments contains two stages. The first one - estimation of the dual curve itself - is very similar to the computation of the Chow polynomial and is simply another case of estimation with heteroscedastic noise. Given the dual curve, two further computations can be performed: (i) computing a set of defining equation of the original curve. (ii) then computing the discrete set of locations where the moving point was at each time instant a picture was made. The first of these computations involves an elimination problem. In that case the sensitivity to noise is a research issue completely open. For this reason, we chose to show, at this stage, only synthetic experiments.

Consider a point moving over the twisted cubic, defined parametrically as follows: $P(t) = [t^3, t^2, t, 1] \in \mathbb{P}^3$. Equivalently it is defined by the following set of equations: $\{Y * T - Z^2 = 0, X * T - Y * Z = 0, X * Z - Y^2 = 0\}$. The point is viewed by a camera moving along a straight line. The tangent to the motion is captured by the camera. Only 34 images are used to recover the trajectory of the point. This is the lower bound given by the theory. Then the dual curve is computed and given by the following polynomial:

$$\Phi = -27A^2D^2 + 18ABDC - 4DB^3 + C^2B^2 - 4AC^3.$$

From the dual curve it is possible to recover the defining equations of the curve generated by the motion of the point. Then one can extract the locations of the point at each time instant an image was made.

6   Summary and Discussion

In this paper we have focused on general algebraic curves as the building blocks from which the camera geometries are to be recovered and as the scene building blocks for the purpose of reconstruction from multiple views, in both static and dynamic cases. The new results derived in this paper include:

1. Extended Kruppa's equations for the recovery of epipolar geometry from two projections of algebraic curves.

2. Dimension analysis for the minimal number of algebraic curves required for a solution of the epipolar geometry.

3. Homography recovery from two views of a general planar curve, when the epipolar geometry is either known or not.

4. The reconstruction from two views of an irreducible curve of degree d is a curve which contains two irreducible components one of degree d and the other of degree $d(d-1)$ — a result that leads to a unique reconstruction of the original curve, for $d > 2$.

5. Formula for the minimal number of views required for the reconstruction of the dual curve.

6. Formula for the minimal number of views required for the reconstruction of the curve representation in $\mathbb{G}(1,3)$.

7. Trajectory triangulation from both tangential and point-based measurements.

8. The number of independent constraints on the trajectory of a mving point provided by a moving camera as a function of the motion the camera.

9. The extraction of practical geometric information on the trajectory of a moving point by simple algorithms.

Some of the algorithms presented in this work lead to solving a system of polynomial equations. There exist two main approaches to handle this problem: (i) computing a Gröbner basis of the ideal defined by the equations, (ii) processing in the dual space via the computation of resultants (see [7, 8] for a detailed presentation). There exists a third method, known as the homotopy method, whose field of applications is broader than the resolution of polynomial systems [1]. However it is generally admitted that the symbolic methods, namely those based on Gröbner basis or resultants, provide better results.

Note that numerical optimization tools like Newton-Raphson or Levenberg-Marquet optimization are not considered here because (i) zero-dimensional polynomial systems which are not overdetermined have more than one root and these optimization methods are designed to extract a single solutions, (ii) the convergence to a solution with these tools is well behaved only when one starts in a small enough neighborhood of the solution.

The use of symbolic tools (either Gröbner basis or resultant) for computer vision applications is not without challenges. First, symbolic computations require large amounts of available computational and memory resources. There is the issue of computational efficiency, scalability to large problems and the questions of effectiveness in the presence of measurement errors. The full answer to these questions is far beyond the scope of this work. The field of symbolic computations for solving polynomial systems is a very active field of research where major progress has been made in the past decade [8, 21, 52]. For example, throughout this paper, the experiments, involving polynomial systems, were performed with one of the latest symbolic tools **Fast GB** developed by Jean-Charles Faugere for efficient and robust Gröbner basis computation. With those latest tools, such as FastGB, one can achieve a high degree of scalability and efficiency in the computations.

Finally the problem of the sensitivity to noise is related to perturbation theory. It is necessary to note that since the computations are symbolic, they do not add any perturbation to the solution. Therefore, as opposed to numerical methods, there is no additional error due to possible truncation during the computations. However, there is very little research on measurement error sensitivity and their propagation throughout the symbolic computations. Such research would be of great interest to the computer vision community, however, this topic is largely open. Nevertheless, a first step in this direction has been done by the introduction of a hybrid of symbolic and numeric computations, especially for the case of zero-dimensional system (which is the case of interest in vision) solved by resultant based methods [52, 41].

References

[1] E. Allgower and K.Georg. Numerical continuation method, An introduction. Number 13 in Computational Mathematics, Springer-Verlag, 1990.

[2] K. Astrom and F. Kahl. Motion Estimation in Image Sequences Using the Deformation of Apparent Contours. In *IEEE Transactions on Pattern Analysis and Machine Intelligence*, **21**(2), February 1999.

[3] S. Avidan and A. Shashua, Trajectory triangulation: 3D reconstruction of moving points from a monocular image sequence. *IEEE Transactions on Pattern Analysis and Machine Intelligence*, **22**(4):348-357, 2000.

[4] M. Barnabei, A. Brini and G.C. Rota, On the exterior calculus of invariant theory. *Journal of Algebra*, **96**, 120-160(1985)

[5] R. Berthilson, K. Astrom and A. Heyden. Reconstruction of curves in R^3, using Factorization and Bundle Adjustment. In *IEEE Transactions on Pattern Analysis and Machine Intelligence*, **21**(2), February 1999.

[6] W. Chojnacki, M. Brooks, A. van den Hengel and D. Gawley, *On the Fitting of Surfaces to Data with Covariances*. PAMI, vol. 22, Nov. 2000.

[7] D. Cox, J. Little and D. O'shea. *Ideals, Varieties and Algorithms, Second Edition*, Springer-Verlag, 1997.

[8] D. Cox, J. Little and D. O'shea. *Using Algebraic Geometry*, Springer-Verlag, 1998.

[9] J. Costeira, T. Kanade, A Multibody Factorization Method for Independent Moving Objects. *International Journal Of Computer Vision*, Kluwer, Vol. 29, Sep. 1998.

[10] Cross and A. Zisserman, *Quadric Reconstruction from Dual-Space Geometry*, 1998.

[11] D. Eisenbud. *Commutative Algebra with a view toward algebraic geometry.* Springer-Verlag, 1995.

[12] D. Eisenbud and J. Harris. *The Geometry of Schemes.* Springer-Verlag, 2000.

[13] O.D. Faugeras *Three-Dimensional Computer Vision, A geometric approach.* MIT Press, 1993.

[14] O. Faugeras and Q.T. Luong, T*he Geometry of Multiple Images*, MIT Press, 2001.

[15] 0.D. Faugeras and T. Papadopoulo. Grassman-Cayley algebra for modeling systems of cameras and the algebraic equations of the manifold of trifocal tensors. *Technical Report - INRIA 3225*, July 1997.

[16] J.C. Faugere. Computing Grobner basis without reduction to zero (F_5). *Technical report*, LIP6, 1998.

[17] J.C. Faugere. *A new efficient algorithm for computing Grobner basis (F_4).*

[18] A.W. Fitzgibbon and A. Zisserman, Multibody Structure and Motion: 3D Reconstruction of Independently Moving Objects. In *Proceedings of European Conference on Computer Vision*, pages 891-906, June 2000.

[19] D. Forsyth, *Recognizing algebraic surfaces from their outlines.*

[20] W. Fulton *Algebraic Curves.*

[21] G.M. Greuel and G. Pfister, *A Singular Introduction to Commutative Algebra.* Springer-Verlag, 2002.

[22] M. Han and T. Kanade, Reconstruction of a Scene with Multiple Linearly Moving Points. In *Proceedings of IEEE Conference on Computer Vision and Pattern recognition*, June 2000.

[23] J. Harris *Algebraic Geometry, a first course.* Springer-Verlag, 1992.

[24] J. Harris and Griffith *Principle of algberaic geometry.*

[25] R. Hartley and A. Zisserman, *Multiple View Geometry in Computer Vision*, Cambridge University Press, 2000.

[26] R. Hartshorne. *Algebraic Geometry.* Springer-Verlag, 1977.

[27] D.P. Huttenlocher, G.A. Klanderman and W.J. Rucklidge Comparing images using the Hausdorff distance. In *IEEE Transactions on Pattern Analysis and Machine Intelligence*, **15**(9), September 1993.

[28] F. Kahl and A. Heyden. Using Conic Correspondence in Two Images to Estimate the Epipolar Geometry. In *Proceedings of the International Conference on Computer Vision*, 1998.

[29] J.Y. Kaminski *Multiple-view Geometry of Algebraic Curves.* Phd dissertation, *The Hebrew University Of Jerusalem*, June 2001.

[30] J.Y. Kaminski and A. Shashua. On Calibration and Reconstruction from Planar Curves. In *Proceedings European Conference on Computer Vision*, 2000.

[31] J.Y. Kaminski, M.Fryers, A.Shashua and M.Teicher. Multiple View Geometry Of (Non-Planar) Algebraic Curves. In *Proceedings of the International Conference on Computer Vision*, 2001.

[32] J.Y. Kaminski and M. Teicher, General Trajectory Triangulation. In *Proceedings of European Conference on Computer Vision*, June 2002.

[33] J.Y. Kaminski and M. Teicher, A General Framework for Trajectory Triangulation. In *Journal of Mathematics Imaging and Vision*, **21**: 27 41, 2004.

[34] S. Lang *Algebra* Addison-Wesley Publishing Company, Inc.

[35] Q.T Luong and T. Vieville. Canonic Representations for the Geometries of Multiple Projective Views. In *Proceedings European Conference on Computer Vision*, 1994.

[36] R.A. Manning C.R. Dyer, Interpolating view and scene motion by dynamic view morphing. In *Proceedings of IEEE Conference on Computer Vision and Pattern recognition*, pages 388-394, June 1999.

[37] B. Matei and P. Meer, A General Method for Errors-in-variables Problems in Computer Vision. In *Proceedings of IEEE Conference on Computer Vision and Pattern recognition*, 2000.

[38] S.J. Maybank and O.D. Faugeras A theory of self-calibration of a moving camera. *International Journal of Computer Vision*, **8**(2):123–151, 1992.

[39] S.D. Ma and X. Chen. Quadric Reconstruction from its Occluding Contours. In *Proceedings International Conference of Pattern Recognition*, 1994.

[40] S.D. Ma and L. Li. Ellipsoid Reconstruction from Three Perspective Views. In *Proceedings International Conference of Pattern Recognition*, 1996.

[41] B. Mourrain and Ph. Trébuchet Algebraic methods for numerical solving. In *Proceedings of the 3rd International Workshop on Symbolic and Numeric Algorithms for Scientific Computing'01*, pp. 42-57, 2002.

[42] T. Papadopoulo and O. Faugeras Computing structure and motion of general 3d curves from monocular sequences of perspective images. In *Proceedings European Conference on Computer Vision*, 1996.

[43] T. Papadopoulo and O. Faugeras Computing structure and motion of general 3d curves from monocular sequences of perspective images. *Technical Report 2765*, INRIA, 1995.

[44] L. Quan. Conic Reconstruction and Correspondence from Two Views. In *IEEE Transactions on Pattern Analysis and Machine Intelligence*, **18**(2), February 1996.

[45] E. Reyssat *Quelques Aspecets des Surfaces de Riemann*. Bikhauser, 1989.

[46] C. Schmid and A. Zisserman. The Geometry and Matching of Curves in Multiple Views. In *Proceedings European Conference on Computer Vision*, 1998.

[47] D. Segal and A. Shashua 3D Reconstruction from Tangent-of-Sight Measurements of a Moving Object Seen from a Moving Camera. In *Proceedings European Conference on Computer Vision*, 2000.

[48] J.G. Semple and G.T. Kneebone. *Algebraic Curves.* Oxford University Press, 1959.

[49] A. Shashua and N. Navab. Relative Affine Structure: *Canonical Model for 3D from 2D Geometry and Applications. IEEE Transactions on Pattern Analysis and Machine Intelligence,* **18**(9):873–883, 1996.

[50] A. Shashua and S. Toelg The Quadric Reference Surface: Theory and Applications. *International Journal of Computer Vision,* **23**(2):185–198, 1997.

[51] A. Shashua and L. Wolf, Homography Tensors: On Algebraic Entities That Represent Three Views of Static or Moving Points. In *Proceedings of European Conference on Computer Vision,* pages 507-521, June 2000.

[52] B. Sturmfels, Solving Systems of Polynomials Equations, *American Mathematical Society,* 2002.

[53] R.J. Walker Algebraic Curves *Princeton University Press,* 1950.

[54] Y. Wexler and A. Shashua, On the synthesis of dynamic scenes from reference view. In *Proceedings of IEEE Conference on Computer Vision and Pattern recognition,* June 2000.

[55] L. Wolf and A. Shashua, On Projection Matrices $\mathbb{P}^k \longrightarrow \mathbb{P}^2$, $k = 3, ..., 6$, and their Applications in Computer Vision. In *Proceedings of IEEE International Conference on Computer Vision,* July 2001.

In: Computer Vision and Robotics
Editor: John X. Liu, pp. 297-310

ISBN 1-59454-357-7
© 2006 Nova Science Publishers, Inc.

Chapter 8

LONG RANGE NAVIGATION OF FLYING VEHICLES WITHOUT GPS RECEIVERS

Yao Jianchao

Centre for Automation Research, University of Maryland,
College Park, MD 20781

Abstract

In this chapter, a new scheme of vision based navigation was proposed for flying vehicles. In this navigation scheme, the main navigation tool is a camera, plus an altimeter. The feasibility of this navigation scheme was carefully studied both from theory and numerical analysis. Unlike most of vision based navigation approaches in which feature trajectories were utilised to compute 3D-platform motion, we use the image geometrical transformation parameters between consecutive frames to infer 3D displacement of camera. Due to this change, the navigation process can be conducted even if there is no salient features that can be extracted from in the image sequence, for example, in the case of flying over the sea. As a result, the long-range navigation becomes possible by use EO sensor. Moreover, the way of improvement navigation accuracy was also addressed. The experiment results demonstrated that the navigation accuracy of this system is compatible to GPS (Global Positioning system), much higher than all kinds of INS (Inertial Navigation System) in terms of position estimation. It is a good alternative choice when the GPS signal is not available

Keyword: Vision Base Navigation, Egomotion estimation, image registration, plane-plus-parallax, Projective model.

1 Introduction

Navigation is the process whereby the measurements, provided by some kind of sensors, are used to determine the position of the vehicle in which they are installed. Among navigation tools, the integration of GPS (global Positioning system) with INS (Inertial Navigation system) is the predominant approach for moving platform navigation, since it provides both the high accuracy of position and long-range stability. Despite of this, the need to use an EO sensor as an alternative navigation instrument is increased, due to its advantage of full-autonomy, no limitation to navigation range, cost-effective and being cable of environment perception.

The crucial issue of vision based navigation is to recover the scene structure and camera motion from image sequence, which has long been an important topic in computer vision and has been approached in many ways. Since the metric structure and less latency are much concerned in navigation, the recursive approach is much popularly used in vision based navigation. Previous application for guidance of flight vehicle includes image-based navigation system for low-altitude flight [1][2], autonomous take-off and landing [3]. Although these excellent works demonstrated the powerful capability of EO sensor in navigation, the role of EO sensor is mainly used for the environment perception. The most part of navigation task during the whole flight is done by the onboard INS/GPS. Another characteristic of their work is the feature-based approach. Since features can only remain in the field of view for certain period of time, the feature based approach is not suitable for long range navigation, in particular the flight path covering scenario of sea view in which the feature is impossible to extract reliably. As a result, EO sensor is still served as an auxiliary instrument for navigation. To the best of our knowledge, there is no previous attempt to take EO sensor as main navigation tool for long range navigation of flight vehicle.

In the chapter, we propose a new navigation scheme for the flight vehicle in which EO sensor serves as the main navigation instrument. The work is inspired by the excellent work of M. Irani al [6] and other previous research on "plan-plus-parallax". The fundamental idea of the scheme is to infer inter-frame 3D incremental motion of camera from image displacement field embedded in two consecutive frames under planar scene assumption. The configuration of the vision based navigation system consists of a camera and an altimeter. The purpose of introducing an altimeter as an aid to the camera is to avoid the issue of uncertainty in scaling factor, so that metric structure of motion can be recovered uniquely. The proposed computational module of navigation includes two components. One is the consecutive registration algorithm based on projective model, the other is the estimation of position and orientation of camera from estimated projective model parameters and height input. In the registration algorithm, since the dominant image motion is induced by the sensor's movement, the robust estimation framework via dynamic weighting is adopted, so as to avoid the perturbation of the secondary or multiple image motion. In the motion estimation module, we implemented both least square approach and Extended Kalman filter approach, and compared their navigation accuracy. The main problem associated with the framework is the accumulated error over the time, in particular for orientation estimation. The navigation accuracy thus decreases dramatically for long range flight path. In order to solve the issue, we introduce orientation sensor as another aid to camera to increase its capability to avoid accumulation error, thus long range navigation via EO sensor becomes feasible. Our experiment demonstrated that navigation error is only 10 meter plus near 10,000 consecutive frames processing. This accuracy is compatible to GPS (Global Positioning system), much higher than all kinds of INS (Inertial Navigation System) in terms of position estimation. It is a good alternative choice when the GPS signal is not available.

The remainder of the chapter is organised as follows. Section 2 gives the description problem and its formulation. Section 3 addresses the implementation issues associated with the navigation scheme. Section 4 presents the experimental evaluation of navigation algorithm, showing its superiority over other navigation tools. Finally, we close the paper with conclusion.

2   Theoretic Analysis of the New Navigation Scheme

The egomotion estimation refers to the problem of determining the relative orientation and translation of camera based on images it captures. In order to formulate the problem, the conformal camera projection model widely used in photogrammetric system was employed for deriving the image transform model, given the sensor's motion parameters. The alignment of the world co-ordinate to the camera centred co-ordinates is conducted in following way. We firstly shift the world co-ordinates by (x_0, y_0, z_0), the location of the camera centre in the world coordintes, and then rotate the resulting coordintes around the x-axis by ω, followed by a rotation by ϕ around the resulting y-axis, and finally, a ratation by κ around the resulting z-axis, as shown in figure 1. Assuming the coordinates of a point in the world coordinates system are (x_w, y_w, z_w), and the coordinate of the point in the camera centered coordinate system are (x_1, y_1, z_1) and its image coordintes (X_1, Y_1). Due to the rigid motion of camera, it becomes $(x_2, y_2, z_2)'$ with respect to second camera coordinate system, and its image coordinates is changed to (X_2, Y_2).

2.1   General Planar Scene

If we assume the general planar scene, we have

$$ax_w + by_w + cz_w = 1 \qquad (1)$$

Based on the assumption, as well as 3D rigid body motion equation and prespective transformation of camera, we can derive the following image transformation]

$$\begin{cases} X_2 = \dfrac{AX_1 + BY_1 + C}{EX_1 + FY_1 + G} \\ Y_2 = \dfrac{HX_1 + IY_1 + J}{EX_1 + FY_1 + G} \end{cases} \qquad (2)$$

Where

$$A = (1 - m_4)r_{11} + \delta x_0 m_1; B = (1 - m_4)r_{12} + \delta x_0 m_2; C = \frac{f}{\varepsilon}[(1 - m_4)r_{13} + \delta x_0 m_3]; E = \frac{\varepsilon}{f}[(1 - m_4)r_{31} + \delta z_0 m_1];$$

$$F = \frac{\varepsilon}{f}[(1 - m_4)r_{32} + \delta z_0 m_2]; G = (1 - m_4)r_{33} + \delta z_0 m_3; H = (1 - m_4)r_{21} + \delta y_0 m_1; I = (1 - m_4)r_{22} + \delta y_0 m_2;$$

$$J = \frac{\varepsilon}{f}[(1 - m_4)r_{23} + \delta y_0 m_3]; \qquad (3)$$

and f represents the focal length of camera, ε the pixel spacing. In addition, other variables involved in (3) can be determined by the sensor's position, orientation and surface equation, as follows:

$$\begin{bmatrix} \delta x_0 \\ \delta y_0 \\ \delta z_0 \end{bmatrix} = R_2 \begin{bmatrix} x_{10} - x_{20} \\ y_{10} - y_{20} \\ z_{10} - z_{20} \end{bmatrix};$$

$$\begin{pmatrix} r_{11} & r_{12} & r_{13} \\ r_{21} & r_{22} & r_{23} \\ r_{31} & r_{32} & r_{33} \end{pmatrix} = \mathbf{R}_2 \mathbf{R}_1^t = \begin{bmatrix} r_{11}^2 & r_{12}^2 & r_{13}^2 \\ r_{21}^2 & r_{22}^2 & r_{23}^2 \\ r_{31}^2 & r_{32}^2 & r_{33}^2 \end{bmatrix} \begin{bmatrix} r_{11}^1 & r_{21}^1 & r_{31}^1 \\ r_{12}^1 & r_{22}^1 & r_{32}^1 \\ r_{13}^1 & r_{23}^1 & r_{33}^1 \end{bmatrix}$$

$$m_1 = ar_{11}^1 + br_{12}^1 + cr_{13}^1; m_2 = ar_{21}^1 + br_{22}^1 + cr_{23}^1;$$
$$m_3 = ar_{31}^1 + br_{32}^1 + cr_{33}^1; m_4 = ax_{10} + by_{10} + cz_{10}$$

The R_1, R_2 are the rotation matrix before and after motion respectively. They are computed from orientation angles (ω, ϕ, κ) (see (5)).

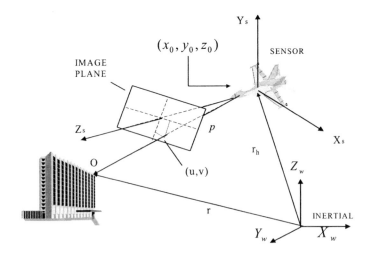

Figure 1. The co-ordinate system involved

For the principle point of first image, we have (X_1, Y_1) =(0,0); its corresponding location in the second image is $(X_2, Y_2) = (\dfrac{C}{G}, \dfrac{J}{G})$. As long as the two cameras are well above the ground, the principal point of first image must be a well-defined point (finite) in the coordintes of second images. Hence $G \neq 0$. The transformation of first image to second image can be determined in terms of eight parameters $a_i, i = 1..8$, as

$$\begin{cases} X_2 = \dfrac{a_3 X_1 + a_5 Y_1 + a_1}{-a_7 X_1 - a_8 Y_1 + 1} \\[3mm] Y_2 = \dfrac{a_4 X_1 + a_6 Y_1 + a_2}{-a_7 X_1 - a_8 Y_1 + 1} \end{cases} \tag{4}$$

Where

$$\begin{aligned} &a_1 = C/G; a_2 = J/G; a_3 = A/G; a_4 = H/G; \\ &a_5 = B/G; a_6 = I/G; a_7 = -E/G; a_8 = -F/G; \end{aligned} \tag{5}$$

The (4) is called the projective model of image transformation. The projective model characterises the image variation induced by camera' motion, under assumption of planar scene.

Given the eight parameters values of a projective model, can we recover uniquely the motion parameters of camera based on these eight values? Unfortunately, the answer is no.

Theorem 1. The relative orientation and translation of the sensor cannot be uniquely recovered from the projective model parameters when general planar scene is assumed.

[Proof] The relative orientation of sensor is described by three independent parameters $\{\delta\psi, \delta\theta, \delta\phi\}$, the relative translation is described by another three independent parameters $\{\delta x_0, \delta y_0, \delta z_0\}$ and general planar scene equation is also expressed by three independent parameters $\{a,b,c\}$. The total independent parameters invloved in (3) is nine. From equation (5), given eight parameters of a projective model $a_i, i = 1..8$, the number of equations that can be constructed is eight. Therefore, the motion parameter of sensor cannot be uniquely determined.

2.2   Ground Scene Case

Note that for the points on the ground plane we have $z_w = D$ (where D is a constant). Based on the assumption, we can also derive the same formulas of image transformation (2), but coeficients of the formulas are different, shown in following

$$A = \frac{f}{\varepsilon}\left(r_{11} + \frac{\delta x_o}{D - z_{lo}} \frac{1}{} r_{13}\right);$$

$$B = \frac{f}{\varepsilon}\left(r_{12} + \frac{\delta x_o}{D - z_{lo}} \frac{1}{} r_{23}\right); \quad C = (\frac{f}{\varepsilon})\left(r_{13} + \frac{\delta x_o}{D - z_{lo}} \frac{1}{} r_{33}\right); \quad E = r_{31} + \frac{\delta z_o}{D - z_{lo}} \frac{1}{} r_{13};$$

$$F = r_{32} + \frac{\delta z_o}{D - z_{1o}} \frac{1}{r_{23}}; \quad G = \frac{f}{\varepsilon}\left(r_{33} + \frac{\delta z_o}{D - z_{1o}} \frac{1}{r_{33}}\right);$$

$$H = \frac{f}{\varepsilon}\left(r_{21} + \frac{\delta y_o}{D - z_{1o}} \frac{1}{r_{13}}\right) \quad I = \frac{f}{\varepsilon}\left(r_{22} + \frac{\delta y_o}{D - z_{1o}} \frac{1}{r_{23}}\right)$$

$$J = (\frac{f}{\varepsilon})^2\left(r_{23} + \frac{\delta y_o}{D - z_{1o}} \frac{1}{r_{33}}\right) \tag{6}$$

If we assume that interframe rotation is small, we have

$$\mathbf{R}_2 \mathbf{R}_1^t = \begin{pmatrix} r_{11} & r_{12} & r_{13} \\ r_{21} & r_{22} & r_{23} \\ r_{31} & r_{32} & r_{33} \end{pmatrix} \approx \begin{bmatrix} 1 & \delta\psi & -\delta\theta \\ -\delta\psi & 1 & \delta\phi \\ \delta\theta & -\delta\phi & 1 \end{bmatrix} \tag{7}$$

Note that $z_{10} - D$ is nothing but the height of aircraft above the ground at time when first image was captured, it can be measured via an altimeter. Thus we let $h = z_{10} - D$. In addition, considering the focal length and the pixel spacing are the intrinsic parameter of camera, which are given beforehand, we let $c = \frac{f}{\varepsilon}$.

For the same reason described in the case of general planar scene, the image transformation can be further reduced from eq. (2) to eq. (4). The parameters of projective model is determined by (5).

Subsituting (6) and (7) into (5), we obtain a linear set of 8 equations

$$\begin{cases} ch\delta\theta + c\delta x_0 r_{33}^1 - a_1 r_{33}^1 \delta z_0 = -a_1 h \\ ch\delta\phi - c\delta y_0 r_{33}^1 + a_2 r_{33}^1 \delta z_0 = a_2 h \\ \delta x_0 r_{13}^1 - a_3 r_{33}^1 \delta z_0 = (1 - a_3)h \\ h\delta\psi + \delta y_0 r_{13}^1 - a_4 r_{33}^1 \delta z_0 = -a_4 h \\ h\delta\psi - \delta x_0 r_{23}^1 + a_5 r_{33}^1 \delta z_0 = -a_5 h \\ \delta y_0 r_{23}^1 - a_6 r_{33}^1 \delta z_0 = (1 - a_6)h \\ h\delta\theta - \delta z_0 (r_{13}^1 + ca_7 r_{33}^1) = -a_7 hc \\ h\delta\phi + \delta z_0 (r_{23}^1 + a_8 r_{33}^1 c) = -a_8 ch \end{cases} \tag{8}$$

Writinng in matrix equation form, we have $\overrightarrow{A} \overrightarrow{x} = \overrightarrow{b}$. Where

$$
A = \begin{bmatrix}
0 & ch & 0 & cr^1_{33} & 0 & -a_1r^1_{33} \\
0 & 0 & ch & 0 & -cr^1_{33} & a_2r^1_{33} \\
0 & 0 & 0 & r^1_{13} & 0 & -a_3r^1_{33} \\
h & 0 & 0 & 0 & r^1_{13} & -a_4r^1_{33} \\
h & 0 & 0 & -r^1_{23} & 0 & -a_5r^1_{33} \\
0 & 0 & 0 & 0 & r^1_{23} & -a_6r^1_{33} \\
0 & h & 0 & 0 & 0 & -(r^1_{13}+ca_7r^1_{33}) \\
0 & 0 & h & 0 & 0 & (r^1_{23}+ca_8r^1_{33})
\end{bmatrix} ; \quad \vec{x} = \begin{bmatrix} \delta\psi \\ \delta\theta \\ \delta\phi \\ \delta x_0 \\ \delta y_0 \\ \delta z_0 \end{bmatrix} ; \quad \vec{b} = \begin{bmatrix} -a_1h \\ a_2h \\ (1-a_5)h \\ -a_4h \\ a_5h \\ (1-a_6)h \\ -a_7hc \\ a_8ch \end{bmatrix} ; \quad (9)
$$

In this set of linear equation, since the number of equation (8) is more than the independent variables (6) , and Matrix A is non-singular matrix, the sensor's motion parameters can be solved uniquely by least sqaure approach.

Theorem 2 : Given the projective model parameters, the relative orientation and translation of the sensor can be uniquely recovered under assumption of ground scene.

Theorem 2 validates our navigation scheme from view of theory. It says with configuration of a on-board camera and an on-board altimeter as a navigation device, the position and oritation of the vehicle in which they are installed can be determined from measurements provided by them.

3 Implementation of Algorithm

The flow chart of implementing above navigation approach is shown in figure 2. There are two functional modules in the system: One is the consecutive frame registration; the other is motion parameter estimation

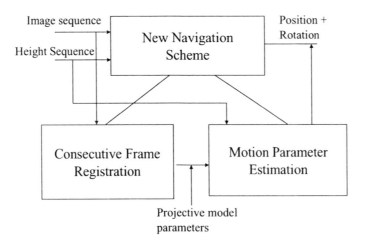

Figure 2. The functional components of the navigation

3.1   Consecutive Frame Registration

The purpose of consecutive frame registration is to estimate projective parameter values, i.e. $\vec{\alpha} = \{a_i\}, i = 1..8$ in equation (4), from two consecutive frames. Since error function between two consecutive frames $f(X_1, Y_1)$ and $g(X_2, Y_2)$ after projective transformation will be represented as:

$$\sum_i | g(X_i', Y_i') - f(X_i, Y_i) |^2 \tag{10}$$

The alignment of two images is to find such optimal parameters $a_i, i = 1..8$ that minimise a weighted sum of errors, i.e.,

$$E = \sum_i w_i (g(X_2, Y_2) - f(X_1, Y_1))^2 = \sum_i w_i r_i^2 \tag{11}$$

Because what we want is the dominant motion induced by the sensor's motion, the algorithm is required to be tolerance of perturbations caused by partial occlusion, multiple motion and noise. Thus robust estimation technique was used. In robust estimation, the quadratic function of residue used in least-squares estimation is replaced by a ρ-function, which assigns small weight for the constraint with larger residue. We used Lorentzion function in the paper, which is given as follows:

$$\rho_{LO}(r_i, \sigma) = \log(1 + \frac{r_i^2}{2\sigma^2})$$

Where r_i is the residue of data constraint and σ is the scale parameter. To use robust estimation in our minimisation framework, we can simply replace the quadratic function in (10) by above ρ-function. This yields the following objective function

$$E(\vec{\alpha}) = \sum_i w_i \rho_{LO}(r_i, \sigma) = \sum_i w_i \rho_{LO}(g(X_2, Y_2) - f(X_2, Y_2), \sigma) \tag{12}$$

Since objective function of (12) is a non-linear, we have to use the iterative non-linear minimisation algorithm, say Levenberg-Marquardt algorithm, to obtain the optimal values of $a_i, i = 1..8$. As we know, in order to make the algorithm to converge to global minimum, the initial estimate of $a_i, i = 1..8$ should be close to the solution. In our algorithm, the initial parameters is obtained by simulated annealing approach incorporated with multiresolution and multiple model strategy, it is then refined via Levenberg-Marquardt algorithm based on intensity-based matching. The detail description of algorithm can be found in [4].

3.2   Motion Parameter Estimation

The goal of motion parameter estimation is to compute the orientation and position of flying vehicle, given the projective parameters $a_i, i = 1..8$ and height of platform. Based on the theoretic derivative in section 2, we implemented two approaches: one is least square approach; the other is Extended Kalman Filter.

3.2.1   Least Square Approach

In terms of equation (8), the incremental position and rotation during period of two consecutive frames are determined via Moore-Penrose pseudo-inverse approach:

$$\vec{x} = (A^T A)^{-1} A^T \vec{b}$$

The current position and orientation is obtained by plus the increment to previous estimate.

3.2.2   Extende Kalman Filter Approach

In order to set up the Kalman-type estimation, we have to specify the state variable, time evolution equation and measurement equation. If we specify the state vector $\vec{s} = [\delta\psi, \delta\theta, \delta\phi, \delta x_0, \delta y_0, \delta z_0]$, which denotes the motion parameter moving from current frame to next frame, we derived the following evolution equation and measurement equation.

(a) Evolution equation

$$\vec{s}_{n+1} = \vec{s}_n + \vec{n}_n$$

where $\vec{n}_n$ is gaussian noise whose covariance is Q_n.

(b) Measurement equation

Measurement equation is used to predict the measurement values (i.e. the projective model parameters), given the current state value estimate. This was formulated in equation (5). This is a set of non-linear equation. The measurement equation of EKF is obtained via the linearization of equation (5), i.e.,

$$\begin{bmatrix} a_1 & a_2 & a_3 & a_4 & a_5 & a_6 & a_7 & a_8 \end{bmatrix}^T = H \vec{s}_n + \vec{n}_M$$

where $\vec{n}_M$ is measurement noise with covariance R and

$$H = \begin{bmatrix} \dfrac{\partial a_1}{\partial \delta\psi} & \dfrac{\partial a_1}{\partial \delta\theta} & \dfrac{\partial a_1}{\partial \delta\phi} & \dfrac{\partial a_1}{\partial \delta x} & \dfrac{\partial a_1}{\partial \delta y} & \dfrac{\partial a_1}{\partial \delta z} \\ \vdots & \vdots & \vdots & \vdots & \vdots & \vdots \\ \dfrac{\partial a_8}{\partial \delta\psi} & \dfrac{\partial a_8}{\partial \delta\theta} & \dfrac{\partial a_8}{\partial \delta\phi} & \dfrac{\partial a_8}{\partial \delta x} & \dfrac{\partial a_8}{\partial \delta y} & \dfrac{\partial a_8}{\partial \delta z} \end{bmatrix}$$

Based on these equations, we can build up EKF for estimation of increment of sensor's parameter.

Via experiment evaluation, we found out that EKF did not demonstrate its superiority over least square method. The fact can be explained as follows: Since the measurement values is computed from image registration algorithm, instead of coming from measurement instrument, the error behaviour of these measurement does not follow the gaussian distribution. Thus violating the basis assumption of using Kalman filter. An more promising research direction is to study and characterise the noise model using non-parametric density distribution, and estimating the motion parameter using a particle filter.

3.2.3   Improving Accuracy of Navigation by an Orientation Sensor

Although the proposed navigation framework was proved theoretically, it is, in fact, subject to the accumulated error. Because the current position and orientation is estimated by summing all the previous increments and initial values, any error arising from estimation of increments in position and orientation leads to inaccuracy of navigation. The fact can also be demonstrated by experimental evaluation of the approach (see first example in next section). From the extensive experiment result analysis, we observe that accuracy of estimation is good for first several hundred frames, it is however degraded dramatically after certain period of time. Furthermore, the large deviation from true value comes first in orientation estimation. It causes subsequently significant error in position estimation due to model error yielded (the error in A matrix of eq. (9)). In order to make the proposed approach work in long range navigation, we introduced an orientation sensor (or INS) as another aid.

By introducing an orientation sensor, the inter-frame angles ($\delta\psi, \delta\theta, \delta\phi$) is measured directly from the sensor, instead of estimation from equation (8). The system of linear equation, i.e. eq. (8), can be simplified to the following

$$\begin{bmatrix} cr_{33}^1 & 0 & -a_1 r_{33}^1 \\ 0 & -cr_{33}^1 & a_2 r_{33}^1 \\ r_{13}^1 & 0 & -a_3 r_{33}^1 \\ 0 & r_{13}^1 & -a_4 r_{33}^1 \\ -r_{23}^1 & 0 & a_5 r_{33}^1 \\ 0 & r_{23}^1 & -a_6 r_{33}^1 \\ 0 & 0 & -(r_{13}^1 + ca_7 r_{33}^1) \\ 0 & 0 & (r_{23}^1 + ca_8 r_{33}^1) \end{bmatrix} \begin{bmatrix} \delta x_0 \\ \delta y_0 \\ \delta z_0 \end{bmatrix} = \begin{bmatrix} -a_1 z_{10} - \delta\theta c z_{10} \\ a_2 z_{10} - \delta\phi c z_{10} \\ (1-a_3) z_{10} \\ -a_4 z_{10} - \delta\psi z_{10} \\ a_5 z_{10} - z_{10}\psi \\ h(1-a_6) \\ -a_7 z_{10} c - \delta\theta z_{10} \\ a_8 c z_{10} - z_{10}\delta\phi \end{bmatrix} \qquad (13)$$

In above equation, we observe that only the last column of data matrix A is affected by the error of the projective model parameters. Therefore, $\delta x_0, \delta y_0$ should be more robust to the error of projective model parameters, compared to δz_0.

The improvement of estimation of δz_0 can be solved in the follow way: from (13) we observe that estimation of δz_0 is an over-constraint problem. If we use equation (13) to estimate δz_0, we find such a solution that satisfying all the constraints as much as possible (in the sense of least mean distance). Since every constraint is affected by the error of projective model, inaccurate constraints lead to actually the wrong solution. Therefore, instead of using all the constraint to find δz_0, we select only one constraint to estimate δz_0. There are two principles of selecting desired constraint for computation of δz_0 among all eight constraints. First of all, the constraint is less affected by the error of projective model; secondly it is better not to include δx_0 or δy_0, since the estimation error of δx_0 or δy_0 will propagate to the estimation of δz_0. Based on this principle, we select the constraint $\delta z_0 (r_{13}^1 + c a_7 r_{33}^1) = z_{10}\theta + a_7 c z_{10}$ (the 7th equation in the (13)) for estimation of δz_0. Therefore, in the algorithm, we first estimate position variable based on (13), subsequently we discard δz_0 solution, and use above constraint for re-compute δz_0.

4   Experimental Evaluation

The experiment was carried by helicopter field trial in which INS, GPS, camera and altimeter were installed. The position and orientation of helicopter at any time are computed via integrated INS and GPS navigation system, which are taken as the true value of navigation. Subsequently, by using our vision based navigation scheme, we estimate the position and orientation of helicopter by using video and altimeter input. After which, we compare it with INS/GPS data. When orientation sensor is required, we utilised the gyroscope reading.

The first example consists of 814 frames, covering both sea and land scene. Figure 3 shows some frames of the sequence (The whole sequence was provided in the supplementary file) Even there is no salient feature in the sea scene, our registration algorithm is still able to produce good result. Figure 4 gives the some projective model parameter estimation for every two consecutive frame. The position and orientation estimation across the whole sequence is shown in figure 5. At final frame, the position error vector is (-27.5289, 0.5024, -17.6844) and orientation error vector is (1.2859°, 1.8153°, 1.0474°). Although the navigation accuracy is acceptable for some application, it degraded fast as the accumulated error about orientation exceeds to certain values, as shown in second example.

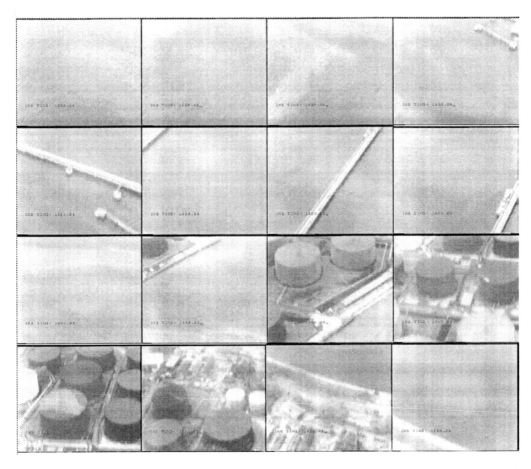

Figure 3. Some frames of the image sequence (1st 50th 100th 150th 200th 250th 350th 400th 450th 500th 550th 600th 650th 700th 750th 800th frame)

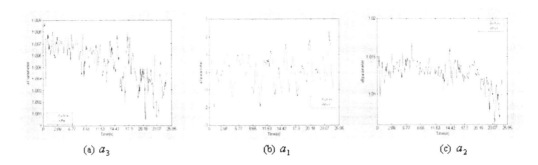

(a) a_3 (b) a_1 (c) a_2

Figure 4. Projective model parameter estimation. The curve in blue indicates the result computed from sesnor's parameter provided by GPS/INS. The curve in red represnts the result obtained y using registration algorithm.

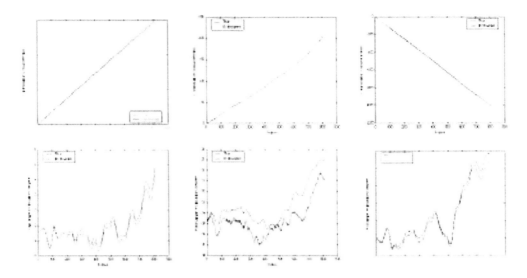

Figure 5. Position (top row) and orientation (bottom row) estimation across the whole sequence

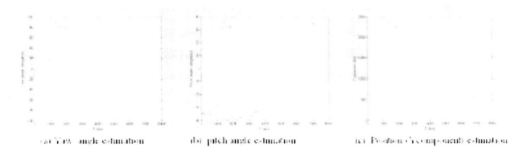

Figure 6. Position and orienttaion estimation based on an camear and an altimeter for second example

The second example is extremely long sequence, consisting of 7500 frames. It covers sea view, urban and wooded area. For such long sequence, the navigation accuracy based on EO and Altimeter is extremely bad after certain period of time, since the accumulated error becomes too large so that estimated value are useless, as shown in figure 6. However, by using an orientation sensor as an aid, the navigation accuracy is extremely improved. Figure 7 shows the position estimation result with help of orientation sensor. The position error vector at final frame is (-0.0329 2.1860 13.4043), thus the navigation error is only 13.58 meter across the whole sequence. The experimental result demonstrated the effective and robust performance of the proposed algorithm. This does be a good alternative navigation scheme when GPS outage for period of time

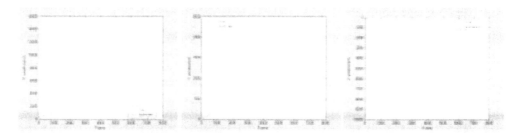

Figure 7. Position estimation by using camera, altimeter and orientation sesnor for second example

Conclusion

The main contribution of the chapter is to develop a new navigation scheme in which the EO sensor is served as the main navigation instrument. We study the feasibility of new navigation scheme theoretically, and also address the implementation issues associated with the scheme. The field trial on the navigation system was conducted; the navigation accuracy was given. The experimental result demonstrated that navigation system consisting of EO, an orientation sensor and altimeter is a robust and high accurate navigation scheme over long period of time. The navigation accuracy of this system is compatible to GPS (Global Positioning system), much higher than all kinds of INS (Inertial Navigation System) in terms of position estimation. It is a good alternative choice when the GPS signal is not available

References

[1] Banavar Sridhar and A.V.Phatak, "Analysis of Image-Based navigation System for Rotocraft Low-Altitude Flight", *IEEE Trans. System, Man, and Cybernetics*, Vol. 22, No. 2, 1992, pp. 290~299

[2] B.K.A.Menon and G.B.Chatterji, B. Sridhar " Electro-Optical Navigation for Aircraft", *IEEE Trans on Aerospace and Electronic System* Vol 29, No.3 July 1993, pp. 825~833

[3] Simon Furst, E.D. Dickmanns " A Vision Based Navigation System for Autonomous Aircraft", *Intelligent Autonomous System* **5**, IAS'98, Sapporo Japan, 01-04 June 1998

[4] Yao Jianchao , Chia Tien Chern "The Practice of Automatic Satellite Image Registration", *Asian Journal of Geoinformatics*, Vol. 3, No.4 June 2003 , pp 11~18

[5] Paul . R.Wolf, Bon A. Dewitt, "Elements of Photogrammetry", 3rd edition, 2000, The McGraw-Hill Companies

[6] M.Irani, B.Rousso and S.Peleg, "Recovery of Ego-Motio using region Alignment*", IEEE Trans. On Pattern Analysis and Machine Intelligence*, Vol. 19, No. 3, March 1997.

In: Computer Vision and Robotics
Editor: John X. Liu, pp. 311-336

ISBN 1-59454-357-7
© 2006 Nova Science Publishers, Inc.

Chapter 9

AN EVALUATION METRIC FOR ADJUSTING PARAMETERS OF SURVEILLANCE VIDEO SYSTEMS

Jesús García, Oscar Pérez,
Antonio Berlanga and José M. Molina[*]
Universidad Carlos III de Madrid. Departamento de Informática
Avda de la Universidad Carlos III, 22. Colmenarejo 28270. Spain

Abstract

In this paper an evaluation metric for calculate the behavior of a video tracking system is proposed. This metric is used for adjusting several parameters of the tracking system in order to improve the performance. The optimization procedure is based on evolutionary computation techniques. The system has been tested in an airport domain where several cameras are deployed for surveillance purposes.

1 Introduction

The application of video cameras for remote surveillance has increased rapidly in the industry for security purposes (Rosin 2003). The installation of many cameras produces a great problem to human operators because the incompatibility of a high analysis of received images with the analysis of the whole information provided for the surveillance video camera net. The solution is the automatic analysis of video frames to represent in a simplify way the video information to be presented to the operator.

A minimal requirement for automatic video surveillance system is the capability of tracking multiple objects or groups of objects in real conditions. A typical video surveillance system is composed of several processes:

---

[*] Funded by CICYT (TIC2002-04491-C02-02)

- A predictive process of the image background, usually Gaussian models are applied to estimate variation in the background
- A detector process of moving targets, detector process works over the previous and actual acquired frames. The detection is carried out by subtracting the two images and analyzing the obtained difference of intensity levels. Detection is directly related with the previous process in order to determine the threshold that defines if a pixel could be considered as a moving target or a variation in the background.
- A grouping pixel process, this process groups correlates adjacent detected pixels to conform detected regions. These regions could be defined by a rectangular area (usually named blob) or by contour shape.
- An association process, this process evaluate which detected blob should be considering as belonging to each existing target.
- A tracking system that maintains a track for each existing target. Usually filters are based on Kalman filter.

Surveillance system depends on many parameters that should be adjusted for a specific implementation. The core of this process is the evaluation of surveillance results. The main point is the definition of a metric to measure the quality of a proposed set of configuration parameters (Black 2003). There are many works (Piater 2003) (Pokrajar 2003) that evaluate video surveillance systems against the ground truth or with synthetic images. In this work we extract the truth values from real images and they are stored in a file. In this file each target is located and positioned in each frame. Targets in the file are defined by six attributes: number of frame, track identifier, min and max value in coordinates x and y of the rectangle that surrounds the target.

Using this metric an evaluation function, we can apply different techniques to evaluate parameters and, then, to optimize them. Classical techniques such as those based on gradient descent are poorly suitable to this optimisation problem. These were discarded due to the high number of local minima presented by the fitness function. We select Evolution Strategies for this problem because they present high robustness and immunity to local extremes and discontinuities in fitness function. In this paper we demonstrate that Evolution Strategies are well matched to this optimisation problem to achieve the desired results (adjust the tracker performance accordingly to all specifications considered).

In this work the surveillance video system is deployed in an airport. The application of video technology in airport areas in a new way to support ground traffic management inside the Advanced Surface Movement, Guidance and Control Systems (A-SMGCS) (FAA 1993) (ECAC 1994) (ICAO 1995) (ASMGCS 2000). Targets moving on the airport are generally commercial aviation aircraft and surface vehicles, such as fuel trucks, luggage convoys, cars, etc. Several aspects should be taken into account for airport surveillance:

- The system will have to deal with fast changes of illumination.
- Cameras should be placed as high as possible to reduce objects overlapping.
- All the interest targets are moving targets, although sometimes they stop.
- The background is not uniform. Steady cameras are used to lower target detection complexity.
- The targets have very dissimilar sizes, orientation dependent shapes, and colors.

- The system has to work in real time with an affordable hardware. Therefore too complex processing algorithms could not be used.
- There are areas with cars, trucks, buses, etc. but without aircraft. Tracking in these areas is not necessary for aircraft control. But, when those vehicles enter aircraft reserved areas, they must be tracked to warn pilots of their potentially dangerous presence.

In the next section, the whole surveillance system is presented, where specific association problems in this application are analyzed. The third section presents a brief resume of the evolution strategies procedure. The proposed metric is presented in section fourth. In section fifth, system output in several scenarios is presented, indicating the response for complex situations, with real image sequences of representative ground operations. Finally, some conclusions are presented.

2   Surveillance Video System

This section describes the structure of an image-based tracking system. It is based on a previously developed prototype, intended to analyse the integration of video technology in A-SMGCS Surveillance function for Madrid/Barajas Airport. This work has been developed jointly by GRPSS group (Grupo de Procesado de Señal y Simulación from Universidad Politécnica de Madrid) and GIAA group (Grupo de Inteligencia Artificial Aplicada from University Carlos III de Madrid). Specifications and details of this video system have appeared in several publications (Besada 2004) (Besada 2001a) (Besada 2001b). The specifications for A-SMGCS require the identification and accurate tracking of all aircraft and vehicles in the airport movement area, in order to improve awareness of surface traffic, conflict monitoring and guidance in a wide range of weather conditions. Basically, camera sensors are being explored as an alternative for surveillance in this area, used as a complementary source of data to conventional sensors such as surface movement radars (Schwabn 1985) (Syletrack 1997) (Ayrsys 1998).

As any multi-target tracking system (Bar-Shalom 1988), the principal points to design are the detection and data association subsystems. A coupled design for detector and tracker blocks has been designed, in order to deal with potential problems such as target splitting and reconnection, occlusions and operation on high-density airport areas (such as inner taxiways and parking zones). The system is oriented to a local tracking using the own camera coordinates. The further practical steps for final integration with other sensors, namely camera calibration, projection to physical coordinates, data fusion, etc, are not covered here.

The system architecture is depicted in figure 1. It is a coupled tracking system where the detected objects are processed to initiate and maintain tracks representing the real targets in the scenario and estimate their location and cinematic state. The tracking feedback over detector allows coherent system behaviour and solves specific problems in this application such as "ghost" targets. The system captures the frames in the video sequence and uses them to compute background estimation. Background statistics are used to detect contrasting pixels corresponding to moving objects. These detected pixels are connected later to form image regions referred to as blobs. Blobs are defined with their spatial borders, generally a rectangular box, centroid location and area. Then, the tracker re-connects these blobs to segment all targets from background and track their motion, applying association and filtering processes.

The association process assigns one or several blobs to each track, while not associated blobs are used to initiate tracks. Map information and masks are used to tune specific aspects such as detection, track initiation, update parameters, etc.

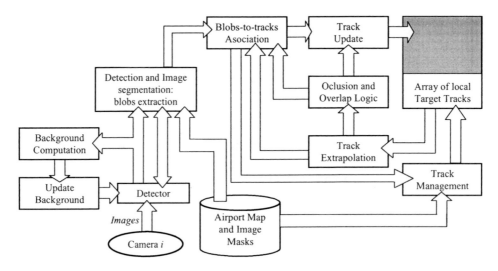

Figure 1. Structure of video surveillance system

To illustrate the process, figure 2 depicts the different levels of information interchanged, from the raw images until the tracks.

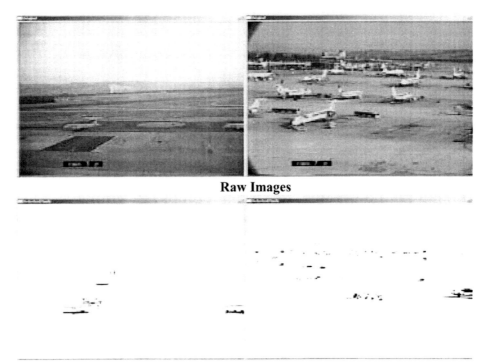

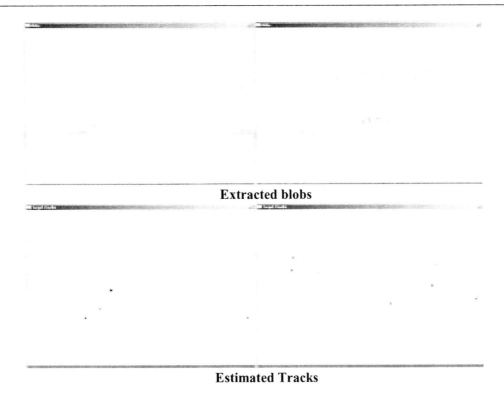

Extracted blobs

Estimated Tracks

Figure 2. Information levels in the processing chain

The most important blocks are described in the following sections.

2.1   Detector and Blobs Extraction

The positioning/tracking algorithm is based on the detection of targets contrasting with local background (Cohen 1998), whose statistics are estimated and updated in an auxiliary image, Background. Then, the pixel level detector is able to extract moving features from this static background, simply comparing the difference with a threshold:

$$Detection(x,y):=[Image(x,y) - Background(x,y)]>THRESHOLD*\sigma \qquad (1)$$

Where σ represents the standard deviation of pixel intensity.

A low threshold would mean a higher sensitivity value, leading to many false detections and higher probability of detection and not corrupting target shape quality. This is one of the key parameters of the system.

The background statistics (mean and variance) for each pixel are estimated, from the sequence of previous images, with a simple iterative process and weights to give higher importance to the most recent frames. Besides, in order to avoid that targets corrupt the background statistics, the update is just performed for pixels not too near of a tracked target, using the tracking information in the detector. So, the statistics for k-th frame are updated as:

$$Background(x,y,k) = \alpha\ Image(x,y,k)+(1-\alpha)\ Background(x,y,k-1) \qquad (2)$$

$$\sigma^2(x,y,k) = \alpha\ [\ Image(x,y,k)\text{-}Background(x,y,k\text{-}1)]^2 + (1\text{-}\alpha)\ \sigma^2\ (x,y,k\text{-}1) \qquad (3)$$

Being x and y pixels out of predicted tracks

If a target is initially stopped within the image and starts moving, its actual background will contrast with the background, calculated from the target image. This leads to the detection of "ghost" targets, effecht than can also be avoided making use of predicted tracks from tracking list in the coupled scheme.

Finally, the algorithm for blobs extraction marks with a unique label all detected pixels connected, by means of a clustering and growing regions algorithm (Snaka 1999). Resulting blobs are then filled, computing the enclosing rectangles, centroids and areas. In order to reduce the number of false detections due to noise, a minimum area, MIN_AREA is required to form blobs.

2.2   Blobs-to-Track Association

This is an essential block for any multi-target tracking system, and it is the focus of interest in this application of video data processing. Its design must take into account the characteristics and quality of data resulting from the detection subsystem. In this case, data are the blobs, resulting from the detection subsystem applied on image sequences of airport surface scenes.

When processing video output in dense airport areas, each available frame presents a set of blob-to-track multi-assignment problems to be solved, where several (or none) blobs may be assigned to the same track and simultaneously several tracks could overlap and share common blobs. So the association problem to solve is the decision of the most proper grouping of blobs and assignation to each track for each frame processed. The characteristics of data to be processed, blobs detected in image sequences of airport surface areas, have been taken into account to develop the image-based tracking system. Due to image irregularities, shadows, occlusions, etc., a first problem of imperfect image segmentation appears, resulting in multiple blobs potentially generated for a single target. This splitting effect occurs with extraneous surface objects, such as luggage convoys, or presence of irregular shadows, and especially when obstacles or other targets appear between the interest target and camera (they are part of background and occlude real targets). So, blobs must be re-connected before track assignment and updating. This problem might be easily solved in single-target scenarios using a blob-grouping algorithm based on the blobs associated to the track in previous frames, defining a spatial gate for each track. However, when multiple targets move closely spaced, their image regions overlap, appearing some targets occluded by other targets or obstacles, so that some blobs can be shared by different tracks. So, a blob-to-track multi-assignment problem has to be solved, where several blobs could be assigned to the same track and simultaneously several tracks could overlap and share common blobs:

- It must group the different blobs representing a single target to avoid track splitting effects. Grouping must adapt to gradual variations in targets sizes and shapes due to changes in distances and orientations of targets.
- When different targets approach, it should avoid mixing their close image regions, since their tracks can be wrongly updated or even one of them discarded, resulting an erroneous single track including more than one target.

The traditional association systems use, together with motion estimation, target position (represented by centroids) extracted from sensor data. Conventional Nearest Neighbor systems (Blackman 1999) deals the assignment between plots and tracks as minimizing a global cost function. This function is computed based on the distance between plots and predicted tracks (residuals) and known statistical models for sensor errors. Bayesian extensions of NN, such as Multiple Hypothesis Tracking (MHT) (Blackman 1999) consider association decisions over several data scans, to ensure track continuity under critical conditions such as presence of false alarms, maneuvers or closely spaced targets. These types of hard-decision systems assume basic constraints of single plot updating each track, and no more than one track updated by the same plot, which are not applicable to the problem dealt.

A possible solution could be the removal of the one-to-one constraints and enumerate all possible grouping and assignment hypothesis, with approaches similar to that suggested in (García 2002). However, these types of solutions could demand excessive computation load to process in real time the frames and it would not ensure solving some problems such as the assignation of corrupted blobs resulting of the mix of several target images. As alternative, an all-neighbors approach, similar to Joint Probabilistic Data Association (Blackman 1999) or PMHT (Gauvrit 1997), seems adequate to this problem, since all blobs potentially gated with each track are used to update it, requiring besides quite lower memory and computation than MHT approaches. Other approaches apply the Expectation-Maximization (Dempster 1977) clustering algorithm for estimating the unknown correspondence among blobs and tracks. The groups of cells representing each target are modelled as a mixture of Gaussian pdfs of unknown parameters, so a likelihood function for those parameters given the measurements are computed at the same time as the unknown correspondence. The application of EM algorithm transforms the hard assignment in a continuous problem, numerically solved with a "hill-climbing" approach. It has been previously applied to data association for computer vision applications (Deallert et al. 2000), and for a probabilistic approach to MHT, PMHT (Gauvrit 1997).

Traditional association systems represent targets with a single position and error parameters. Using a Video Surveillance System, an explicit representation of target shape and dimensions is more adequate to select the set of blobs gated by each track. Track-state vectors with position and cinematic estimates (2D location and velocity referred to the camera plane) are complemented with attributes defining a spatial representation of target extension and shape. So, the predicted target contour is used to gate blobs extracted in next frame. For the sake of simplicity, first a rectangular box has been used to represent the target, as indicated in figure 3. Around the predicted position, $(\hat{x}_p, \hat{y}_p)$, a rectangular box is defined, $(x_{min}, x_{max}, y_{min}, y_{max})$, with the estimated target dimensions. Then, an outer gate, computed with a parameter defined as margin gate, Δ is defined. It represents a permissible area for searching blobs, allowing some freedom to adapta target size and shape with new information. This outer gate allows the system track dynamic variations in target shape along the sequence, for targets not perfectly matching to predictions due to variations in projected shape (changes of orientation, distance, etc.), or maneuvers. This parameter will affect clearly to the tracking performance, since a high margin may increase conflicts and interaction among tracks, while a tight value can produce lose of blobs.

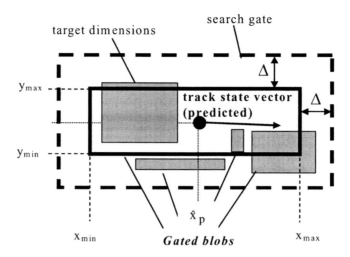

Figure 3. Target segmentation with estimated box

The association algorithm analyzes the track-to-blob correspondence. It firsts checks if the blob and track search rectangular gates are compatible (overlap), and marks those blobs compatible with two or more different tracks as conflictive. Then, all tracks with any blob in conflict are marked as tracks in conflict. Those tracks and blobs in conflict have a different processing to enable conflict resolution. After gating, a grouping algorithm is used to obtain one "pseudoblob" for each track, which will be used to update track state. If there is only one blob associated to the track and the track is not in conflict, the pseudoblob used to update the local track will be this blob. Otherwise, several cases may occur:

When a track is not in conflict, and it has several plots associated with it, they will be merged on a pseudoblob whose bounding limits are the outer limits of all associated blobs. However, in order to avoid problems with grouping different targets and background elements, two constraints are required in the grouping algorithm: density and size tests. If the group of compatible blobs is too big and not dense enough, some blobs (those further from the centroid) are removed from the list until density and size constraints are hold. The details of the grouping algorithm are in the next paragraph.

A conflict situation arises when there are overlapping regions for several targets (conflicting tracks). In this case, the system discards those blobs gated by several tracks, and extrapolates the affected tracks. There are open other alternatives such as solving the conflict using image correlation methods to solve the conflict, or giving priority to greater tracks (if they can be considered as more important, since they represent aircraft on surface)

2.3   Blob Grouping

The grouping system decides if a blobs´ set gated by a certain track can be re-connected to form a single region, the "pseudoblob". This system is needed since targets images often appear split in several blobs. Blob re-connection is strongly related to the other blocks in the tracking system, as indicated in figure 4. After the gating is applied, the compatible blobs with every track are marked. Then, conflict analysis is performed with the available tracks, to decide if they can be used to update them. Finally, non-conflict blobs are analysed previously to track update. As indicated in the figure, track initialization is performed only with blobs not gated by any track, in order to avoid track instability induced by potential tracks around the real ones.

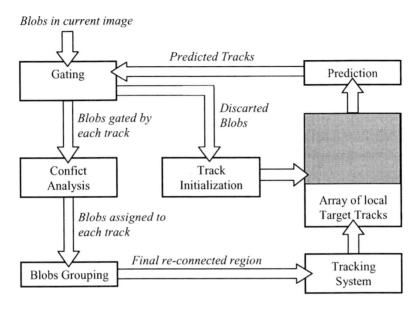

Figure 4. Interaction of grouping and other blocks of tracking system

The grouping criteria avoid bad re-connections of different objects in a single track, a complementary process with the conflict resolution. In these cases, a simple analysis is performed on the posible groups to test two attributes:

- *Maximum allowed size*: this parameter takes into account real aircraft dimensions as the maximum bound for a region updating a track. Since the size is assessed in pixels, this parameter can be adjusted depending on the camera location and image transformation, or simply tuned by analysis of samples with aircraft images in different locations.
- *Minimum group density*: In order to avoid grouping scattered blobs, probably originated by different sources, this criterion assesses the density of the resulting region as a ratio between detected regions and non-detected areas (holes) in the finally reconnected pseudo-blob. So, in the case that this number is bellow a certain threshold, the pseudo-blob is split back in the original blobs, removing elements from the group until the criterion is accomplished. A direct method could be to compare the sum of areas of the connected blobs with the area of the box enclosing them (Figure 5). However, this method depends strongly on the relative situation of the blobs to be tested, since the area of the enclosing rectangular box, aligned with XY axis, changes deeply from the case of blobs aligned with the axis with respect to diagonal configurations. It could be assessed as the ratio of areas computed from the rectangle including all regions and the sum of blobs, but this definition has problems with geometry, as indicated in Figure 5. The situation in the right hand side

represents a probable case of bad grouping and left hand side a probable group, while the density computed with the areas is the same, 50%. So, the selected criterion to assess the density is different, it is the sum of projection over both axes, as displayed in figure 6.

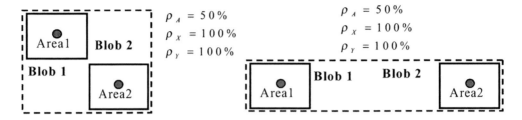

Figure 5. Density and axis density in two situations

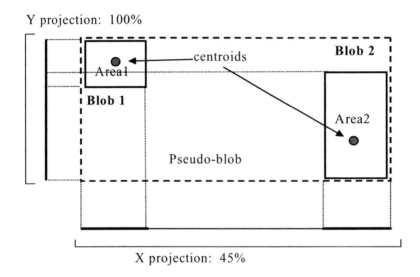

Figure 6. Density of axes covering

The horizontal and vertical densities are computed and then compared with a threshold: $(\rho_X > \rho_{min})$, $(\rho_Y > \rho_{min})$. This criterion imposes a minimum density in the resulting grouped blobs to split the scatter groups until a compact region is left. So the density is defined over the occupied extension of axis, covering of X and Y, as indicated in figure 6, defined as:

$$\rho_X = \frac{occupied\ horizontal\ axis}{horizontal\ length} \qquad \rho_Y = \frac{occupied\ vertical\ axis}{vertical\ length} \tag{4}$$

Where the occupied extensions are computed with a division of both axes for the grouped blob and counting the bins covered by at least the projection of one blob.

2.4   Tracks Filtering, Initiation and Deletion

A recursive filter is used to update both centroid position and velocity for each track from the sequence of assigned values, by means of a decoupled Kalman filter for each cartesian coordinate, with a piecewise constant white acceleration model (Blackman 1999). So, the association of blobs to tracks determines the evolution of tracks representing the targets. The acceleration variance must be enough to allow manoeuvres and projection changes. The predicted rectangular gate, with its margin, is used for gating, so it is important that the filter is "locked" to real trajectory, otherwise track would lose its real blobs to finally drop.

Tracking initialization and management takes non-associated blobs to any previous track. It requires that not-gated blobs extracted in successive frames accomplish certain properties such as a maximum velocity, and similar sizes. As mentioned above, blobs for initialization are those not gated for any other pre-after a potential track is initiated, it is required a third assigned blob to confirm it, otherwise it is automatically erased.

3   Evolution Strategies

In numeric optimization problems, when f is a smooth, low-dimensional function, there are a number of classic optimization methods available. The best case is for low-dimensional analytical functions, where solutions can be analytically determined, or found with simple sampling methods. If partial derivatives of function with respect to input parameters are available, gradient-descent methods could be used to find the directions leading to minimum. However, these gradient-descent methods quickly converge and stop at local minima, so additional steps must be added to find the global minimum. For instance, with a moderated number of global minima we could run several gradient-descent solvers to find the best solution. The problem is that the number of similar local minima increases exponentially with dimensionality, making these types of solvers unfeasible.

For complex domains, Evolutionary Algorithms (EA), (Back 2000 a,b), has proven to be robust and efficient stochastic optimization methods, combining properties of volume and path-oriented searching techniques. The label "Evolutionary Algorithms" groups a set of techniques that are inspired in a metaphor of the biological evolution. Otherwise, the EA, neural nets and fuzzy systems constitute the soft computing paradigm. The great popularity achieved for this kind of techniques should fundamentally to that they have acceptable performance at acceptable costs on a wide range of problems and they obtain better result than other techniques when handle complex problems with lots of data and parameters, exist many local minimums and exist complex or unknown relationships between parameters. However, always will be necessary to take his lacks into account when apply to a problem of optimization. They are computationally expensive, therefore are not adequate for optimizations that take place in real-time applications, require adjusting intrinsic parameters and most of all, the optimal solution in finite time is not guaranteed.

A new terminology was introduced in EA paradigm, taken from the Biology, to make reference to basic concepts of the numerical classical optimization's field. Thus, the candidate solution is named "individual", the quality function is called "fitness function" and the problem that must be resolved is the "environment" that the evolution takes place. The process of evolution happens in equilibrium among two antagonistic tendencies, on the one hand some operators's tendency is to increase the genetic diversity in the population and on the other hand the mechanism of selection decreases this diversity.

In the following figure the evolutionary process is shown. The cycle starts generating a random population, and then the selection operator chooses the fittest solutions to the problem. The application of some genetic operators produces, "breeds", new solutions and the replacement process generates a new population.

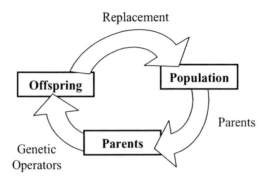

Figure 7. Evolutionary Wheel

Evolution Strategies (ES) (Rechenberg 1994), (Schwefel 1977, 1993) are the evolutionary algorithms specifically conceived for numerical optimization, and have been successfully applied to engineering optimization problems with real-valued vector representations (Bäck 1996). They combine a search process which randomly scans the feasible region (exploration) and local optimization along certain paths (exploitation), achieving very acceptable rates of robustness and efficiency. Each solution to the problem is defined as an individual in a population, being codified each individual with a couple of real valued vectors: the searched parameters and a standard deviation of each parameter used in the search process.

A general ES is defined as an 8-tuple (Bäck 1996): ES= (I, Φ, Ω, Ψ, s, ι, μ, λ)

Where $I = (\vec{x}, \vec{\sigma}, \vec{\alpha}) = \Re^n \times \Re_+^{n_\sigma} \times [-\pi, \pi]^{n_\alpha}$ is the space of individuals, $n_\sigma \in \{1, ..., n\}$ and $n_\alpha \in \{0, (2n-n_\sigma)(n_\sigma-1)/2\}$, $\Phi : I \rightarrow \Re = f$, is the fitness function, $\Omega = \{m_{\{\tau, \tau', \beta\}}: I^\lambda \rightarrow I^\lambda\} \cup \{r_{\{rx, r\sigma, r\alpha\}}: I^\mu \rightarrow I^\lambda\}$ are the genetic operators, mutation and crossover operators. $\Psi(P) = s(P \cup m_{\{\tau, \tau', \beta\}}(r_{\{rx, r\sigma, r\alpha\}}(P)))$ is the process to generate a new set of individuals, s is the selection operator and ι is the termination criterion. In this work, the definition of the individual has been simplified: the rotation angles n_α have not been taken into account, $n_\alpha=0$.

The mutation operator generates new individuals as follows:

$$\sigma_i' = \sigma_i \cdot \exp(\tau' \cdot N(0,1) + \tau \cdot N_i(0,1)) \tag{5}$$

$$\vec{x}' = \vec{x} + \sigma_i' \cdot \vec{N}(0,1) \tag{6}$$

In the following figure, the general outline of ES is showed.

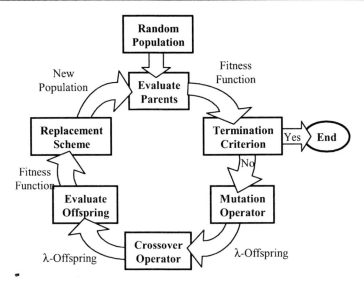

Figure 8: General outline of an ES

ES has several formulations, but the most common form is (μ, λ)-ES, where $\lambda > \mu \geq 1$, (μ, λ) means that μ-parents generate λ-offspring through crossover and mutation in each generation. The best μ offspring are selected deterministically from the λ offspring and replace the current parents. ES considers that strategy parameters, which roughly define the size of mutations, are controlled by a "self-adaptive" property of their own. An extension of the selection scheme is the use of elitism; this formulation is called $(\mu+\lambda)$-ES. In each generation, the best μ-offspring of the set μ-parents and λ-offspring replace current parents. Thus, the best solutions are maintained through generation. The computational cost of (μ, λ)-ES and $(\mu+\lambda)$-ES formulation is the same.

4   Evaluation System

4.1   Need of an Evaluator

One of the most important aims of our study is to calculate some parameters which allow the evaluation of the performance of our tracking system. To achieve this goal, the measurements given for the tracking system are compared with the ideal output. This ground truth is the result of a careful study from pre-recorded video sequences and a subsequent process in which a human operator selects points for each target.

This process can be explained for each video in several steps as follows:

1)   The most interesting objectives are selected in order to analyze their trajectories. The criterion for selection is the size and position of the different targets in the videos. The bigger target the better, and the more difficult to be distinguished from a close object the more interesting for our study.

2)   The coordinates of the targets are selected frame by frame by surrounding them with rectangles and taking the upper left corner and lower right corner as location of our objectives at this moment. This location is referred to the upper left corner of the

complete image which represents the pixel (0, 0). Then, the range of values varies from 0 to 767 pixels in the x-axis and from 0 to 575 in y-axis. Thus, the ground truth can be defined as a set of rectangles that define the trajectory of each target.

3) Finally, the ground truth data are stored in a table which will be used to compare to the result tracks of the tracking system.

In order to have a clear idea of the steps described above a flow chart is shown in figure 9.

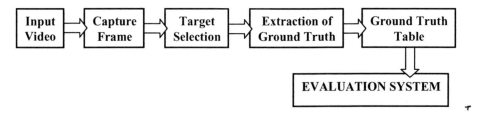

Figure 9. Steps to get the ground truth tracks of the selected targets

The results trajectories have to be as similar as possible the ground truth tracks. Thus, the next step is the comparison of the ideal trajectories with the detected ones so that a group of parameters can be obtained to analyze the results and determine the quality of our detections. Furthermore, the quality will be quantified by means of different weights to each parameter and some other calculations in order to have an outcome of the analysis part which will be the input of the evolutionary strategy program. All this process is shown in figure 10.

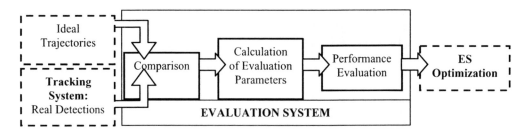

Figure 10. Evaluation System modules

Next paragraphs try to explain how the ideas which have been explained before are carried out to build our evaluator. The description will be done by following the order of the flow chart showed above.

4.2   Ideal Trajectories

As stated above, each target is located and positioned in each frame and these data are recorded in a table in seven columns:

1) Number of frame which has been analyzed. There are so many lines a frame as selected targets in this specific frame.

2) Track identifier, which will be a number between 1 and the number of tracks, given to a specific trajectory.
3) Value of the minimum x coordinates of the rectangle which surrounds the target.
4) Value of the maximum x coordinates of the rectangle which surrounds the target.
5) Value of the minimum y coordinates of the rectangle which surrounds the target.
6) Value of the maximum y coordinates of the rectangle which surrounds the target.
7) Number of line in the text file.

Figure 11 below shows the format of the data for the ideal values stored after a careful extraction process which is done for each video.

```
...
30 2 489 629 354 403 22
30 3 0 13 351 363 23
31 2 468 607 354 404 24
31 3 0 27 351 368 25
32 2 449 587 357 404 26
32 3 0 40 347 367 27
...
```

Figure 11. Example of the ground truth table

4.3   Evaluation Metrics

The next process (see Figure 12) is carried out as many times as estimated tracks a frame are returned by the tracking system. The information that is needed are the necessary data to evaluate the performance of our tracking system: time of prediction, track identifier, value of the minimum x and y and maximum x and y which surround the target (shape of a rectangle), the codes or identifiers of each ideal trajectories and a matrix to store the results. These data have been estimated by the tracking system explained in the former chapters.

First of all, the result tracks are checked to see if they match with the ground truth tracks registered in the ground truth table. For example, as we see in the next pictures (figure 13), the real image shows two aircrafts in the parallel taxiways while the tracking system displays three targets. Then, the target which is in the middle of the screen (the 'ghost target' mentioned in the former chapter) would not pass the test and it would be marked as a mismatched track.

Thus, if the test is passed, the next task of the evaluator is to find couples among all the tracks estimated for our system and the ground truth tracks. In figure 4, it is very clear to check that the plane that is on the left side, whose ground truth tracks have been extracted and stored previously, matches with the track displayed on the left side (dark blue). The same process could be followed for the aircraft that is on the right side.

Once the couples are matched, the parameters to evaluate the quality of our system can be computed and stored in a matrix. The next list describes the quantities and how they are estimated. They parameters are divided in 'accuracy metrics' and 'continuity metrics'.

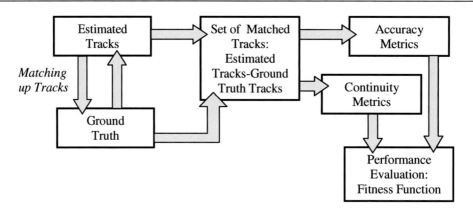

Figure 12. Calculation of Evaluation Metrics

Figure 13. Example of mismatch for the dark red track in the center of the image. Above, the real image shot by the cameras is shown. Below, it can be observed the tracks that our system estimates.

Accuracy Metrics:

1) *Error in area (in percentage)*:
   To begin with the first parameter, the difference between the ideal area and the estimated area is computed

2) *X-Error and Y-Error*:
   The difference among the x and y coordinates of the bounding box of an object estimated by the tracking system and the ground truth.

3) *Overlap between the real and the detected area of the rectangles (in percentage)*:
   The inside overlap and the outside overlap between the areas are computed and the program takes the biggest one. The next example shows a clearer ideal of this concept. Figure 14 shows the bigger rectangle which represents the real location of the target and, the small one represents the location detected by our system. The two overlaps are computed and the program selects the highest value.

Figure 14. Calculation of the overlap between the ground truth (big rectangle) and the estimated location (small rectangle)

Continuity Metrics:

4) *Commutation*:

One of the most important parameters to measure by the evaluator is the commutation. It is defined as follow: the first time the track is estimated, the tracking system marks it with an identifier. If this identifier changes in subsequent frames, the track is considered a commuted track. There could be several reasons why the identifier changes, but the most common is the loss of the track for a short time and subsequent recovery.

Figure 15 shows a good example of a commutation. The track of the aircraft that is going from left side to the right side of the screen is lost by the tracking system and recovered three frames later. The identifier that the tracking system gives to the recovered track is different than the former one. The evaluator realizes that the new track is in fact the old one and marks it with a commutation flag.

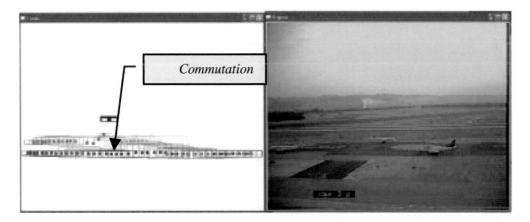

Figure 15. Example of commutation for the aircraft which goes from the left side to the right side of the screen. It is seen that the color of the track changes since it is considered a different track for our tracking system. The evaluator realizes that both detections match with the same ideal track

5) *Number of tracks*:

It is checked if more than one detected track is matched with the same ideal track. If this happens, the program keeps the detected track which has a bigger overlapped area value, removes the other one and marks the frame with a flag that indicates the number of detected tracks associated to this ideal one. The next graph is obtained for the last case in order to show the value of the parameter explained above (figure 16).

As it is observed, only one detected track is associated to one ideal track except for five frames where the value is 0 or 2. The quality of the system will become increasingly better if the value of this last parameter is equal to one.

If the behavior of our system is checked for a more difficult case, it is seen that the evaluator part is a key tool to improve the performance.

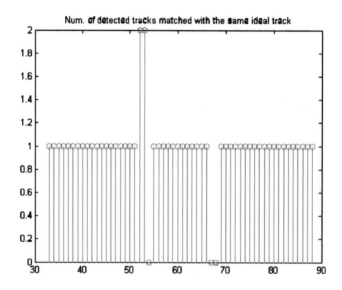

Figure 16. Number of tracks detected by our system which match with the same ideal track

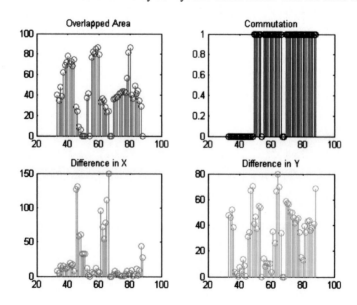

Figure 17. Values of the parameters for the evaluation

Figure 17 displays the values of the X-Error,Y-Error, overlapped area and commutation parameters for this track under study. It is observed that the first detection occurs in the 32nd frame and disappears in the 49th frame. It is recovered by the tracking system in the 52nd

frame where the commutation flag is switched on. The different values of the error in the y axis from the center of the ground truth rectangle tell us that the parameters of the Kalman filter must be adjusted in order to have a better performance. Between the 60th and the 80th frames, the values of overlapped area and errors in x and y coordinates get worse, just in the moment where the aircraft of the study crosses with another airplane. The conclusion is that some other arguments that rule the tracking system should be improved.

The second video under study presents more targets and movements which increase the difficulty of detection. Thus, the next study is focused on the big aircraft that heads from the right side to the left side of the screen. The first detection shows the next parameters with which the quality of the system is questioned.

Figure 18. Background of the second video under study

Figure 19. Result of the tracking system for the big airplane

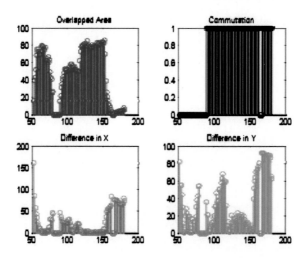

Figure 20. Values of the parameters to measure the quality of the aircraft's track

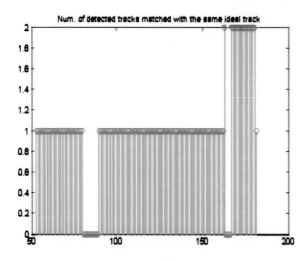

Figure 21. Number of tracks estimated by our system which match with the same ideal track for the big aircraft of the video

Some conclusions can be inferred by the careful study of figures 19, 20 and 21. First of all, the commutation graph and the screen of tracks show how a commutation is produced at frame 90th. Then, some parameters must be changed to avoid this loss of tracking. Furthermore, the Y-Error is higher than X-Error which indicates that the filter arguments should be adjusted better. Finally, the last graph shows that two tracks are estimated to mach up with the same ideal one from the frame number 160.

5   Adjusting Parameters of Surveillance System

In this specific problem, one individual will represent a set of parameters of the whole tracking system. A parameter affects to the detector, modifying the regions detection threshold of the image that they will be considered to update the current tracks, the remaining parameters modify the behavior of some stages of the tracking system, such as the initialization of tracks and mainly the association procedure.

The three parameters which are going to be selected are the following:

- The "threshold" that defines if a pixel could be considered as moving target or a variation in the background.
- The "minimum area of blob" defines a minimum area of blob in order to reduce false detections due to noise.
- The "margin gate" or outer gate defines a permissible area for searching blobs around the defined rectangular box (x_{min}, x_{max}, y_{min}, y_{max}), allowing some freedom to adapt target size and shape with new information.

We have implemented ES for this problem with a size of 6+6 individuals and mutation factor $\Delta\sigma=0.5$. The fitness function will be based on the metric function previously described in section 4. The type of crossover used in this work is the discrete crossover and the two standard types of ES replacement schemes, $(\mu+\lambda)$-ES and (μ,λ)-ES, were used to select the individual to the next generation.

It is important to notice that simulations are carried out using common random numbers to evaluate all individuals in all generations, enhancing system comparison within the optimization loop. In other words, the noise samples used to simulate all scenarios in the RMS evaluation are the same for each individual, in order to exploit the advantages coming from the use of a deterministic fitness function. Besides, the number of iterations was selected to guarantee that confidence intervals of estimated figures were short in relation to the estimated values.

A basic aspect to achieve successful optimization in any evolutionary algorithm is the control of diversity, but this appropriateness will depend on the problem's landscape. If a population converges to a particular point in a search space too fast in relation to the roughness of its landscape it is very probable that it will end in a local minimum. On the contrary, a too slow convergence will require from a large computational effort to find the solution. ES gives the higher importance to the mutation operator, achieving the interesting property of being "self-adaptive" in the sizes of steps carried out during mutation, as indicated in step 3 of algorithm above. Before selecting an algorithm for optimization it is interesting considering the point of view of the "No Free Lunch" (NFL) theorem [8], which asserts that no optimization procedure is better than a random search if the performance measurement consists in averaging arbitrary fitness functions. The performance of ES has been widely analyzed under a set of well-known test functions [9], [10]. They are artificial analytical functions used as benchmarks for comparison of representative properties of optimization techniques, such as convergence velocity under unimodal landscapes, robustness with multimodality, non-linearity, constraints, presence of flat plateaus at different heights, etc. However, the performance on these test functions can not be directly extrapolated to real engineering applications.

5.1   Fitness Function

This section calculates a number by giving a specific weight to each of the next parameters and computing the sum of all of them. The result constitutes the measurement of the quality level for the tracking system and is the input for the next part of the system (the evolutionary strategies programs).

Here it will be explained how the final result is obtained by means of a sum of different factors:

- The first one is a counter which stores how many times the ground truth and the tracked object data do not match up. Furthermore, this counter is normalized by the difference between the last and first frame in which the ideal track disappears and appears.

- The three next terms are the total sum of the overlapped areas and the central errors of x and y axes which are normalized by a number which indicates how many times these values are different from zero.

- The next two factors are two counters:

    ☐ Counter number one: how many times the ground truth track is matched with more than one tracked object data
    ☐ Counter number two: how many times the ground truth track is not matched with any track at all.

- Finally, the last term is the number of commutations in the track under study.

The three last factors are normalized by the same value of normalization as the first one.

5.2   Performance of Optimization

This section of the study shows how the analysis of the evaluation system, and the subsequent use of ES, improve considerably the performance of our tracking system. Three parameters explained above are going to be studied in order to see the effects of them in the optimization of the tracking system: the threshold, the minimum area and margin gate.

The first example shows the video of the three aircrafts used in the former chapters to explain how the tracking system works. The study is focused on the airplane that moves from the left side to the right side of the screen. The adjusting parameters of the system are randomly selected by the program:

- Threshold: 39
- Minimum Area of blob: 7
- Margin gate: 1.0838

The result of the first execution can be seen in figure 22.

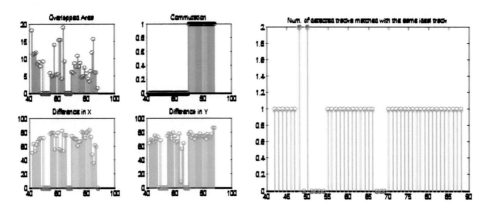

Figure 22. Performance of example 1 before ES optimization

After using the ES program, the performance of our system improves. The values of the three parameters under study are:

- Threshold: 16.7
- Minimum Area of blob: 3.15
- Margin gate: 10.95

The two first parameters have lower values and the last one is higher. That means, for example, that the criterion for a pixel to be considered as a moving target is less restricted. Then, the sensitivity value and the probability of detection are higher. Moreover, the value of the minimum area that defines a blob is also lower so that much more blobs are considered by the system to likely form future rectangles. And finally, the higher value of margin gate permitted the search of new valuable information around the rectangle to adapt the target size and shape with new information.

Thus, the result is a better performance of our system that can be observed in figure 23.

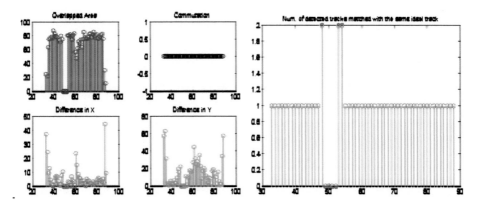

Figure 23. Performance of example 1 after ES optimization

After the optimization of the adjusting parameters, the first estimation track is done earlier (34th frame) than in the previous example (42nd frame) and the track is lost once instead of twice.

The second example takes the second video considered in former chapters and the aircraft that goes from the left side to the right side of the screen as the main target to study. The values of the three parameters on which our study is focus:

- Threshold: 16.7
- Minimum Area of blob: 2
- Margin gate: 2

The surveillance system performance is shown in figure 24.

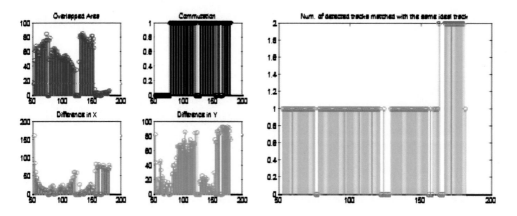

Figure 24. Performance of example 2 before ES optimization

The ES program optimizes these three parameters, giving as a result:

- Threshold: 38.94
- Minimum Area of blob: 6.9885
- Margin gate: 1.0838

The surveillance system performance is shown in figure 25.

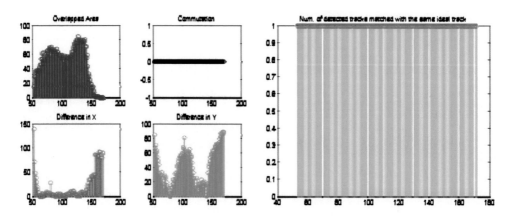

Figure 25. Performance of example 2 after ES optimization

The new values show the opposite situation that we had in the previous example. The threshold and the minimum area of blob are higher and the margin gate is lower. That means that these values decrease conflicts and interaction among tracks (no commutation), at the same time that the detection probably and the criterion threshold to form a blob are lower.

6 Conclusions

We have presented a novel process to evaluate the performance of a tracking system based on the extraction of information from images filmed by a camera.

The ground truth tracks, which have been previously selected and stored by a human operator, are compared to the estimated tracks. The comparison is carried out by means of a set of evaluation metrics which are used to compute a number that represents the quality of the system. Then, the proposed metric constitutes the argument to be introduced to the evolutionary strategy (ES) whose function is the optimization of the parameters that rule the tracking system. This process is repeated until the result and the parameters are good enough to assure that the system will do a proper performance. The study tests several videos and shows the improvement of the results for the optimization of three parameters of the tracking system.

In future works we will implement the optimization of the whole set of parameters using the results of this paper as valuable background. Furthermore, we plan the evaluation over a high value of videos which present very different number of targets and weather conditions.

References

(Airsys 1998) "Cooperative Area Precision Tracking System (CAPTS). Final report of test results". Frankfurt Airport. Airsys ATM Gmbh. January 1998

(ASMGCS 2000) "European Manual of Advanced Surface Movement and Control Systems (ASMGCS)". Draft. Volume 1: *Operational Requirements*. Draft Version 04. 08/24/2000

(Bäck 1996) T. Bäck, "Evolutionary Algorithms in Theory and Practice", *Oxford University Press*, New York, 1996

(Bäck 2000a) T. Bäck, D.B. Fogel, Z.Michalewicz. "Evolutionary Computation: Advanced Algorithms and Operators", Institute of Physics, London, 2000

(Bäck 2000b) T. Bäck, D.B. Fogel, Z.Michalewicz. "Evolutionary Computation: Basic Algorithms and Operators", Institute of Physics, London, 2000

(Bar-Shalom 1988) Y. Bar-Shalom, T. E. Fortmann. "Tracking and Data Association". *Mathematics in science and engineering. Academic Press*, Inc. 1988

(Besada 2001a) Juan A. Besada, Javier Portillo, Jesús García, José M. Molina, Ángeles Varona, Germán Gonzalez. "Image-Based Automatic Surveillance for Airport Surface". FUSION 2001Conference. Montreal, Canada. August 2001

(Besada 2001b) J. A. Besada, J. Portillo, J. García, J. M. Molina. "Image-Based Automatic Surveillance for Airport Surface". Fourth International Conference on Information Fusion. FUSION 2001. Montreal, Canada. August 2001

(Besada 2004) J. A. Besada, J. M. Molina, J. García, A. Berlanga, J. Portillo. "Aircraft Identification integrated in an Airport Surface Surveillance Video System". *Machine Vision & Applications*, Vol 15, No 3. July 2004

(Black 2003) J. Black, T. Ellis and P. Rosin. "A Novel Method for Video Tracking Performance Evaluation", *Joint IEEE Int. Workshop on Visual Surveillance and Performance Evaluation of Tracking and Surveillance* (VS-PETS), 2003

(Blackman 1999) S. Blackman, R. Popoli. "Design and Analysis of Modern Tracking Systems". Artech House. 1999

(Cohen 1998) I. Cohen and G. Medioni. "Detecting and Tracking Moving Objects in Video from an Airborne Observer", *Proc. IEEE Image Understanding Workshop*, pp. 217-222, 1998

(Dempster 1977) A. Dempster, N. Laird, D. Rubin. "Maximum likelihood from incomplete data via the {EM} algorithm". *Journal of the Royal Statistical Society 39 (Series B)*, 1-38, 1977

(ECAC 1994) *Proceedings of the ECAC APATSI and EC Workshop on A-SMGCS*. Frankfurt, Germany. April 1994

(FAA 1993) U.S. Department of Transportation. FAA. "The Future Airport Surface Movement Safety, Guidance and Control Systems: A vision for Transition into the 21st Century". Washington. November 1993

(García 2002) J. García, J. A. Besada, J. M. Molina, J. Portillo. "Fuzzy data association for image-based tracking in dense scenarios". *IEEE International Conference on Fuzzy Systems*. Honolulu, Hawaii. May 2002

(Gauvrit 1997) H. Gauvrit, J.P le Cadre, C. Jauffret. "A Formulation of Multitarget Tracking as an Incomplete Data Problem". *IEEE Transactions on Aerospace and Electronic Systems*. October 1997

(ICAO 1995) "Manual of SMGCS. ICAO". Doc. 9476-AN/927

(Piater 2003) J. H. Piater, J. L. Crowley. "Multi-Modal Tracking of Interacting Targets Using Gaussian Approximations". *IEEE International Workshop on Performance Evaluation of Tracking and Surveillance* (PETS), 2001

(Pokrajac 2003) D. Pokrajac and L. J. Latecki. "Spatiotemporal Blocks-Based Moving Objects Identification and Tracking", *IEEE Int. W. Visual Surveillance and Performance Evaluation of Tracking and Surveillance* (VS-PETS), October 2003

(Rechenberg 1994) I. Rechenberg. "Evolutionsstrategie'94". frommannholzboog, Stuttgart, 1994.

(Rosin 2003) P.L. Rosin and E. Ioannidis, "Evaluation of global image thresholding for change detection", *Pattern Recognition Letters*, vol. 24, no. 14, pp. 2345-2356, 2003

(Sanka 1999) M Sanka, V. Hlavac, R. Boyle, "Image Processing, Analysis and Machine Vision", Brooks/Cole Publishing Company, 1999

(Schwabn 1985) C. E. Schwabn D. P. Rost. "Airport Surface Detection Equipment" *Proceedings of the IEEE*, No. 2, February 1985

(Schwefel 1977) H.-P. Schwefel. "Numerische Optimierung von Computer-Modellen mittels der Evolutionsstrategie". *Birkh¨auser*, Basel, 1977

(Schwefel 1995) H.-P. Schwefel . "Evolution and Optimum Seeking: The Sixth Generation". John Wiley & Sons, Inc. New York, NY, USA, 1995

INDEX